Effects of Accumulation of Air Pollutants
in Forest Ecosystems

Effects of Accumulation
of
Air Pollutants
in Forest Ecosystems

Proceedings of a Workshop held at Göttingen, West Germany,
May 16-18, 1982

edited by

B. ULRICH
Institut für Bodenkunde und Waldernährung der Universität Göttingen, West Germany

and

J. PANKRATH
Umweltbundesamt, Berlin (West), Germany

D. Reidel Publishing Company
Dordrecht, Holland / Boston, U.S.A. / London, England

Library of Congress Cataloging in Publication Data

Main entry under title:

Effects of accumulation of air pollutants in forest ecosystems.

    Sponsored by the Environmental Agency of the Federal Republic
of Germany.
    Includes index.
    1. Air–Pollution–Environmental aspects–Congresses. 2. Forest
ecology–Congresses. 3. Forest soils–Congresses. 4. Soil ecology–
Congresses. I. Ulrich, B. (Bernard), 1926–   II. Pankrath,
Jürgen. III. Germany (West). Umweltbundesamt.
QH545.A3E33   1983   581.5'2642   83-3203
ISBN 90-277-1476-2

Published by D. Reidel Publishing Company,
P.O. Box 17, 3300 AA Dordrecht, Holland.

Sold and distributed in the U.S.A. and Canada
by Kluwer Boston Inc.,
190 Old Derby Street, Hingham, MA 02043, U.S.A.

In all other countries, sold and distributed
by Kluwer Academic Publishers Group,
P.O. Box 322, 3300 AH Dordrecht, Holland.

D. Reidel Publishing Company is a member of the Kluwer Group.

Printed in The Netherlands

CONTENTS

Preface   ix

B. Ulrich: A concept of forest ecosystem stability and
        of acid deposition as driving force for
        destabilization   1

TOPIC 1: PROCESSES AND RATES OF DEPOSITION, STORAGE PLACES
        OF DEPOSITED AIR POLLUTANTS

B. Ulrich: Interaction of forest canopies with atmospheric
        constituents: SO_2, alkali and earth alkali
        cations and chloride   33

R. Mayer: Interaction of forest canopies with atmospheric
        constituents: Aluminum and heavy metals   47

K.D. Höfken: Input of acidifiers and heavy metals to a
        German forest area due to dry and wet deposition   57

W. Thomas, W. Riess and R. Herrmann: Processes and rates
        of deposition of air pollutants in different
        ecosystems   65

L. Granat: Measurements of surface resistance during dry
        deposition of SO_2 to wet and dry coniferous
        forest   83

TOPIC 2: PROCESSES AND RATES OF PROTON PRODUCTION BY
        DISCOUPLING OF THE ION CYCLE, AND OF PROTON
        CONSUMPTION BY SILICATE WEATHERING

E. Matzner and B. Ulrich: The turnover of protons by
        mineralization and ion uptake in a beech
        (Fagus Silvatica) and a Norway spruce ecosystem   93

S.I. Nillson: Effects on soil chemistry as a consequence
        of proton input   105

M.J. Mazzarino, H. Heinrichs and H. Fölster: Holocene
        versus accelerated actual proton consumption in
        German forest soils   113

TOPIC 3: EFFECTS ON CHEMICAL SOIL STATE

B. Ulrich: Soil acidity and its relations to acid
        deposition   127

E. Matzner: Balances of element fluxes within different
        ecosystems impacted by acid rain   147

J. Prenzel: A mechanism for storage and retrieval of acid
        in acid soils   157

N. v. Breemen and E.R. Jordens: Effects of atmospheric
        ammonium sulfate on calcareous and non-calcareous
        soils of woodlands in the Netherlands   171

H.G. Miller: Studies of proton flux in forests and heaths
        in Scotland   183

I.K. Morrison: Composition of percolate from reconstructed
        profiles of two Jack Pine Forest soils as
        influenced by acid input   195

G. Abrahamsen: Sulphur pollution: Ca, Mg and Al in soil
        and soil water and possible effects on forest
        trees   207

R.A. Skeffington: Soil properties under three species of
        tree in southern England in relation to acid
        deposition in throughfall   219

G. Brümmer and U. Herms: Influence of soil reaction and
        organic matter on the solubility of heavy metals
        in soils   233

TOPIC 4: EFFECTS ON BIOLOGICAL SOIL STATE AND ON ANIMALS

M-W. v. Buch: Micro-morphological characteristics of humus
        forms as indicators of increased environmental
        stress in Hamburg's forests   247

A. Hüttermann, B. Fedderau-Himme and K. Rosenplänter:
        Biochemical reactivity in forest soils as
        indicators for environmental pollution   257

S. Bombosch: Mercury - Accumulation in game   271

TOPIC 5: EFFECTS OF SOIL ACIDIFICATION AND ACCUMULATION
        OF AIR POLLUTANTS ON PLANTS

T. Keller: Air pollutant deposition and effects on plants   285

K.F. Wentzel: IUFRO-Studies on maximal SO_2 emissions
        standards to protect forests   295

H. Flühler: Longtermed fluoride pollution of a forest
        ecosystem: Time, the dimension of pitfalls and
        limitations   303

S. Athari and H. Kramer: The problem of determining growth
          losses in Norway Spruce stands caused by
          environmental factors   319

J.B. Reemtsma: First information about inventory of
          emission depending damages on Norway Spruce in
          Lower Saxony/Fed. Rep. Germany   327

G.H. Tomlinson, II: Die-back of Red Spruce, acid deposition
          and changes in soil nutrient status - a review   331

K. Kreutzer, A. Knorr, F. Brosinger and P. Kretzschmar:
          Scots Pine-dying within the neighbourhood of an
          industrial area   343

K.E. Rehfuess, H. Flurl, F. Franz and E. Raunecker: Growth
          patterns, phloem nutrient contents and root
          characteristics of beech (Fagus sylv.L.) on
          soils of different reaction   359

J. Bauch: Biological alterations in the stem and root of
          fir and spruce due to pollution influence   377

Index   387

PREFACE

This volume is based on a workshop on "Effects of accumulation of air pollutants in forest ecosystems", held in Göttingen, Federal Republic of Germany, from May 16-18, 1982. This workshop was initiated and sponsored by the Environmental Agency of the Federal Republic of Germany (project officer: Dr. J. Pankrath) as part of a research contract (project leader: Dr. B. Ulrich).

THE PROBLEM SEEN UNDER THE ASPECT OF ADMINISTRATION

The problem of forest damage caused by air pollution is not new in Europe. Already in 1983 a comprehensive report from Schroeder and Reuss about vegetation damages by fume in the Harz mountains was published. In 1923, Prof. Dr. Julius Stocklasa of the Bohemian Technical Highschool in Prague was concerned with research of toxical effects of sulphur dioxide in his publication "The damage of vegetation by flue gas and exhalations of facilities". This comprehensive and instructive work concludes with the sentence: "It is already high time for the governments of all cultural states to take legal, police and private measures in order to prevent damage by flue gases".
In the neighbourhood of industries with high gaseous and dust emissions damages have been shown to occur for a long time; these deleterious effects have influenced the growth of trees and in extreme cases have even caused their early death.
All over Central Europe a scattered but immense forest damage is confirmed. In the last years a reduction in the vitality of forests has been observed especially in areas remote from industrialized regions. This exhaustion of forest ecosystems is the result of an environmental stress which has lasted already for more than several decades.
The new complex disease of coniferous stands especially is obviously pointing towards more serious impairment. Trees, being the final link of the ecological hierarchy of a forest are disturbed in their ecological equilibrium. The biological and/or

B. Ulrich and J. Pankrath (eds.), Effects of Accumulation of Air Pollutants in Forest Ecosystems, ix–xvii.

soil buffering capacity of forest stands which normally supports
the resistance of its species to different kinds of air pollution
is obviously exhausted in several areas. In the Federal Republic
of Germany by 10 percent of the forest area exhibit deleterious
impacts mainly caused by a supply of acidifying air pollutants.
Whereas the symptoms of forest damage are uncontested, it is
only with regard to the causes of forest injury that the opinions
of forest experts strikingly disagree.
At the moment it is a matter of fact that the impairment of air
pollutants on vegetation as well as the acidification of soils
is one of the main causes in the reduction of vitality of large
forest areas. For instance, acidification of soil signifies,
according to the current state of knowledge, that toxic and
root injuring aluminium ions are released.
In addition, forest areas are very effectively filtering pollutants out of the air. Thus, the uptake of harmful airborne substances is increased by a significant amount if compared to the
open field. This filtering effect of the canopy, called interception, works the whole year for coniferous stands; for
deciduous stands it is limited to the foliage period.

A few decades ago the impairment of regions remote from air
pollution sources was hidden by growth stimulating effects of
Nitrogen and Sulphur compounds. The increase of the source heights
has made the deleterious effects gradually become more evident
on a growing scale.
It is not intented to discuss the different opinions on possible
causes of forest damage, nor are investigations on the contribution of air pollutants to be presented at this workshop anticipated; however, it is necessary to call upon the deposition of
acidifying air pollutants which is to be particular to the
widespread forest disease, and to consider technical and legal
instruments for an air quality control which are also beneficial
to larger regions.
Samples of precipitation in areas remote from industrialized
regions of the Federal Republic of Germany show a diluted acid
with a mean pH value of 4.2, corresponding to a hydrogen ion
concentration of 63 μg H^+/l. Analyses of cations and anions in
precipitation indicate that H^+ ions, characterizing acidity, are
due to dissociation of sulphuric (about 2/3) and nitric acid
(about 1/3) originating in the oxidation of sulphur dioxide and
nitrogen oxides, respectively.
In comparison to sulphuric and nitric acids other strong acids
derived from hydrogen fluoride and hydrogen chloride are mainly
of local concern.

In the Federal Republic of Germany for instance 3,55 Mio t SO_2
had been released annually into the atmosphere, an amount that
was shared by the following emission source categories:

- Power plants, heating plants and industrial
  power plants   56 %
- Industry   28 %
- Small industry and domestic consumption   13 %
- Traffic   3 %

The corresponding NO_x-emissions resulted in about 3 Mio t NO_2,
shared by the following source categories:
- Power plants, heating plants and industrial
  power plants   31 %
- Industry   19 %
- Small industry and domestic consumption   5 %
- Traffic   45 %

The SO_2 emissions are due to the utilization of fossil fuels; in
the last few years a stabilisation of the SO_2 emission level was
reached by substitution of coal, supply of fuels with low
sulphur content and application of flue gas desulphurization
techniques. In contrast to SO_2, the NO_x emissions due to firing
processes had increased according to the growth of consumption
of primary energy; this evidently illustrates the fact that
measures for reduction had hardly been applied.
A successful environmental policy has to limit the release of
air pollutants at the source. In demanding the current state of
technology one merely obeys the legal rules for protection of
human health and for conservation of natural resources. It is
to be regarded with great care, however, that in the past
decreasing pollution levels in densely populated areas have often
been accompanied by increasing pollution in more remote areas
with low pollution levels. Thus the building of tall stacks to
unburden the neighbourhood of high level emittors clearly
exhibits its partial disadvantage.
The Federal Emission Law in Germany contains legal instruments
to ensure clean air. Installations which need a licence, have to
obey the emission and air quality standards of the Technical
Regulation of Air Protection. Technical reduction measures like
-   low emission firing technologies
-   low emission fuels
-   advanced gas purification processes
have to be used.
In the Federal Republic of Germany the necessary legal instruments
for an effective clean air policy have been elaborated. Priority
for further action is given to the amendment of the Technical
Regulation of Air Protection, to an Ordinance for large firing
installations limiting especially precursors of acids like SO_2,
NO_x, HF and HCl in operating new and - with a certain delay in
time - in operating old installations, and to diminish emissions
from motor vehicles.
The ecological air pollution issue is, moreover, an international

problem which cannot be solved by national measures alone. It is
well known that the transboundary transport of air pollution in
Europe causes a manifold of mutual interferences. For instance,
the annual sulphur deposition in the Federal Republic of Germany
is due half to deposition from indigeneous sources and half due
to emissions from other European countries. Within the framework
of the ECE Convention on Long Range Transboundary Air Pollution
it is the common objective to strengthen the co-operation of
the contracting countries in order to combat harmful air pollution. The contracting countries shall endeavour to diminish and
to prevent air pollution as far as possible.
The 1982 Stockholm Conference on Acidification of the Environment
demonstrated that the problems of air pollution have international
dimensions.
Any reduction of acidifying emissions of sulphur and nitrogen
compounds will be beneficial to the environment. Control
techniques are already available today.
The demand for further scientifically and technically research
must not obscure, however, the unmediate and unquestionable need
for measures for combating the acidification of the environment.
In implementing the 1979 ECE Convention on Long Range Transboundary Air Pollution priority must be given to internationally
harmonited measures taken at the source of air pollution in
accordance with the   state of technology.

THE PROBLEM SEEN UNDER THE ASPECT OF SCIENCE

From the scientific point of view, the forest damage and die-back
now occuring in Central Europe raises several questions. Two of
them may be put here: Why was science not able to foresee this
development? Why is science, even after the large scale outbreak
of forest damages, not able to give rapidly clear proof of what
is going on ?
These questions touch on the philosophical background of science.
To answer them, it is necessary to consider the procedure of how
objective knowledge is gained in general. We do so by following
Karl Popper who stated that any increase in objective knowledge
is attained by passing through the following stages:
  1. condensing of the available knowledge and the observations
     into a hypothesis;
  2. trying to discard the hypothesis with the help of new
     experiments and observations;
  3. and as soon as it becomes impossible to find any contra-
     diction, accepting the hypothesis as proven theory.

In the light of this science theory, a condensing hypothesis
forms the starting point to gain objective knowledge. In a
forest ecosystem, all system components which may be affected
have to be included into the hypothesis.

In the last decades of the 19th century the hypothesis was
developed and accepted that forest damages due to air pollution
are caused by the direct action of gases like SO_2 and HF on the
leaves. By short term gas chamber experiments with young vital
plants and by observations in the field, concentrations of SO_2
and HF in the air have been fixed below which no injury should
occur to defined organisms. This hypothesis is the base of the
clean air acts existing in the different countries. A consequence
of this hypothesis was the strategy to build large stacks in
order to dilute increasing emissions and to remain below the
accepted treshhold levels of SO_2 concentration in air.

This hypothesis has neglected the acidity being formed from SO_2
and NO_x. It has neglected that other air pollutants like heavy
metals and organic compounds including oxidants may be involved
(this was considered later to some extent in addition). The
hypothesis further neglected other parts of the ecosystem like
bark, wood, roots, ground vegetation, the soil and the organisms
living in it. By neglecting all these components in air and in
the ecosystem, the hypothesis was implicitely expressing that
acidity and other air pollutants have no effects, and that other
parts of the ecosystem are not affected. The hypothesis even
prevented research in these respects.
This example shows the implications of scientific hypotheses.
Scientists carry a high responsibility just by building hypotheses. There are some disadvantages if pieces are included in
the hypothesis which are not necessary components. But it can
include a high risk if pieces are excluded which in fact are
necessary components of the hypothesis.
We must assume that man is now executing an experiment with
forest ecosystems by producing air pollution. This experiment
may have started with industrialisation, but it was the rapid
increase in energy consumption after the Second World War that
gave it a global character. We are witness of a unique and
irreversible process which is inaccessible in principle to
experimental proof. This does not mean that experiments are not
necessary; but these experiments can only check individual elements of the hypothesis; proof of the hypothesis as a whole is
impossible.
It was the aim of the workshop to throw light on the connexions
and processes which should be integral parts of a hypothesis
describing the present situation of forest ecosystems. It is
hoped that the papers presented in this volume contribute
alltogether to the formation of a condensing hypothesis. The
objective of such a hypothesis cannot be the present ongoing
forest damage, it must be the understanding and quantification
of forest ecosystem stability and resilience. The ongoing large
scale experiment with forest ecosystems allows to reach this
knowledge and forces us to gain it.

STATEMENTS AND RECOMMENDATIONS

The final discussion at the workshop was dedicated to the consequences which should be drawn from the present state of knowledge. The content of this discussion is given in the following
statements and recommendations. In addition, many participants
have given during and (in written form) after the workshop their
comments to special scientific questions. These comments have
been incorporated in the papers of Ulrich. We thank all
participants for their cooperation.

A.   EMISSION CONTROL

1. It is agreed that there is convincing evidence for widespread
   and increasing damage to forests in Germany and other Central
   European countries. The participants recommend that urgents
   steps be taken to reduce emission at the sources of air
   pollution which are contributing to this pattern.

2. The air pollutants contributing include gaseous compounds
   like SO_2, NO_x and HF, acids formed mainly from SO_2 and NO_x,
   inorganic compounds of varying toxicity like heavy metals,
   organic micropollutants like hydrocarbons, and oxidants. Due
   to the interrelationships between groups of pollutants and
   between single chemical compounds, it is difficult, if at all
   possible, to attribute special damages, to special air pollutants. Thus, for example, in presence of acids the toxicity
   of heavy metals and of SO_2 is increased. Measures which limit
   at once the emission of several pollutants, like fume gas
   washing, may be therefore especially efficient.

3. An efficient emission control requires detailed knowledge
   about the contribution of the various emission sources.
   Existing lack of information, e.g. about the gaseous emission
   of chemical elements from combustion plants, can have
   unforeseen consequences, due to the potential toxicity of some
   of the elements involved. It is therefore recommended to
   demand by law annual mass balances of chemical elements for
   power and garbage combustion plants.

4. The influence of air pollutants on forest ecosystems cannot
   be deduced from the concentration of a few components like
   SO_2 in the air. It is therefore recommended to install a
   net-work for the measurement of deposition rates. The measurement should include acid, the main cations and anions, and
   heavy metals and other inorganic and organic compounds of
   potential toxicity. On the base of the existing methods, this
   can only be achieved by use of the flux balance method.

B.   FOREST ECOSYSTEM CONTROL

5. The late detection of the consequences of accumulation of air
   pollutants in forest ecosystems is a consequence of missing
   forest ecosystem monitoring. It is therefore recommended to
   continue and to complete existing ecosystem monitoring (e.g.
   Solling), and to establish a net-work of continuously moni-
   tored ecosystems, covering different areas as well as
   different types of ecosystems. The monitoring has to be based
   on flux balances and inventories of species and of significant
   chemical compounds. It must be suited to detect long-term
   changes in the ecosystem at their beginning.

6. As a consequence of long range transport of air pollutants,
   no "zero plot" can be found anymore. This indicates the
   necessity of specimen banking as a means to evaluate lateron
   any changes which may have occured over a period of time.
   This activity may be connected with the ecosystem monitoring
   program.

7. The intensive ecosystem monitoring on a few places must be
   supplemented by a more dense net-work of forest inventories.
   A method of forest inventory should designed which is based
   on fixed sample plots, allowing permanent or consecutive
   taking of inventories. An inventory based on forest management plans cannot meet the requirement. In case where older
   inventories are missing old aerial photographs can be used.
   They may be available as far back as about 1930. Their information on forest stands can be transferred to dendrometric
   data by means of modern photogrammetry. It is revommended to
   stress application of this tool for inventory purposes.

C.   GAPS IN KNOWLEDGE, RESEARCH NEEDS

Deposition:

- Methods for measuring deposition rates for all ecologically
  relevant air pollutants. Application of micrometeorological
  and flux balance methods at the same sites for comparison is
  strongly recommended.
- Chemistry (acidity, dissolved inorganic and organic compounds
  of potential toxicity) of cloud droplets, mist, and fog, and
  their deposition.
- Importance of episodic events of high acid deposition on
  canopy surfaces for erosion of cuticula, mobilization of
  deposited heavy metals, and for toxicity of SO_2

Effect of Air Pollutants on Plants:

- With the combustion of fossil fuel and of garbage, many chemical elements and organic micropollutants are widely distributed
  in the ecosphere and their effects on plants are not known.
  Even if a correlation has been found between the deposition of
  a pollutant and the appearance of a damage, it cannot be
  excluded that other toxicants may be involved. In order to
  make our knowledge more complete, it is therefore necessary to
  include as many air pollutants into investigations as possible.

- There is an urgent need to attract specialized physiologists
  and biochemists to contribute for an explanation of the many
  definite findings which have also been demonstrated during the
  workshop. Altered mechanisms and reactions in a diseased tree
  have to be interpreted. This is also true for the mycorrhiza,
  often forming the interphase between plant root and soil.

Effects of deposited air pollutants on soils, decomposers and
roots

- Soils are natural objects. Their acidification leads to the
  appearance of toxic ions like Al ions. This induces the tendency that living beings reduce their contact with and finally
  retract from the mineral soil where the toxic substance is
  being produced. In the past, investigations of these effects
  have been mainly limited to agricultural crops. There is a
  great need to investigate:
  - the nature of soil acidity under the condition of high
    sulfuric acid input
  - the kind of ion species, especially of Al, existing in soil
    solutions
  - the action of different ion species on living cells
    (bacteria, fungi, mycorrhiza, plant roots, soil animals).

- Soils are formed by the interaction of lithosphere, atmosphere,
  hydrosphere and biosphere. The retraction of life from mineral
  soil is bound to have consequences for soil development. The
  long-term perspective of the present ongoing soil acidification
  should attract the attention of soil biology as well as of
  soil genetics.
- Balance studies of chemicals in soils are bound to the knowledge of the water balance of the soil and of the ecosystem.
  Soil water modeling must be developed further to be able to
  provide the necessary information.

'orest Ecosystems as Links in the Ecosphere

n respect to the water cycle, forest ecosystems represent the
ink between atmosphere and hydrosphere. It has been demonstrated

that, due to strong soil acidification, some forest ecosystems
act already not as sinks but as sources for some heavy metals of
potential toxicity. There is an urgent need to investigate:

- the transport processes of chemicals through the soil and
  underlying geological layers to the water body
- the risk of future ground water acidification in forested
  areas, due to increasing soil acidity.

Forest Ecosystems as Indicators

It is agreed that the problem of increasing damages to forests
may be a problem of long-living species.

                                        Jürgen Pankrath
                                        Bernhard Ulrich

A CONCEPT OF FOREST ECOSYSTEM STABILITY AND OF ACID DEPOSITION
AS DRIVING FORCE FOR DESTABILIZATION

B. Ulrich

Institut für Bodenkunde und Waldernährung der
Universität, D-3400 Göttingen, Büsgenweg 2

ABSTRACT

A theory is proposed which explains the acidification and
alcalinization of soils, respectively, as consequence of the
discoupling of the ion cycle in the ecosystem. Under the
assumption that the ecosystem tends to minimize net proton
production or consumption in order to keep the chemical soil
state in optimal conditions for growth, the characteristic
features of stable forest ecosystems showing high resilience are
deduced. A sequence of ecosystem states is described; aggradation
phase, stability range I (high resilience), destabilization
phase I (humus disintegration), stability range II (low resilience),
destabilization phase II (build up of decomposer refuge, podzolization). A continuous input of acidity exceeding the rate of
base cation release by silicate weathering within the root zone
forces forest ecosystems from the stability ranges into the
transition states (destabilization phases). The concept of stress
and strain is used to deduce how acid deposition superimposes
natural stress factors and may trigger forest damages connected
with climatical extremes and pests.

INTRODUCTION

Many forest soils are acid. This was accepted as a fact. In the
northern hemisphere, acid forest soils have been found the more
often the farer north and the higher (altitude) the location has
been. Since precipitation increases with increasing altitude it
was assumed that stronger soil acidification in higher altitudes
is caused by stronger leaching through higher precipitation. But

1

B. Ulrich and J. Pankrath (eds.), Effects of Accumulation of Air Pollutants in Forest Ecosystems, 1–29.
Copyright © 1983 by D. Reidel Publishing Company.

in a soil which contains no cation acids, percolation with
water or with a very diluted NaCl solution (NaCl from sea spray),
being in equilibrium with the CO_2 content of air, will not cause
the pH to drop below 5. This means that precipitation alone will
not cause the formation and accumulation of stronger organic
acids and cation acids in soils.

It was further found that acid forest soils usually carry an
acid top organic layer and may show bleached A horizons as well
as iron enriched B horizons. The process leading to such a soil
profile was called podzolization. Many investigations showed
that the vegetation plays a deciding role in podzolization,
mainly by the kind of leaf litter produced. This made clear that
vegetation and decomposers are involved in the process of soil
acidification called podzolization. The presence and the role of
organic acids in this process was clearly demonstrated, but the
cause for their formation remained hidden.

In Central Europe, podzolization has been found to be related to
a change from forest to heath ecosystems. The disappearance of
trees has been explained by utilization of biomass by man and
by grazing, prohibiting the regeneration of trees. This may not
be the sole reason. What does soil acidification mean in respect
to forest ecosystem stability and resilience?

ROSENQUIST (1) was one of the first who tried to compare the
role of natural soil acidification and of acid deposition. He
made clear that the role of acid deposition has to be weighted
with natural acidification.

How can such a weighing be done? Since we consider processes
acting in time, the adequate measures will be rates. Thus we
have to compare rates of natural soil acidification with rates
of acid deposition. No data existed a few years ago on rates of
natural production of strong acids in soils.

In the following a theory, first presented in 1978 (2,3,8) is
shortly outlined. This theory explains the acidification and
alcalinization of soils, respectively, as consequence of the
discoupling of the ion cycle within the ecosystem. Similar
approaches have been proposed by DRISCOLL (4), ANDERSSON et al.
(5) and SOLLINS et al. (6). The theory shows further how the
internal proton production or consumption in soils can be measured
and calculated (from the cation/anion balance of the storage
changes in soil compartments). Corresponding experimental data
are given by MATZNER et al (this volume). The theory is expanded
to define aggradation phases, stability ranges of different
resilience, and destabilization phases of forest ecosystems.
Stability ranges were characterized by steady states, aggradation
and destabilization phases (transition states) are connected with
increase and decrease of storages (7).

Evolution is not limited to species, also ecosystems can pass
through evolutionary stages by adjusting to changing conditions
during geological periods. The stability ranges and transition
states described are typical for the geological and climatical

conditions of Central Europe. Under different conditions, other stability ranges can occur as a result of ecosystem evolution.

The element cycles in forest ecosystems are subjected to two general rules: the rule of continuation of mass and the principle of electrical neutrality. The last one has especially to be considered if the elements pass from one phase to another one as ions. Precondition for the application of these rules on data sets is the simultaneous measurement of the total ion turnover in the ecosystem. This is methodical difficult, but can be achieved with enough accuracy (8,9). By considering possible errors, the measuring approach can be limited to the quantitatively most important ions: all those cations and anions can be neglected whose contribution to cation sum and anion sum, respectively, is less than the error of this sum.

THE ION CYCLE IN THE ECOSYSTEM IN STEADY STATE

By defining the ecosystem in a proper way, all processes included in ion turnover can be reduced to four fluxes: input (e.g. from atmosphere, or by weathering of soil minerals), output (e.g. leaching, denitrification), uptake from soil solution into biomass (by plants and microorganisms), and release of ions during decomposition (mineralization). We consider an ecosystem in which input and output are equal for each element. For such an ecosystem, quasi-steady state can be achieved: in the temporal and spatial mean all stores in the system are constant. This covers phytomass, total biomass, soil organic matter, the various chemical elements and their various binding forms. In such an ecosystem the rate of assimilation must be equal to the rate of respiration, or, if one includes also the turnover of elements other than C, H and O, the rate of phytomass production must be equal to the rate of mineralization:

$$CO_2 + H_2O + X^+Y^- \underset{\substack{\text{energy} \\ \text{mineralization}}}{\overset{\text{phytomass production}}{\rightleftharpoons}} CH_2OX^+Y^- + O_2$$

In eq. 1, X^+Y^- symbolyzes the cations and anions of the elements other than C, H and O; they are taken up from the soil solution and return during mineralization again to the soil solution. For the ion cycles coupled in soil, the principle of electroneutrality says that in each compartment, soil solution or plant, the cation/anion balance must be zero. If during phytomass production more anions Y^- are taken up than cations X^+ (or vice versa), the resulting charge inequilibria must be balanced. If this is not possible, ion uptake must cease. The same holds for mineralization. The charge balancing can occur by proton transfer. If anion uptake exceeds cation uptake, protons must be taken up in

addition (or OH⁻ must be transferred from plant to soil; this
can happen by transport of HCO_3^- which, by being transformed to
CO_2 and H_2O, transfers one proton from the soil solution into
H_2O). If cation uptake exceeds anion uptake, the proton transfer
must occur in the opposite direction. If in the steady state the
process of phytomass production is exactly balanced by mineralization, than the proton transfers connected with uptake and
mineralization are of opposite direction and balance each other.

It is known from many investigations that the ion turnover
occuring by mineralization and uptake is connected with proton
production or consumption and causes changes in the pH value in
the medium (for plant uptake see 10). Hitherto the single processes have been looked at, e.g. the uptake of nitrogen as ammonium
or nitrate. The effect in the ecosystem,however, can be judged
only if the total ion turnover is considered simultaneously, and
if the ionic status (uncharged, cation, anion) in each compartment (soil solid phase, soil solution, plant) is taken into
consideration (MATZNER et al. this volume). Calculations of the
proton production in ecosystems which are based on single processes only and not on complete ion balances (11) are therefore
meaningless.

THE ION CYCLE OF AN EVEN AGED BEECH STAND - AN EXAMPLE

The turnover rate in the ion cycle can be approached by measuring
and calculating the rate of ion uptake by the vegetation. In
table 1 data are given for a beech (Fagus silvatica) stand,
120 years old, on a podzolic brown earth developed within a
loess layer overlying red sandstone; the site is located in
Central Europe in the Solling district in 500 m altitude. The
total uptake rate is calculated as the sum of transport rates in
litter and canopy leaching, the accumulation rate in the increme-
ment, and the turnover rate of fine roots.

Since the form of nitrogen uptake is unknown, the both extreme
cases are considered separately:
    hypothesis A: N uptake exclusively as NO_3^-
    hypothesis B: N uptake exclusively as NH_4^+
From the data of table 1 it follows that the rate of cation
uptake amounts to 5.29 keq, the rate of anion uptake to 1.01 keq,
and the rate of N uptake to 5.80 keq. In the case of hypothesis
A, N uptake occurs as anion. In this case the anion uptake
amounts to 6.81 and exceeds the cation uptake by 1.52 keq. This
anion surplus would have been balanced by proton uptake from
soil. In the case of hypothesis B, the N uptake occurs as cation.
In this case the surplus is in cation uptake and amounts to
10.08 keq; it would have to be balanced by a corresponding proton
transfer from the plant into the soil.

Table 1: Ion uptake in a beech forest on podzolic brown earth in the Solling

way of turnover	Na	K	$\frac{1}{2}$ Ca	$\frac{1}{2}$ Mg	$\frac{1}{2}$ Mn	$\frac{1}{3}$ Al	$\frac{1}{3}$ Fe	Cl	$\frac{1}{2}$ SO$_4$	H$_2$PO$_4$	N$_t$
					keq . ha^{-1} . yr^{-1}						
litter fall	0,03	0,41	0,81	0,13	0,19	0,00	0,07	0,02	0,28	0,13	3,5
canopy leaching	0,03	0,48	0,33	0,04	0,07	0,00	0,00	0,06	0,18	0,00	0,00
storage in forest increment	0,00	0,17	0,39	0,14	0,12	0,01	0,04	0,00	0,09	0,07	0,93
storage in stocks and roots > 5mm (increment)	0,03	0,03	0,05	0,02	0,01	0,01	0,01	n.d.	0,01	0,01	0,16
turnover of roots < 2 mm	0,04	0,14	0,29	0,19	0,03	0,60	0,38	n.d.	0,08	0,06	1,21
sum = plant uptake	0,13	1,23	1,87	0,52	0,42	0,62	0,50	0,1	0,64	0,27	5,80
sum in kg element	3,0	48	37	6,3	12	5,6	9,3		10	8,4	81
uptake in above-ground plant parts in kg element	1,4	41	31	3,8	10	0,1	2,0	2,8	7,5	6,2	62

To account for the ionic status, the precursor of the ions present in soil solution and the binding form in the plant must be known. As binding forms in the soil matrix and in the plant are assumed:

N: only organic binding forms with charge zero; fixed or exchangeable NH_4^+ in soil as well as NO_3^- in plant are neglected

Na, K, Mg, Ca, Mn, Al, Fe: as cations, the charge is balanced by inorganic or organic anions

S, P, Cl: as anions balanced by cations. This is an oversimplification in case of S (S containing amino acids), but the error introduced can be neglected for the following consideration. In case of P, the pyrophsphate bound creates dissociating OH groups during its breakdown.

With the exception of N assimilation and denitrification, the input and output of all elements occurs in ionic form. The precursor of assimilated N is uncharged N_2, the reaction products of denitrification are also uncharged compounds (N_2, N_2O).

If the N accumulated in the plants stems from N_2 (after assimilation) or soil organic N (after mineralization), its transfer to plant organic N is not connected with any net change in charge neither in soil nor in plant, that means it is not connected with any net proton transfer. It is therefore not possible to calculate net proton production in soil from NH_4/NO_3 ratio during ion uptake. In the example given in table 1, the net proton source term connected with uptake is equivalent to the difference between cation uptake (5.29 keq) and anion uptake (1.01), giving a proton production in soil of 4.28 keq $ha^{-1}yr^{-1}$.

From the point of view of the ion cycle, ion uptake is controlled by two principles:
-   maintenance of electrical neutrality (cation/anion balance) in the flux
-   maintenance of H^+/OH^- balance in the compartments (plant and soil)

The maintenance of H^+/OH^- balance requires that a NH_4/NO_3 ratio in uptake differing from 1 is accompanied by an equivalent net transfer of protons from soil to plant (NO_3^- surplus) or reverse (NH_4^+ surplus). If uptake and mineralization occur at the same place and time with the same rate, the proton flux connected with uptake would, under steady state conditions, be supplied by the process of mineralization. It will be discussed later that the resilience of the ecosystem is the greater the better this coupling is achieved.

Plants and soils contain buffer systems which are able to produce or consume protons without great changes in the chemical potential of the proton in the solution (pH). In the case of N transfer from charge zero in soil to charge zero in plant, the buffer systems in soil and plant counteract each other. This is demonstrated by the following reaction chain, illustrating the principle:

mineralization: $\quad\quad\quad\quad N_{org}^{o} \longrightarrow H^+ + NO_3^-$

soil buffering: $\quad \underline{\quad}\!\!-K + H^+ \longrightarrow \underline{\quad}\!\!-H + K^+$

plant buffering: $\quad CO_2 + H_2O + (K^+NO_3^-) \longrightarrow (K^+HCO_3^-)_{comp.I} +$

$$(H^+NO_3^-)_{comp.II}$$

protein formation: $(H^+NO_3^-) \longrightarrow N_{org}^{o}$

In case of a surplus in the uptake by NO_3^-, a possible proton consumption in soil during the transfer of organic bound N from soil to plant results in the formation of basicity in the plant. As example the formation of $KHCO_3$ has been postulated, but the basicity may appear in any other compound. This means that a plant alcalinization occurs which is equivalent to the previous soil acidification. In case of a NH_4^+ surplus (surplus exceeding NO_3^- uptake), a plant acidification occurs which is equivalent to the soil alcalinization due to NH_4^+ formation ($NH_3+H^+ \longrightarrow NH_4^+$). The buffer (pH-stat) operating in plants may than limit N uptake.

This seems possible especially in soils with extreme nitrification conditions: in neutral soils poor in organic matter due to an extreme high rate of nitrification (NH_4 uptake tends to zero), and in soils with acute podzolization due to an extreme low rate of nitrification (NO_3 uptake tends to zero). For both conditions, special adaptations seem to have developed during evolution (e.g. mycorrhiza).
There are several mechanisms to transfer alcalinity from plant to soil: leaf leaching of bicarbonates, accumulation of Ca salts in leafs before leaf fall, diffusion of bicarbonates from root to soil. The transfer of acidity from plant to soil seems to be limited mainly to the root. The compartmentation of the soil/root system should allow proton transfer reactions of the following type:

$$(NH_4^+Cl^-)_{soil} + H_2O + (CO_2)_{plant} \longrightarrow (NH_4^+HCO_3^-)_{plant} + (H^+Cl^-)_{soil}$$

The exudation of organic acids (R-COOH) does'nt fulfill this purpose since it is not connected with a change in the H^+/OH^- balance of the plant. Only if the organic molecule is taken up again as salt and the cation is transferred to an (insoluble) hydroxide, there would be a proton transfer from plant to soil:

$$(R\text{-}COOH)_{plant} \longrightarrow (R\text{-}COOH)_{soil}$$

soil: $\quad\quad R\text{-}COOH + X^+ \longrightarrow R\text{-}COOX + H^+$

$$(R\text{-}COOX)_{soil} \longrightarrow (R\text{-}COOX)_{plant}$$

plant: $R\text{-}COOX + H_2O \longrightarrow R\text{-}COOH + XOH$

The cation X must be a cation acid like Al, Mn, Fe ion species. The
process may be responsible for the accumulation of cation acids,
especially Al, in roots. As a consequence of this process, long
living plant like trees would have to renew regularly those
roots, where the accumulation occurs. This should increase the
root turnover. A higher root turnover needs more photosynthates
and will therefore reduce the growth of the overstory (bole
increment) even if the net photosynthesis remains at the same
level. This process may be therefore the main reason for the
decrease in forest yield on acid soils.
From this point of view, Al toxicity in trees may be connected
with the cation species X^+ existing in soil solution. If X^+
is a highly charged cation like Al^{3+}, it may be transferred from
soil to plant as cationic complex $(R\text{-}COOX)^+$. By the release of
X in the plant, a soluble instead of an insoluble ion species is
formed.
Cell metabolism should play a major role in all of these transfer
reactions, as driving force as well as providing sinks and
sources for the reactants. The base cations (Ca), bound in
exchangeable form in the free space, can act as an intermediate
buffer system. The capacity of this buffer system is limited, but
it can be regenerated by metabolic activity.
If one looks at the size of the balance as well as at the processes transferring alcalinity or acidity from plant to soil,
pure nitrate nutrition should be much easier balanced by the
plant than pure ammonium nutrition. Balance A (only NO_3 uptake,
1.5 keq) is much lower than balance B (only NH_4 uptake, 10 keq).
Exclusive NO_3 uptake should be much easier compensated in the
plant by formation and disposal of equivalent alcalinity, than
exclusive NH_4 uptake by formation and disposal of equivalent
acidity. This means that exclusive NH_4 uptake should limit
phytomass production much more than exclusive NO_3 uptake. This
may be a main reason for the low productivity of strongly acidified soils where nitrification is prohibited.

AGGRADATION PHASE

The term "aggradation phase" is used here for ecosystems in
which the organic matter storage in mineral soil is increasing
through accumulation of organic matter with narrow C/N ratio
($\sim 10/1$). The humusform is any subtype of mull. With this phase,
soil and ecosystem development starts at a freshly exposed
surface of loose sedimentary rocks, e.g. after glaciation. As
an example a soil may be considered where an amount of 100,000
kg organic matter per ha has been accumulated, the carbon content

of organic matter = 50 %, C/N ratio = 10, CEC (cation exchange capacity) = 4 eq/kg C. The CEC indicates a production of 200 kmol H^+ha^-1.

The N accumulated during the aggradation phase originates in untouched ecosystems mainly from N_2 (legumes), to a small fraction from nitrate input with rain. If the nitrate input is balanced by base cations, its accumulation in the ecosystem in form of organic N is connected with a proton consumption. This proton consumption occurs in plants, but is finally transferred to the soil (e.g. by root exudation, canopy leaching, litter fall). Assuming a nitrate input of 4 kg N $ha^{-1}yr^{-1}$, it will take 700 years to balance the acidification in the soil caused by the formation of the CEC of soil organic matter. The accumulation of soil organic matter with low C/N ratio in the mineral soil need therefore not to lead to soil acidification.

A forest plantation may swing back into the aggradation phase after clear cutting, under the influence of the freshly developed herb layer providing easily decomposable litter. In such a case the aggradation phase may end already after a few years, when the forest canopy closes, the herbs are suppressed, and the microclimate is changing again. In this case, the N accumulated stems from the N mineralized in the organic top layer. If the HNO_3 formed by nitrification is transported to the A horizon and taken up there, no change in soil acidity occurs. If the uptake is suppressed, e.g. by the use of herbicides, and the nitrate leached, the soil acidifies. It is not known whether during the mineralization of the organic top layer a transfer reaction of the following type is possible:

$$2 \ HNO_3 + Ca(HCO_3)_2 \longrightarrow Ca(NO_3)_2 + 2 \ H_2CO_3$$

It seems probable that the microcompartmentation existing in the soil allows such a transfer of acidity from the strong nitric acid to the weak carbonic acid (which passes over into $CO_2 + H_2O$). In such a case, the uptake of nitrate from the A horizon would be connected with a proton consumption, that means with a decrease in acidity.

STABILITY WITH HIGH RESILIENCE (STABILITY RANGE I)

Till now the theory rests on two general accepted physico-chemical rules and a hypothetical state of the ecosystem, the steady state. One can say that each ecosystem tends towards the steady state but will never reach it (12). If the ecosystem is not in the steady state, phytomass production and minerali-zation does not compensate each other and a net production or consumption of protons in the soil will usually be the conse-quence. The fact that the steady state can **never be achieved**

has three reasons:
- the climate, acting as regulator and determining the aim of
  ecosystem development, is not constant, but exhibits
  statistical variations. The ecosystem as a whole should tend
  to follow this variations.
- elements of the ecosystem like microorganisms, plants, animals,
  but also structural units, have a limited life span and have
  to be replaced continuously. The fallout of elements of the
  system represents a stress.
- ion uptake and mineralization may not happen in the same
  microcompartment of the soil. Thus there is always spatial
  discoupling of the ion cycle.

The seasonal variation of climate results in the seasonal discoupling of the ion cycle. After what has been said, anticipation
of nitrification ahead of nitrate uptake will cause an acidification push by the net production of HNO_3. In a humid climate,
acidification pushes of this kind are mainly regulated by
temperature; dryness plays a role especially as a delaying
factor. Mineralization begins at 0^oC, increases from 5° on and
reaches its optimum at 30° (13). After the warming up of soils
in spring the nitrification may start earlier than nitrate uptake,
thus leading to a seasonal acidification push. As soon as nitrate
uptake follows, the deacidification phase starts. Also the wetting
of a dried soil in summer or autumn may cause a seasonal acidification push.

In stable ecosystems, temporal discouplings occur in opposite
direction (e.g. proton consumption following proton production),
their effects on the chemical soil state do therefore compensate
in the temporal mean. Spatial discouplings are compensated by
the mixing activity of soil burrowing animals. The temporal or
spatial discoupling may, however, already lead to cell injury
in microorganisms and roots as consequence of the appearance of
toxic cation acids like Al ions. There is therefore a need to
buffer a net proton production for the time interval of discoupling by basic cations. This intermediate buffering is taken
over by exchangeable Ca (and Mg) (see ULRICH, this volume).

If one considers cation acid (e.g. Al) toxicity as the main
factor of chemical stress in natural, untouched forest ecosystems
(nutrients being well balanced), than the capacity of the cation
exchange buffer existing in soil is one major factor determining
the resilience of the ecosystem. Parameters of this buffer system
are the cation exchange capacity (CEC in keq $ha^{-1}z^{-1}$, z = soil
depth) and the percentage of exchangeable basic cations (Ca+Mg).
Another major factor of resilience is the mixing activity of
soil burrowing animals.

The absence of toxic cation acids allows all species of primary
producers and of secondary producers (decomposers and consumers)
to exist and thus to compete with each other. For the climatical
and hydrological conditions given one should therefore expect

the greatest diversity, compared with ecosystems on acidified
soils. The absence of toxic cation acids means further that
roots tend to maximize the contact with soil material where
water and nutrients are stored, and that bacteria play the
dominant role in the decomposers.
A stable forest ecosystem with high resilience (stability
range I) should therefore, under the conditions of Central
Europe, being characterized by the following properties:
- it is composed of relatively many species, which are
  structured in layers
- the soil is deeply rooted, the roots are homogeniously
  distributed
- the decomposers are characterized by the activity of earth
  worms
- the soil stays throughout in the silicate (or carbonate)
  buffer range and shows no depth gradient in the chemical
  soil state
- soil organic matter is accumulated throughout the whole
  rooting zone, due to the activity of soil burrowing animals,
  the soil is of crumby structure.
Examples can be found which demonstrate that untouched forest
ecosystems in Central Europe would exhibit these properties.

In the higher altitudes of the subalpine mountains it is
possible that a biological depth gradient existed in soil due
to frost action. This should result in the formation of an
organic top layer which is colonized mainly by arthropods,
whereas the mineral soil can still be characterized by earthworm activity. The biological depth gradient implies a chemical
depth gradient, that is the organic top layer is more acid than
the mineral soil. The climatical limitations to decomposers
should thus lower the resilience of the ecosystem – forest ecosystems in higher altitudes or farer north are more easily
subjected to plastic strain by climatic or man-made stress.

Even in ecosystems with the soil staying in the silicate buffer
range, there is some loss of cations (mainly Ca^{2+}), accompanied
by HCO_3^-. This leaching from soil is responsible for the salt
content of soft groundwater. In order to keep the ecosystem
stable, this loss has to be balanced by silicate weathering.
A long-lasting discoupling of the ion cycle is to be expected
if, in the course of natural ecosystem development, the
weatherable silicates in the root zone are exhausted. If the
output (leaching) of Ca and Mg becomes larger than the input
(from atmosphere and by weathering), the ecosystem passes over
into a nonstationary transition state which is characterized by
decreasing stores of exchangeable Ca and Mg. The soil acidifies,
the chemical soil state passes over from the silicate into the
cation exchange buffer range. The time span needed for this
process can be estimated from the buffer capacity and the buffer

rate in the silicate buffer range of the soil. With the values
given elsewhere (ULRICH, this volume), the buffer capacity
amounts to 750 keq ha^{-1}m^{-1} per 1 % silicate content in the soil.
With the buffer rate of 0.4 kmol H$^+$ ha^{-1}yr^{-1} for 1 m soil depth,
found by MAZZARINO (this volume) for a soil of the last interglaciation period (soil "Dasburg"), a silicate content of 1 % would
last for around 2000 years. If this buffer rate is accepted as
typical for the natural development of untouched ecosystems on
comparable soils in an interglaciation period, it means that
soils with silicate contents above 5 % would still be in the
silicate buffer range or swinging between silicate and cation
exchange buffer range, provided that there was no acidifying
influence by man's activities.

CLIMATIC ACIDIFICATION PUSHES, PHASE OF HUMUS DISINTEGRATION
(DESTABILIZATION PHASE I)

Soils in the silicate buffer range without exchangeable Al at
clay surfaces buffer acidification pushes by H/Ca exchange back
and forth without any drastic change in chemical soil state;
even the pH value may remain almost stable. The higher the
percentage of exchangeable Al in the soil, the more pH can drop
during an acidification push and recover during deacidification
phase. Such seasonal pH variations have been recorded for forest
ecosystems (14), they are typical for soil horizons swinging
between silicate and cation exchange buffer range.

Especially in the higher and cooler regions of the subalpine
mountains, soil temperature in the deeper rooting zone may be
so far from the optimum of mineralization that in cool, humid
years residues from root decomposition rich in N may be accumulated. The increase in soil temperature in and following warm,
dry years can result in a strong increase in the mineralization
rate as soon as the soil layer is rewetted again. Such climate
fluctuations are typical for Central and North Europe. The
strong and deep reeching acidification pushes in or following
warm, dry years are called climatic acidification pushes. They
are thus separated from the seasonal acidification pushes which
are weaker and restricted closer to soil surface. It is of
fundamental ecological importance that warm, dry years are a
heavy load to ecosystems not only due to water stress and dryness,
but also due to climatic acidification pushes. The lower the
exchangeable Ca+Mg percentage, the more often toxic cation acids
may be formed during an acidification push and cause injury to
decomposers (bacteria) and roots.
A very long lasting discoupling of the ion cycle should result
in the loss of the organic matter which was accumulated in the
whole rooting zone, including the subsoil, during the aggradation
phase and which was kept on a constant level during the

stability range I. This type of discoupling is called humus
disintegration (3,15). The ecosystem is in a non-stationary
transition state which is necessarily limited in time. The time
period under question may be decades to centuries, however.
Under natural conditions this process seems to be bound to
strong climate changes which leads to the destruction of the
forest ecosystem. Examples are the changes from interglacial to
glacial periods. The destruction of the forest ecosystem by man
may trigger the same process.
Humus disintegration is the consequence of the discoupling
between formation and breakdown of stable soil organic matter (3).
Like with lignin decomposition (20), the rate of breakdown of
stable soil organic matter should increase with decreasing pH
and reach a maximum at pH 4.0 to 4.5. The formation of stable
soil organic matter is thought to be connected with the autolysis
of bacterial cultures (16, biological humification). This process
may happen during gut passage in soil animals, in faeces etc.
During the autolysis, phenolic rings can be formed and polymerize
to water insoluble humic acids. The polymerization of the
phenolic rings should be influenced by the presence of cation
acids. If cation acids like Al ions are present in solution or
at clay surfaces, they can form organic complexes of a relatively
low degree of polymerization, which are stable against autoxi-
dation. Polymerization is than stopped. Thus, the newly formed
organic matter is easily dispersed or water soluble. It will
therefore being decomposed at a higher rate. This means that the
rate of formation of stable soil organic matter is decreased, at
the same time the rate of breakdown is increased. The consequence
is a loss of soil organic matter.
As an example we consider the same soil as before. We assume
that the total soil organic matter is mineralized, the nitrogen
being nitrified. Denitrification is assumed to be zero. In such
a case, there would be a proton consumption by mineralization
of the Ca saturated acidic groups amounting to 200 kmol/ha. On
the other hand, nitrification produces 360 kmol H^+/ha. The
balance of both partial processes yields a proton production of
160 kmol H^+/ha.
If one assumes that this process takes place within 100 years,
the mean annual rate would be 1.6 kmol H^+/ha. This exceeds for
many soils the possible buffer rate in the silicate buffer range,
thus increasing soil acidity. Existing data indicate that in the
early stages of humus disintegration N rich compounds are
mineralized preferably. This leads to an increase in the C/N
ratio of the remaining soil organic matter. During each acidification push, the NH_4/NO_3 ratio in ion uptake will be shifted to
a high nitrate surplus. This will tend to limit the uptake rate
of NO_3, thus prolonging the period of discoupling. In a system
not being close to steady state, the chemical reactions in soil
can occur far from equilibrium. Such soils can, for example, be
more acid (iron buffer range) than their final state would be

(aluminium buffer range, as long as clay minerals are present).

Sustained damage to primary producers is also caused by man.
Clear cutting, grazing and shifting agriculture, may result in
sustained reduced primary production and, even more, in sustained
reduction of understory root and overstory leaf etc. litter.
It further changes the microclimate of the soil, that is the
climatical conditions for the decomposers. Both effects, the
reduction in litter production and the increase in soil temperature, operate in the direction of lowering organic matter stores
in the mineral soil. They may thus initiate humus disintegration.
The process can be started by the acidity produced (alcalinity
not returned) by the discoupling of the ion cycle due to biomass
utilization.

ACIDIFICATION EFFECTS OF BIOMASS UTILIZATION

In table 2 the annual uptake rates and ion balances are given
for different tree compartments of the same stand as shown in
table 1. The tree compartments correspond to different degrees
of phytomass utilization: bole (wood + bark), bole + branches,
total overstory, total overstory and understory. The cation/
anion balance shows that as long as only the woody parts of the
overstory are used and exported from the ecosystem, the soil
acidification is in a range (0.6 kmol H^+ $ha^{-1}yr^{-1}$) which can
be buffered by "better" soils by base cation release during
silicate weathering. If the nitrogen rich parts are continuously
exported from the ecosystem (e.g. by litter utilization, which
played a role during the last centuries), soil acidification
increases strongly. Even the utilization of total overstory and
understory represents a realistic picture of some forest areas
in Central Europe during the last centuries. All forest soils,
where such utilizations have been practisized, have been
strongly acidified when modern forestry started. The consequences
of this acidification will be discussed later (destabilization
phase II).
One should assume that all forest ecosystems in Central Europe
have passed after the last glaciation period through the aggradation phase and reached the stability range I. With the exception
of parent material very low in silicate content, and of shallow
soils, most forest ecosystems should still be in stability range
I. This is not the case. Most forest subsoils are in the cation
exchange or aluminium buffer range, they show no activity of
soil burrowing animals, and the roots are inhomogeneously
distributed. These soils must be assumed to have passed through
the phase of humus disintegration centuries or millenniums ago.
It must further be assumed that in most cases this process has
been initiated by biomass utilization through man.

Table 2:   Annual rates of ion uptake in various tree
           compartments (the same stand as in table 1)

	bole	bole + branches	bole + branches + leaves	total overstory + understory
	------	keq per ha	and year	--------------
cation sum	0.69	0.83	2.44	4.34
anion sum	0.11	0.16	0.59	0.83
cation/anion balance	+0.58	+0.67	+1.85	+3.51

INITIATION OF HUMUS DISINTEGRATION BY ACID DEPOSITION

It seems that in Central Europe all forest ecosystems which have
not been subjected to humus disintegration before, have now
switched over into this phase. There is no actual reason for
this than acid deposition. As discussed later, acid deposition
buffered at the leaf surface is transferred to the soil close
to the roots via the regulation of ion uptake (cation/anion
balance). Acid deposition is therefore especially suited for
acidifying the soil in the deeper rooting zone. After a reduction
of exchangeable Ca, this may in combination with a climatic
acidification push initiate humus disintegration and prevent the
process to be stopped in a deacidification phase.

In the beginning stages of humus disintegration, the ecosystem
resembles the stable ecosystem in the silicate buffer range with
the exception of low pH values especially in the deeper rooting
zone: the humusform can be still mull, the rooting deep reaching
and being homogeneous, the soil structure crumb-like. Due to the
continuous surplus of nitrification compared to nitrate uptake,
tree growth may be excellent and in the shrub and herb layer
plants indicating high nitrogen supply appear. The Al ions
released from clay minerals are bound to soil organic matter;
this reduces their toxicity. The feature typical for the process
running in the deeper rooting zone is the leaching of nitrate in
combination with Ca and Mg. The leaching of nitrate indicates the
continuous net nitrification (i.e. formation of HNO_3) in the
rooting zone. The leaching of Ca and Mg indicates the acidification, i.e. the loss of basic cations and their replacement by
Mn and Al ions (i.e. cation acids). The process of humus disintegration can be stopped at any point, if the rate of proton load
(HNO_3 formed exceeding HNO_3 uptake + acid deposition) becomes
smaller than the rate of proton consumption by base cation
release during silicate weathering. All external influences
(climatic effects, changes in plant cover, biomass utilization
etc.) which lower nitrification rate and increase rate of nitrate
uptake will lower the rate of proton production and thus tend to

stop the process. That this is reached under the influence of
acid deposition is very improbable, however. Also liming may
be ineffective, if not the earthworm activity is high enough to
transport substantial amounts of buffering substances down to
the subsoil.

Due to its long duration and the excellent growth of Al tolerant
tree species, the process is very difficult to recognize and
nobody is aware of its dramatic end: the change of the chemical
soil state into the aluminium and iron buffer range with all of
its consequences for plant growth and decomposer activity. Many
stands showing fir die-back seem to be subjected to humus
disintegration (17). The same seems to be the case with some
forest stands in the subalpine mountains. If in the final phase
the root system becomes shallow, soils may become compacted on
the long run with the possible consequences of erosion and
waterlogging.

At the present stage of knowledge, the best measure of humus
disintegration seems to be the monitoring of net losses (surplus
of output above input) of nitrate and base cations from the
rooting zone. To avoid changes in the seepage due to denitrification, the measurements have to be done immediately below the
rooting zone by means of suction lysimeters for collection of
the percolating soil solution (18,19).

MAZZARINO (this volume) gives for soils developed after the last
glaciation proton consumption rates by base cation release
during silicate weathering between 0.6 and 1.1 keq $ha^{-1}yr^{-1}$ for
around 1 m soil depth. The soils Spanbeck 4 and Hof 3 (each 1.1
kmol H^+) have passed through the humus disintegration phase,
the soil West 1 (0.6 kmol H^+) only partially. These data indicate
that during humus disintegration, the rate of silicate weathering
may be increased substantially (more than doubled).

DESTABILIZATION PHASE II: BUILDUP OF A DECOMPOSER REFUGE

Soils which have passed through the phase of humus disintegration
more or less completely and are staying in the cation exchange
or aluminium buffer range, have lost the soil burrowing animals
and tend therefore to carry through litter decomposition in a
top organic layer separated from mineral soil. The lower part of
top organic layer may be rooted. In this case not only leaf but
also root litter contributes to its formation. The accumulation
of a top organic layer means in its essence the buildup of a
decomposer refuge after the mineral soil has become toxic by the
presence of cation acids. It is accompanied by the loss of
ground vegetation and started by the retardation of leaf litter
decomposition. As a consequence, a fermentation layer (OF horizon)
is forming (humusform: F-mull). If the OF increases and roots
are stretching between OF and A horizon, the accumulation of a
well decomposed OH horizon starts (humusform: mull-like moder).

Most dead roots seem to provide a compartment for decomposers
which is free of toxins (like cation acids or water soluble
phenols). Root decomposition leads therefore usually to the formation of highly polymerized humin material. The accumulation
can be stopped at any stage, provided that the conditions for
decomposers allow to carry through decomposition and mineralization at the same rate as litter production (reaching of a
steady state). This is usually achieved after developing a full
OL-OF-OH profile (humusform: moder).
Almost all of the nitrogen accumulated in the top organic layer
stems from mineral soil organic N which has been transferred via
plant uptake and litter fall. Its transfer was not connected
with a net proton production or consumption. As in plant uptake,
the acidification caused by the accumulation of the organic
matter is given by its cation/anion balance. Since the anions P,
S and Cl can be neglected, the acidification is equivalent to the
cation accumulation. In the beginning stages of the process, the
cations accumulated are mainly Ca and Mg, and the relative amount
accumulated is low, resulting in cation contents of about
1 eq/kg C. If one assumes that 1/3 of the annual litter production
of 3000 kg dry matter corresponding to 1500 kg C, is withdrawn
from mineralization and being accumulated, the acidification rate
is 0.5 keq H^+ $ha^{-1}yr^{-1}$. These conditions apply to soils where
the A horizon stays in the cation exchange buffer range. As
discussed in detail by MATZNER et al. (this volume), the acidity
appears partly as soluble organic acids, which are leached from
the O to the A horizon. Another part can be due to NH_4^+ uptake.
Both processes become effective in the A horizon. Even a low
proton production rate as the one mentioned above is thus
generated in a very limited soil volume and will there exceed the
buffer rate due to base cation release in silicate weathering.
This means that the A horizon acidifies and may pass through the
cation buffer range into the aluminium and further into the iron
buffer range. This process continues as long as the accumulation
of the top organic layer continues. If the A horizon reaches the
Al buffer range, Al accumulation in the OH horizon starts; if
it reaches the Fe buffer range, podzolization including Fe
accumulation in the OH horizon starts. Parallel with the transfer
of Al and Fe ions from the A into the O horizon the cation
carrying capacity of the O horizon is increasing and may reach
values of 2 eq/kg C (table 3).
Podzolization can be seen at the soil profile by the appearance
of a greyish Ae horizon. During the last 3 decades, this has
become a widespread feature in forests of Central Europe. The
investigations in the Solling yielded for this stage between 1966
and 1979 mean annual rates of organic matter accumulation in the
top organic layer of 1200 kg (beech) to 1500 kg (spruce) per ha.
The net proton production has been calculated from the ion flux
balance and shows values varying between 1.2 and 3.2 (mean: 2.2)
kmol H^+ $ha^{-1}yr^{-1}$ for Fagus silvatica, and between 1.2 and 5.6

Table 3:   Cation equivalent sum and its composition in
           needles, litter, and organic top layer
           (Picea abies, Solling plot Fl, 1973)

	cation sum eq/kg organic d.m.	1/2 Ca	1/3 Fe	1/3 Al
		-----% of cation sum----		
needles		33	1.1	1.3
litter		46	8	9
OL horizon	0.49	18	28	32
OF horizon	0.88	11	44	37
OH horizon	1.0	9	39	46

(mean: 3.7) for Picea abies (MATZNER et al., this volume). One
can compare the amount of acidity produced during 30 years at a
rate of 3 kmol H^+ $ha^{-1}yr^{-1}$ with the buffer capacity in the cation
exchange buffer range. This comparison yields that the amount of
acidity is sufficient to leach the base cations from a 30 cm
thick soil layer containing 5 % clay, or a 7.5 cm thick soil
layer containing 20 % clay. The first example corresponds to
typical podzols on sandy soils, the second to the type of
podzolization as it is now widespread on loamy soils in Central
Europe.

The typical podzols are heath podzols which developed centuries
ago under the influence of extreme strong biomass utilization on
sandy soils. As the development of heath ecosystems shows (22,23),
podzolization is a cyclic process, an acidification phase is
followed by a deacidification phase. The acidification phase is
connected with the accumulation of the organic top layer. As a
consequence of the strong soil acidification, the plant cover
decays, the accumulated organic matter starts to decompose, and
the ecosystem passes over in the deacidification phase. In this
phase other plants can immigrate and repress the vegetation
which has been lead to podzolization. Such plants are very often
acid tolerant grasses which do not form woody roots and are thus
not subjected to accumulation of cation acids inside the plant
body. A succession may develop which leads back to aluminium or
even cation exchange buffer range, if weatherable silicates are
still available.

Also evenaged forest plantations can exert a strong tendency for
podzolization. The tendency is the higher, the less decomposable
the leaf litter is. It is known that phenols diminish the
decomposibility (21), it seems that the presence of this toxins
in the liquid phase is the deciding factor. There exists probably
a feedback mechanism: Al (cation acid) tolerant shrub and tree
species often produce water soluble phenols, which are able to
complex cation acids and are a tool to increase NH_4 uptake and
to decrease Al toxicity. On the other hand, these phenols may be
toxic to soil bacteria which are responsible for the chemical
breakdown of dead organic matter. Thus, adaption to NH_4 nutrition,
cation acid tolerance and low litter decomposability may often be

coupled. To reach a steady state between litter production and
mineralization may than require that the concentration of
phenols in microcompartments of the fermentation layer falls
below a treshhold level. If the canopy of an evenaged young
stand of spruce or pine closes, the heavily decomposable needle
litter becomes the only source for the decomposers. Soils in the
Al and Fe buffer range show a low tendency to oxidize and polymerize soluble phenols, in contrary they stabilize them by the
formation of metal-organic complexes with Al and Fe ions. Under
these conditions a strong decrease in needle litter mineralization can occur, which leads to the accumulation of an organic
top layer and a strong proton production. The adverse effects
of evenaged spruce plantations on soils were discussed in
Central Europe already in the 2nd and 3rd decade of this
century (24,25). The expectation that the growth of Norway
spruce will decline in the 2nd and 3rd tree generation did not
fulfill, however, in the 4th, 5th and 6th decade of this
century. It seems that the delay in the expected growth decline
has an external reason, being caused by the increase in deposition
of nutrients like N and Ca from air pollution.
The phase where this nutrient input stimulated tree growth in
Central Europe is now passed through, and the fate of the
forests is now determined to a great degree by the podzolization
caused by acid precipitation.
The example of heath ecosystems shows that actual podzolization
lasts only for a limited period of time. The development of
evenaged coniferous plantations growing on soils in Al and Fe
buffer range leads to the same conclusion. 20 to 30 years after
canopy closure the canopy is opened again by eliminating trees.
After having accumulated a decomposer refuge, the change in
microclimate caused by the canopy opening may be sufficient to
reach steady state between litter production and organic matter
mineralization, and the proton production stops. On soils in the
silicate or cation exchange buffer range, the elimination of a
limited number of trees may be enough to reach steady state. On
soils being deeply acidified to the Al and Fe buffer range, all
trees may be eliminated.
In past, the development of destabilization phase II has been
mainly triggered by biomass utilization. It has been the natural
end of the phase of humus disintegration. Modern forestry has
triggered destabilization phase II by planting even-aged dense
forests with tree species which posess heavily decomposable
litter. Modern forestry has such contributed to soil acidification in Central Europe. On the other hand, the theory
developed allows the conclusion that under the conditions of
modern forestry in Central Europe, podzolization should only
occur in even-aged dense coniferous forests. The widespread
existence of actual podzolization in almost all forests with an
organic top layer can therefore only be due to acid deposition.
As stated above there are reasons to assume that the rates of

proton production connected with podzolization can be endured
only for a few decades. All information available points to the
fact that species with long-lived roots like trees tolerate a
continuous production of cation acids in the rooted soil only
for a limited period of time. For many forests in Central Europe,
half of this time span may already be over.

STABILITY WITH LOW RESILIENCE: STABILITY RANGE II

As already stated, the destabilization phase II can be inter-
rupted at any stage. As soon as the mineralization rate approaches
the rate of litter production, proton production ceases and the
ecosystem can again reach a quasi-steady state. The chemical
soil state may then slowly pass back to the cation exchange
buffer range. The rate of this recovery process depends upon the
rate of proton consumption by silicate weathering (base cation
release): by this process cation acids are transferred to
uncharged compounds and base cations produced, so the percentage
of exchangeable Ca+Mg can slowly increase. Finally a steady state
may be achieved where the base cation percentage fluctuates
around a mean value corresponding to the effects of climatic
acidification pushes and deacidification phases.

This stability range is characterized by a much lower resilience
than stability range I with the soil staying in the silicate
buffer range. The missing of soil burrowing animals has the
consequence that within the soil compartments develop with
different chemical state: Between these compartments chemical
gradients exist. Such compartments are the organic top layer and
the mineral soil, the different horizons within these layers,
and the interior and surface of aggregates within these horizons.
In any of these compartments, the cation/anion balance is kept
at zero, and the H^+/OH^- balance must be kept close to zero, in
order to maintain steady state. Deviations from these requirements, caused by the processes of mineralization and uptake,
must be compensated by the transport of ions from one compartment to another. Such transports can be mediated by water flow,
by transport within roots, and by diffusion. The extention of
roots into the organic top layer may be of great advantage to
avoid spatial discoupling of proton production and proton
consumption, but makes the plants susceptible to drought periods.
Within the mineral soil, the roots are growing along the surface
of aggregates. The humic substances formed during root decomposition are therefore accumulating at the aggregate surfaces
close to the roots and are not mixed within the total soil mass.
This increases considerably the risk for cation acid (Al)
toxicity in case of a strong seasonal acidification push: The
nitric acid forms in a very small soil volume only which is close
to the roots. In this small soil volume the buffer capacity is

very limited. The frequency of the formation of toxic cation
acids injuring microorganisms and roots is in this stability
range II therefore much greater than in stability range I.

Soils with a high rate of base cation release by silicate
weathering may have been able to return during decades or centuries to the silicate buffer range. Suche ecosystems swing back
to the aggradation phase and may finally return into stability
range I. Only the presence of interlayer Al in the clay
minerals may than remember to the acidification phase which has
passed through before. For most soils which developed on sedimentary rocks, glacial deposits and wind deposits in Central
Europe, however, the recovery after ceasing of humus disintegration ended in stability range II. This applies also for soils
where humus disintegration ceased centuries ago, and where a
forest ecosystem with beech as almost only tree species
developed again (Luzulo-Fagetum). If the silicate buffer rate of
these soils was not sufficient to bring the chemical soil state
back to the silicate buffer range, it will also not be sufficient
to buffer the additional acid entering the ecosystem as acid
deposition. The present knowledge allows to state that all these
soils will acidify under the influence of acid deposition. These
forest ecosystems have already or will in near future switch
over again into the destabilization phase II. The morphological
feature are the greyish A horizons indicating podzolization.

THE SUPERIMPOSEMENT OF NATURAL STRESS AND STRAIN BY ACID
DEPOSITION

LEWITT (27) introduced the concept of stress and strain, developed in physics and applied e.g. in geology, into biology. The
action of a stress factor on a system results in a strain. This
strain may be elastic (reversible) or plastic (irreversible).
Plastic strain means that the system has changed persistent
some of its properties. Plastic strain can result in invisible
(latent) or in visible changes. Visible plastic strain is usually
called injury.
This concept helps in realizing possible interactions of various
stress factors. If the action of any stress factor results in
a plastic strain, this may alter the reaction of the system on
the influence of a second stress factor: the range of elastic
strain may become smaller, plastic strain may begin at a lower
treshhold.
Fig. 1 gives an overview about stress and strain in forest ecosystems. The strain is related to ecosystem processes, the
physiological reactions of the organisms behind are not
considered.
The climate is the driving force for the ecosystem. Its variability causes stress. The variability in the heat climate and

Fig. 1: Stress and strain in forest ecosystems -
        an overview

stress factor	causal relationship	plastic strain
CLIMATE		
heat climate		
warm	acidification push	root injury
cool	deacidification phase	recovery
humidity climate		
wet	O_2 deficiency in soil	root injury
dry	water deficiency in soil	root injury
mechanical climate		
no wind	-	-
storm, snow, ice	mechanical stress to roots and canopies	
chemical climate		
normal	low nutrient input	-
close to sea		
coast	NaCl salt damage	crown deformance
air pollution	manyfold	leaf injury
		soil acidification
		bark(cambium)injury
		crown deformance
		root injury
		damage to decom- posers: destabilization
CONSUMERS		
pests		
virosis		
bacteria		diseases
fungi		wood rot
insects		feeding damages
man		
biomass utili- zation	diminishing nutrient stocks changing microlimate diminishing feeding source for decom- posers	complex: destabilization

NATURAL FALLING OUT OF ELEMENTS OF THE ECOSYSTEM

death of indi- viduals	e.g. change in microclimate	non in stable ecosystems
deterioration of structural units		

in the humidity climate can cause root injury. One should assume that forest ecosystems and trees as their main components are well adapted to the stress caused by the mechanical components of the climate(wind, snow cover, ice) . Since man has changed the provenience and the species composition in most areas of Central Europe, elastic strain due to these stress factors may be reduced, and plastic strain may appear earlier, in man-made forests.

The only component of the climate which has been changed drastically during the last century is the chemical climate, its change is due to air pollution. In Central Europe the input of acidity and many chemical elements into forest ecosystems has been increased by a factor between 10 and 100. The input of nutrients like sulfur, nitrogen, calcium and magnesium has increased to such a degree that it may amount to more than half of the annual uptake of these nutrients by the trees (8).

Also pests can be looked at as stress factors. Plants being subjected to plastic strain by other stress factors can be expected to be attacked easier by pests. Damages at the outer surface of the plant (leaves, bark, roots) may allow pathogenic organisms to invade a tissue. Invisible plastic strain may influence the production of secondary plant substances which may control pathogenic organisms (28, 38, 39, 40). Changes in cell metabolism may attract insects.

Biomass utilization by man plays a role in Central Europe since 5000 years. In Germany it has lead to the destruction of forest ecosystems on large areas, to soil acidification, and it has brought into existence heath and grass ecosystems of low productivity (14). Restriction of biomass utilization in forests to timber by modern forestry has lead to the recovery of many forests. This recovery was supported by the deposition of nutrients from air pollution.

Also the natural falling out of elements of the ecosystem must be considered as a stress factor. An ecosystem can continue its existence only if the natural replacement of these elements (e.g. trees in the forest ecosystem) is possible. This would be the case if the strain caused by the falling out remains elastic.

From the concept of stress and strain it follows that we would need to know the whole stress situation in order to evaluate the effect of a special stress factor. This knowledge is not available since we do'nt know the invisible strain which air pollutants and other stress factors are causing in the ecosystem as a whole and in individual organisms. We must expect that the action of many stress factors on a low level of intensity at the same time (e.g. acidity and heavy metals) can limit the elastic strain for any other stress factor so that plastic strain occurs. It is to be expected that the visible plastic strain (injury) follows after a period of invisible plastic strain. WALLACE (29) has shown that various heavy metals behave additiv in causing

injury: concentrations below treshhold level may lead to injury
in case of various heavy metals being present.
If one accepts that all stress factors may interact, the problem
of finding the cause for an injury must be defined as the search
for the chain of effects; that is for the sequence of strains
caused by various stress factors. It seems that in forest ecosystems the effects start with damages to soil microorganisms,
that is to the decomposer chain, and to the short roots including
their mycorrhiza. To the plants, the damages of the decomposer
chain operate indirect by discoupling the ion cycle. This causes
imbalanced nutrient supply and the release of toxic cation acids
like Al ions which in turn may cause root injury.

Taking into consideration the many stress factors causing root
injury, one may look at root injury as a second stress and ask
for the strain it causes in the plant as a whole. During the
evolution, the chemical stress in the soil has forced to develop
adaption strategies which allow the plant to survive after a root
injury. Especially plants growing on acid soils seem to possess
a high capability to cut off damaged parts of the root system and
to renew those parts rapidly. It is obvious that the renewal is
bound to the condition that the stress causing the injury is
diminished to a level where it causes no plastic strain. Depending
upon soil conditions, the recovery time may therefore vary
considerably. The longer recovery needs, the stronger the strain
to the plant of a root injury should be. Injury caused by
toxicity of cation acids like Al ions should very much depend
upon the degree of soil acidification. The more acids are accumulated in soil, and the stronger these acids are, the longer a
toxic concentration level may persist in the soil solution after
an acidification push. Under the conditions of deposition of
heavy metals from air pollution, a further effect comes into play.
Under all soil conditions these heavy metals seem to be accumulated in the roots (30), probably in the cell wall space of the
long roots and of lignified roots. The accumulation of Al and
heavy metals in this tissue means the accumulation of potential
toxins close to living cell surfaces. If after the injury of
mycorrhiza and short roots an acid soil solution can penetrate
into these tissues, it can mobilize the Al and heavy metals
present and transfer them into toxic cation acids as result of
the buffering of protons. This will especially be the case if the
Ca content in the root is already low as in strongly acidified
soils (cf. ULRICH, this volume, BAUCH, this volume). It must be
expected therefore that the accumulation of Al and heavy metals
in roots increases the extent of root injury after it has started
as consequence of an acidification push, and it prolongs the time
necessary for recovery.
A plant with damaged roots may continue water uptake as long as
no air enters the vessels and causes embolism. Without doubt
plastic strain due to drought is increased. Therefore, if

drought is causing injury, one can nothing say about the primary
cause without knowing whether there have been root damages before.
If this is the case, than the primary cause of the drought damage
may be the stress factor which caused the root injury.

In a plant with damaged roots, part of the water entering the
xylem vessels may not have passed through living cells of the
endodermis. During this passage the ionic composition of the
transpiration stream is controlled. The avoidance of this passage
means that the nutrient composition of the transpiration stream
may become unbalanced and that even cation acids can become components of the transpiration stream. Probability and extent of
these effects increase with increasing soil acidification. As a
consequence different strains may develop: nutrient deficiency
(especially of Mg and B), increased leaching of nutrients from
leaves (strengthening nutrient deficiencies), reduced basicity
transported to leaves, disturbances in cell metabolism (invisible
plastic strain, reducing production of secondary plant substances,
thus lowering the resistance against pathogens). A further
possibility is that pathogens invade the plants through the
injured roots and are transported passively with the transpiration
stream inside the root system and further into other tissues.
This is one link in the chain which leads to fir dye-back (31,32,
33);the primary causes of the fir dye-back are therefore the soil
conditions which allow the climatic variations to cause root
injury and which prevent rapid recovery. If these soil conditions
are due to acid deposition, than acid deposition is the primary
cause.
The same must be expected to be applicable to root rotting fungi.
Since many wind-thrown trees show uncured root damages (34),
acid deposition can be the primary cause for forest damage by
wind-throw. This seems in fact to be the case for the heavy wind-
throw occuring in November 1972 which caused forest damage to more
than 100000 ha in Lower Saxony. The root damage was induced by
the acidification push following the warm-dry years 1967/69, but
due to soil acidification by acid deposition. The extinction of
the forests having root damage by this wind-throw may be the
reason that the forest damages which developed during the last
years (as consequence of the acidification push induced by the
warm-dry years 1975/76) have not shown up already in 1973 and
1974.
Tree leaves can buffer acidity deposited on the leaf by exchange
with Ca; it is assumed that Ca is exchangeable bound on acidic
groups in the cell wall and that the buffering occurs by H/Ca
exchange. During stomata closure this exchange may be reversed
forced by an input of $Ca(HCO_3)_2$; the carbonic acid formed may be
assimilated or transferred to H_2O and CO_2:

$$\text{buffering:} \quad \boxed{\begin{matrix}\text{cell}\\\text{wall}\end{matrix}} \begin{matrix}-O-\\-O-\end{matrix} Ca^{2+} + 2H^+ + SO_4^{2-} \rightarrow \boxed{\begin{matrix}\text{cell}\\\text{wall}\end{matrix}} \begin{matrix}-OH\\-OH\end{matrix} + Ca^{2+} + SO_4^{2-}$$
$$\text{leached}$$

recharging: $\boxed{\begin{array}{c}\text{cell}\\\text{wall}\end{array}}\!\!\begin{array}{c}-\text{OH}\\-\text{OH}\end{array} + Ca^{2+} + 2\ HCO_3^- \rightarrow \boxed{\begin{array}{c}\text{cell}\\\text{wall}\end{array}}\!\!\begin{array}{c}-O-\\-O-\end{array}Ca^{2+} + H_2O + CO_2$

transpiration
stream input

If the Ca is supplied from the exchangeable pool in the soil, the amount of basicity fed in the transpiration stream is balanced by an equivalent amount of acidity formed in the soil:

soil acidi-fication : $\boxed{\begin{array}{c}\text{soil ex-}\\\text{changer}\end{array}}\!Ca^{2+} + H_2CO_3 \rightarrow \boxed{\begin{array}{c}\text{soil ex-}\\\text{changer}\end{array}}\!\!\begin{array}{c}-H^+\\-H^+\end{array} + Ca^{2+} + 2HCO_3^-$

from root   input to trans-
and decom-   piration stream
poser respiration

Each living cell acts as a biochemical pH-stat. Within cells organic acids and their salts form a buffer system. To be effective outside cells, the effects of the biochemical reactions controlling pH must be transferred to the cell wall space (free space, apoplast) and through the xylem or phloem from one tissue to another. According to the pH range encountered, carbonic acid and its salts has to take over this task within the transpiration stream and the free space in roots and leaves connected with it.

Proton buffering occuring in leaves is thus immediately transferred to the soil. From a physico-chemical point of view it needs a pH around or above 5 and an appreciable amount of exchangeable Ca ions present at the acidic groups in the Free Space of the roots to charge the transpiration stream continuously with $Ca(HCO_3)_2$. Decreasing Ca content in the root indicates that the pH in the Free Space is dropping, the formation of $Ca(HCO_3)_2$ becomes limited, and the transport of basicity to the leaves is diminished. As shown elsewhere(ULRICH, this volume), the reduction of the Ca content in roots proceeds proportional to the reduction of exchangeable basic cations in the soil. This means that the pH-stat in root cells cannot maintain optimal conditions in the free space of the root. Low Ca contents and Ca/Al ratios in roots thus indicate invisible plastic strain. If after a root injury, acid soil solution containing cation acids flows directly into the transpiration stream, this acidity may balance the basicity charged in from intact roots more or less.

pH measurements in throughfall of forest canopies show that the pH in the water films covering leaf surfaces may drop to 2.75 in the monthly mean (35), which means that within episodes it may be substantial lower.

The reduction in the amount of basicity transported to the leaves should have several consequences:
-   pH values below 4 and especially below 3 may, depending upon
    their duration, lead to cuticula erosion (36). This may
    increase the leaching of nutrients and decrease the resistance

against drought (cuticular transpiration) and pathogens.
- There is convincing evidence (37) that cell injury by SO_2 is bound to the transport of physical dissolved SO_2 ($SO_2 \cdot H_2O$, H_2SO_3) through the cell membrane. The existence of physical dissolved SO_2 depends upon pH, the ratio of SO_{2aq} to HSO_3^- is 0.0 1 at pH 3.8 and 0.1 at pH 2.8. This means that with pH values approaching 3 within the cell wall space of the stomata, very low concentrations of SO_2 should already be enough to cause cell injury. Under conditions of low pH in throughfall (as indicator for pH in water films covering plant surfaces), SO_2 injury may occur at SO_2 concentrations of 20 $\mu g/m^3$ (WENTZEL, this volume). Direct SO_2 injury to leaves seems to be connected with many of the forest damages reported.
- As shown by MAYER (this volume), heavy metals from deposition are accumulated on leaves and bark. That means that there is a tendency for increasing concentrations of potential toxins during the life span of these plant organs. The solubility of the compounds accumulated increases with decreasing pH, at pH below 3 many of the heavy metals exist as ions (cation acids) in the solution. The longer the accumulation of these elements continues, and the weaker the buffering action of leaves and bark becomes, the greater is the probability that cation acids pass over into cells and influence cell metabolism. The lowest degree of strain should be connected with disturbances of the ion balance in the cell, especially the balance of micronutrients. This may result in disturbances of growth hormons. Such effects are in fact part of the damages which can be observed. The highest degree of strain is cell death and the formation of necrotic tissue. Also this seems to play a considerable role and may explain bark necrosis as well as necrosis at the twig base leading to premature abscission of green leaves and short twigs throughout the whole vegetation period, provided the plant surfaces are wetted. Both effects are very widespread and can be observed at many tree species. Any stage of increased leaf and twig losses and dyeback can be found in the forests, from the very beginning up to tree death. The symptoms are now occuring at trees of any age, provided they are exposed and can accumulate heavy metals from deposition. The acidity, whether direct or in form of cation acids, may also injure the aperture mechanism of the stomata and thus decrease drought resistance.

Many of the possible strains caused by root damage as stress factor are at present of hypothetical nature and need further detailed investigation. On the other hand, the possible strains mentioned rest on accepted theories and on observations and can at present not be falsified. They should therefore be accepted as building stones in a general hypothesis which includes all possibilities to explain the forest damage. Such a general hypothesis should be the guiding principle for further activities.

LITERATURE

1. Rosenquist, I.Th. (1977): Sur jord - surt vann.
   Ingenioerforlaget A/S, Oslo, 123 p.
2. Ulrich, B. (1980): In T.C. Hutchinson and M.Havas (eds.):
   Effects of Acid Precipitation on Terrestrial Ecosystems,
   Plenum Press New York, pp. 255-282.
3. Ulrich, B. (1981): Z.Pflanzenernähr. Bodenk. 144, 647-659
4. Driscoll, Ch.T. (1980): Chemical characterization of some
   dilute acidified lakes and streams in the Adirondack
   region of New York State. Ph.D.Thesis Cornell Univ.
5. Andersson, F.T., T. Fagerström and S.I. Nilsson (1980): see
   2., pp. 319-334
6. Sollins, P., C.C. Grier, F.M. McCarison, K.Cormack and
   R.Royel (1980): Ecological Monographs 50, 261-285
7. Bormann, F.H. and G.E. Likens (1979): Pattern and Process in
   a forested ecosystem. Springer Verlag New York
8. Ulrich, B., R. Mayer und P.K. Khanna (1979): Schriften Forstl.
   Fak.,Univ. Göttingen 58, 291 p., Sauerländer Verlag
   Frankfurt
9. Matzner, E. und B. Ulrich (1981): Z.Pflanzenernähr.Bodenk.
   144, 660-681
10. Nye, P.H. and P.B. Tinker (1977): Solute movement in the
    soil-root system. Blackwell Oxford
11. Straughan, I.R., A.A. Elseewi and A.L. Page (1981): An over-
    view of acid deposition and its impacts on soils. Paper
    given at the Conference on Heavy metals in the environ-
    ment, Amsterdam, Sept. 1981
12. Ulrich, B. (1981): Forstarchiv 52, 165-170
13. Alexander, M. (1961): Introduction to soil microbiology.
    Wiley New York
14. Ellenberg, H. (1978): Vegetation Mitteleuropas mit den Alpen.
    Ulmer Stuttgart
15. Ulrich, B. (1980): Forstwiss.Centralbl. 99,376-384
16. Scheffer, F. und B. Ulrich (1960): Morphologie, Biologie,
    Chemie und Dynamik des Humus. Enke Verlag Stuttgart
17. Ulrich, B. (1981): Forstwiss.Centralbl. 100, 228-296
18. Cole, D.W. (1958): Soil Sci.85, 293-296
19. Hetsch, W., F. Beese und B. Ulrich (1979): Z.Pflanzenernähr.
    Bodenk. 142, 69-85
20. Crawford, R.L. (1981): Lignin biodegradation and transfor-
    mation. Wiley New York
21. Swift, M.J., O.W. Heal and J.M. Anderson (1979): Decompo-
    sition in terrestrial ecosystems. Blackwell Oxford,
    chapter 4.3.3
22. Gimmingham, C.A. (1972): Ecology of heathlands. Chapmann and
    Hall London
23. Matzner, E. and B. Ulrich (1980): Z. Pflanzenernähr.Bodenk.
    143, 666-678

24. Wiedemann, E. (1923): Zuwachsrückgang und Wuchsstockungen
       der Fichte. Laux Tharandt
25. Kraus, G., F. Müller, G. Gärtner und H. Schanz (1939):
       Thar.Forstl.Jahrb. 20, 483-709
26. Ulrich, B., R. Mayer and P.K. Khanna (1980): Soil Sci. 130,
       193-199
27. Levitt, J. (1980): Responses of plants to environmental
       stresses. II. Water, radiation, salt and other stresses.
       Academic Press New York
28. Lunderstädt, J. (1980): Z.Pflanzenernähr.Bodenk. 143,412-421
29. Wallace, A. (1982): Soil Sci. 133, 319-323
30. Mayer, R. und H. Heinrichs (1981): Z.Pflanzenernähr.Bodenk.
       144, 637-646
31. Schütt, P. (1981): Forstwiss.Centralbl. 100, 174-179
32. Blaschke, H. (1981): Forstwiss.Centralbl. 100, 190-195
33. Brill, H., E. Bock und J. Bauch (1981): Forstwiss.Centralbl.
       100, 195-206
34. Bazzigher, G. und P.Schmidt (1969): Schweiz.Zeitschr.f.
       Forstwesen 120, Nr. 10
35. Matzner, E., P.K. Khanna, K.J. Meiwes, M.Lindheim, J.Prenzel
       und B. Ulrich (1982): Elementflüsse in Waldökosystemen im
       Solling - Datendokumentation. Göttinger Bodenkundl.Ber.71,
       1-267
36. Jacobsen, J.S. (1980): see 2., pp. 151-160
37. Spedding, D.J., I.Ziegler, R.Hampp and H.Ziegler (1980):
       Z.Pflanzenphysiol. 96 351-364
38. Cates, R.G. (1980): Oecologia 46, pp. 22-31
39. Haukioja, E. (1980): Oikos 35, pp. 202-213
40. Schoonhoven, L.M. (1982): Ent.exp. & appl. 31, pp. 57-69

Topic 1:

Processes and Rates of Deposition, Storage Places of Deposited Air Pollutants

INTERACTION OF FOREST CANOPIES WITH ATMOSPHERIC CONSTITUENTS:
SO_2, ALKALI AND EARTH ALKALI CATIONS AND CHLORIDE

Bernhard Ulrich

Institut für Bodenkunde und Waldernährung der
Universität, Büsgenweg 2, D3400 Göttingen

ABSTRACT

From the view point of the receiving ecosystem, a distinction is
made between precipitation deposition and interception deposition. Precipitation deposition occurs by gravity and is the
sum of deposition with rain and snow and as dust particles; the
receiving surface has no influence on the deposition rate.
Interception deposition is the sum of impaction of aerosols and
droplets (fog, cloud droplets), and of gas absorption; the
receiving surface influences the deposition rate greatly by its
size, kind and chemistry. An approach is presented to calculate
the rate of interception deposition on forest canopies and to
account for the various processes. The approach rests on the
measurement of precipitation (wet) deposition and throughfall.
Flux balance equations are used to account for sink and source
terms of the canopy. The approach is demonstrated with data from
a beech and a spruce forest covering the period from 1969 to
1981. The data indicate that the SO_2 absorption in the canopy
(dry deposition) is limited by the ability of the trees to buffer
the protons produced during SO_2 absorption in water films at
inner and outer plant surfaces. Under conditions of high soil
acidity and low tree vitality, the buffer ability of the trees
can cease. This limits dry deposition of SO_2 to very low rates.

TYPES OF AIR POLLUTANTS

Most of the elements supplied from the air to forest ecosystems
have a natural source which may be outweighted by anthropogenic
sources. The following grouping is relevant in Central Europe

33

B. Ulrich and J. Pankrath (eds.), Effects of Accumulation of Air Pollutants in Forest Ecosystems, 33–45.

which is under strong influence of many anthropogenic sources
with contributions from soil dust and sea salt:
- neutral salts: e.g. Na^+, Cl^- (from sea salt)
- nutrients: Mg^{2+}, Ca^{2+}, NH_4^+, NO_3^-, SO_4^{2-}, micronutrients
- acid formers: SO_2, NO_x, Cl_2
- potential toxins: SO_2, HF, heavy metals, As, Se, hydrocarbons
- oxidant formers: NO_x, gaseous organic compounds.
This list is by far not complete. To understand the reaction of
the forest ecosystem on air pollution, it is necessary to consider (and therefore: to measure) all air pollutants from each
of these groups, which can be expected to exert effects and
where the deposition rates are substantially higher than
natural background.

DEPOSITION PROCESSES

The following deposition processes can be distinguished:
- wet deposition by rain or snow (A)
- sedimentation (gravimetric fall) of particles (B)
- impaction of aerosols (C)
- impaction of mist, fog, cloud droplets (C)
- absorption of gases (example: SO_2) (D)
  - on wet surfaces like foliage, bark, wet snow
  - inside stomata
    - on cell walls
    - on to mesophyll and palisade cell surfaces
- reemission
Wet deposition of rain or snow is not influenced quantitatively
by the receiving surface. Deposition of particles (sedimentation,
impaction) is strongly dependent on particle size (deposition
increases substantially for particles larger than approximately
2 μm diameter and also depends on wind speed and the structure
of the receiving surface). Fog, mist, cloud droplet could be
regarded as large particles and are very efficiently captured by
a canopy making the rate of deposition dependent mostly on the
wind speed. Deposition of gases varies with kind and state of
the receiving surface.
The deposition can be looked at from two different points of
view: either from the depositing compound or from the receiving
surface. From the point of view of depositing compounds wet and
dry deposition are distinguished. From the point of view of the
receiving surface it makes sense to distinguish between precipitation deposition and interception deposition. These terms
are shortly defined in the following comparison:

Process	view point of depositing compound	view point of receiving surface
A precipitation of rain or snow "particles" with dissolved (soluble) or undissolved (unsoluble) content	wet deposition	precipitation deposition
B precipitation of particles other than rain or snow according to gravity	dry	precipitation deposition
C impaction of aerosols including fog and cloud droplets according to air or Brownian movement[+]	deposition	interception deposition
D dissolution of gases on wet surfaces (with subsequent chemical reactions)	deposition	interception deposition

[+] the impaction of fog and cloud droplets is sometimes grouped under wet deposition.

It is evident that the processes summarized under the term interception deposition depend upon size, kind and (chemical) state of the receiving surface. This has important consequences regarding the measuring techniques.
After deposition chemical reactions can occur. The following reaction types can be distinguished:
- acid production (e.g. SO$_2$)
- proton buffering (e.g. cation exchange on cell walls)
- precipitation or dissolution, depending on pH (e.g. heavy metals)
- reactions with living plant cells (reactions with cell membranes, uptake into cells, assimilation in the cell metabolism)
- formation of gaseous compounds (e.g. sulfides)
After deposition a translocation could also occur within the canopy or by leaching by rain from the outside as well as from the inside of leaves and needles. Much of the deposited material will by this process be transferred from the canopy to the ground.

MEASURING APPROACHES

For the measurements of dry deposition of gases and particles, several approaches exist in principle but with regard to a forest canopy the majority of estimates rely on 1) theoretical estimates based on laboratory studies, 2) field measurements with chamber technique or 3) estimates based on throughfall measurements, in some cases supplemented with 4) watershed budgets.
With regard to deposition processes involved no artificial

collector surfaces can be expected to give realistic values of
dry (interception) deposition especially for the uptake of gases.
The surface of the unchanged receiving system seems to be the
only acceptable sampling device, which can be assumed to give
realistic dry (interception) deposition values. This includes
even the unchanged metabolism of plant canopies, since the dry
deposition of gases depend upon the chemical reactions following
absorption (installing infinite sinks).
Dry deposition of aerosols is usually estimated from aerosol
composition and concentration (preferably for several size
classes) coupled with the typical deposition rate for each size
particle interval.
The importance of dry deposition (for instance as compared with
wet deposition) varies very significantly between areas. Deposition on wet canopy may be very rapid in the simultaneous
presence of SO_2 and ammonia while it may be unimportant in a
remote area where the throughfall is quite acid. Areas where fog
or low clouds are frequent may receive large additional input by
the interception of droplets.
Measurements based on throughfall measurements use the living
forest canopy with its variation in surface and metabolism as
receiving surface. In this paper the approach of MAYER and ULRICH
(1) taking a forest canopy as measuring device for determining
interception deposition is further developed. The approach
is based on the flux balance of the forest canopy or of the
whole ecosystem (MAYER, this volume), depending upon the mobility
of the air pollutant considered. This approach is in principle
applicable to any air pollutant, including organic micropollutants. In the flux balance equation used, the function of the
canopy or the whole vegetation cover to act as sink or source
for the air pollutant considered has to be expressed quantitatively as annual rate.

FOREST CANOPIES AS MEASURING DEVICES FOR INTERCEPTION DEPOSITION

A forest canopy can be looked at as a compartment which is connected with its environment by inputs and outputs.

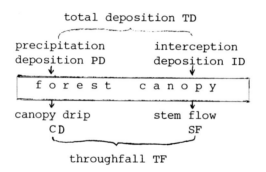

The input, total deposition TD, consists of precipitation deposition PD and interception deposition ID, where PD can be
measured with simple precipitation samplers and chemical analysis. The output occurs as canopy drip and stem flow, which
yield together the throughfall TF. Also TF can be measured by
collecting the precipitation beneath the canopy. The forest
canopy may itself exert a sink or source function Q. If X denotes
a distinct chemical compound or ion and if annual rates are considered, the following balance equation applies:

$$TF_X = PD_X + ID_X + Q_X \tag{1}$$

If PD_X and TF_X are measured, knowledge of Q_X is necessary in
order to solve equation 1 for the interception deposition of
X, ID_X.
The canopy can act as a source in the following cases:
- BA: leaching of ions from senescent leaves mainly in autumn
       $X \in \{Na, Mg, Ca, Cl, SO_4\}$
- KA: cation exchange in the leaf tissue (see also below)
       $X \in \{Ca, Mg\}$
- PA: leaching of ions throughout the growing season due to
       metabolic processes
       $X \in \{K, Mn\}$
-      dissolution of undissolved matter in deposited particles
       $X \in \{Al, heavy metals\}$
For the last case see MAYER, this volume. The other cases
reflect ecosystem internal cycling processes which have to be
subtracted from TF in order not to overestimate TD.

The canopy can act as a sink in the following cases:
-      assimilation, i.e. uptake into cell metabolism
       $X = NH_4^+$. At high rates of sulfur deposition the fraction
       of S assimilated can be neglected
- KA: cation exchange in the leaf tissue (see also above,
       exchange of H$^+$ against Ca and Mg)
       $X = H^+$
-      storage of particles
       $X \in \{Al, heavy metals\}$
-      precipitation of dissolved ions
       $X \in \{Al, heavy metals\}$
For the elements considered in this paper only the exchange of
protons has to be taken into account.

FLOW RATES OF IONS IN PRECIPITATION AND THROUGHFALL

The approach has been applied to two forest ecosystems, a 120
yrs. old beech (Fagus silvatica) stand and a 85 yrs. old Norway
spruce (Picea abies) stand. Both stands lie close together in
the Solling mountains, 9°30' east of Greenwich and 51°40' north,

on a plateau 500 m above sea level. The site is described in
detail by ELLENBERG (5).
The mean monthly flow rates between 1969 and 1979 in precipitation and throughfall of water and the ions H, Na, K, NH4, Mg,
Ca, Al, Mn, Fe, Cl, NO3, SO4 and organic bounds N are published
elsewhere (3,4,6,7).

LEACHING OF IONS FROM SENESCENT LEAVES

Inspection of the monthly flow rates in the beech stand shows
that there is an increase in the flow rates of some ions in
September and October compared to the earlier part of the vegetation period from May to August. This increase is attributed to
the leaching of ions from senescent leaves. For the ions Na, Mg,
Ca, Fe, Cl, NO3 and SO4, this increase amounts to 2 to 7 % of
the flow rate in throughfall (3). This means that the leaching
of ions contributes only a small percentage to the flow in
throughfall. Nevertheless this increase was subtracted from the
annual value of the transport rate in throughfall before further
data evaluation, thus correcting for the source term BA.

INTERCEPTION DEPOSITION OF Na, Cl AND S

After correcting for the leaching of ions from senescent leaves,
no other source or sink terms exist for Na, Cl and SO4. Thus the
interception deposition can be calculated from equation 1 by
setting Q_X = BA:

$$ID_X = TF_X - PD_X - BA_X \qquad X \in \{Na, Cl, S\} \qquad (2)$$

In table 1 the values of the various flow rates are presented
as mean ($\bar{x}$, $\tilde{x}$) of the years 1969 to 1981 together with their
variance (s_x, range R) between years. This calculation rests
on the fact that the rate in throughfall is strongly increased
outside the vegetation period in winter (for data see 3), when
especially for the leafless beech stand leaching from the canopy
cannot occur. This fact as well as the common knowledge about
ion uptake allows to set PA, the leaching of ions throughout
the growing season due to metabolic processes, for these ions
to zero.
On the base of the data presented, two possibilities exist to
estimate interception deposition of particulate sulfur (sulfate).
One can either assume that the ID of Na (and Cl) reflects the
ability of the canopy to capture aerosols and droplets. This
assumption leads to

$$ID_X^{spruce} = (ID_{Na}^{spruce}/ID_{Na}^{beech}) \cdot ID_X^{beech} \qquad (3)$$

Table 1: Calculation of interception deposition of Na, Cl and S
          Mean values 1969 to 1981

		---- beech ---- fluxes in kg			---- spruce ---- per ha and year		
		Na	Cl	S	Na	Cl	S
precipitation deposition PD	x̄	7.85	16.9	23.6	7.85	16.9	23.6
	s$_x$	2.16	4.7	2.0			
throughfall TF	x̄	14.3	33.1	52.0	17.3	39.0	87.4
	s$_x$	2.6	5.6	7.2	4.4	8.3	12.5
leaching from senescent leaves BA	x̃	0.8	2.0	1.4	–	–	–
	R	0 -3.3	0.6-6.5	0 -5.1			
interception deposition ID	x̄	5.7	13.8	25.9	8.88	18.9	61.7
	s$_x$	1.4	2.8	5.3	3.15	7.4	7.9
ID/PD	x̃	0.8	0.9	1.1	1.0	1.1	2.6
	R	0.4-1.2	0.5-1.3	0.8-1.4	0.4-1.9	0.3-2.1	1.7-3.3
ID$_{spruce}$/ID$_{beech}$	x̃				1.6	1.3	2.3
	R		spruce/beech		0.7-2.3	0.6-1.9	1.7-3.6

With this assumption, the ID of particulate sulfate in the
spruce stand is calculated on the base of the ratio of the ID
of Na (or NaCl) between spruce and beech, and on the base of the
ID of sulfur in the beech stand. This assumption underlies the
data presented earlier (3). Inherent to this assumption is the
neglection of the ID of gaseous sulfur (SO_2) in the beech stand
which is certainly not true.
Another assumption which can be made is that the ratio of ID to
PD for particulate sulfate is the same as for Na. It is obvious
that the ID of Na, which originates from sea spray, occurs in
particulate form.With this assumption, the following equation
may be used for calculation of the interception deposition of
particulate sulfur:

$$ID_{S, particulate} = (ID/PD)_{Na} \cdot PD_S \qquad (4)$$

mean values of the ratios used in equation 3 and 4 are given in
table 1. The result of the calculation using equation 4 is given
in table 2.

Table 2:   Median values and range of interception depo-
           sition of particulate and gaseous sulfur

		beech	spruce
		kg S ha^{-1}	yr^{-1}
Interception deposition (total)	$\tilde{x}$	25	61
	R	17-33	44-82
Interception of particulate sulfur (SO_4^{2-})	$\tilde{x}$	17	25
	R	10-24	9-38
Interception of gaseous sulfur (SO_2)	$\tilde{x}$	5	32
	R	0-16	20-46

INTERCEPTION DEPOSITION OF H, Mg and Ca

Whereas Mg and Ca are, like Na, deposited only in aerosols or
droplets (particulate), protons may either be deposited in
particulate form or they are formed as a consequence of SO_2
absorption. NO_x can be neglected since its ID is comparatively
low. For the calculation of the ID of particulate H, Mg and Ca,
the reasoning in the last section can be applied and equations
3 or 4 can be used. As with sulfur, the data published earlier
(3,4) have been calculated using equation 3. In the following,
equation 4 is used with X $\in$ {H, Ca, Mg} .
Part of the acid produced after absorption of SO_2 can be buffered
by exchange with Ca and to a lesser degree Mg ions bound on
acidic groups of the cell wall. This occurs probably mainly
inside the stomata. After stomata closure the plant can restore

exchangeable Ca by the influx of a solution of higher pH with the transpiration stream (ULRICH, this volume). The exchanged cations diffuse out of the stomata and are leached from the canopy with the throughfall. This process is listed in the previous sections as cation exchange. Its rate can be calculated using the following balance equation:

$$KA_X = TD_X - (PD + BA + ID_{particulate})_X \qquad X \in \{Ca, Mg\} \quad (5)$$

The results of these calculations are presented in fig. 1 and table 3 together with the data for interception deposition of sulfur compounds.
The ID of particulate sulfate is probably mainly due to the interception of fog and cloud droplets. The situation where this happens exists more often during the leafless period, where the intercepting surface of the beech is much smaller compared to spruce. This may explain the higher values in spruce. Nevertheless, there is a positive correlation between the beech and spruce data (r = 0.71). The ID of gaseous SO_2 depends also on weather conditions (wet surfaces, stomata opening) and shows therefore a similar positive correlation between beech and spruce (r = 0.71). It depends further upon the pH in the solution according to the following equations:

$$SO_2 + H_2O = H^+ + HSO_3^-$$

$$\frac{(H^+)(HSO_3^-)}{(SO_{2aq}, H_2SO_3)} = 10^{-1.81} \qquad (6)$$

and $\qquad HSO_3^- + \frac{1}{2}O_2 = HSO_4^- \qquad (7)$

At pH 4.8 the molar ratio of HSO_3^- to dissolved SO_2 (SO_{2aq} and H_2SO_3) is 1000, at pH 3.8 it is 100. The oxidation of SO_2 (HSO_3^-) to SO_3 (HSO_4^-) occurs with a rate of some percent per hour (6). If pH is kept high (pH above $\sim$ 4.5) by some buffering reaction and the oxidation continues, the surface behaves as an infinite sink: the absorption of SO_2 can continue with a steady rate as long as the surface is wet and accessible for gas exchange. Plants growing on soils with pH > 5 (soils in the carbonate or silicate buffer range) behave this way. They get enough basicity with the transpiration stream to keep the pH at wet leaf surfaces and in the stomata above $\sim$ 4.5. This is also true for beech trees on lime stone (7). pH values approaching 4 or less indicate that the solution acts not as an infinite sink, but the rate of absorption of SO_2 is kinetically limited, e.g. by the rate of proton buffering.
In the throughfall the absorbed SO_2 appears as sulfate ion SO_4^{2-}, this must be accompanied by either protons or, after buffer reactions, by Ca and Mg cations. The measured (H^+) and calculated

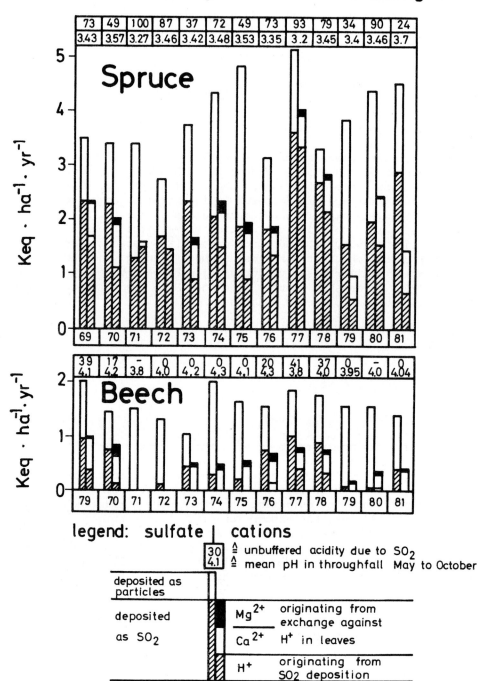

Fig.1 Interception deposition of S and H
cation exchange in leaves of Ca and Mg

Table 3:   Annual mean deposition data of H, Mg, Ca, and S for the period 1969 to 1981

		beech				spruce			
		H	Mg	Ca	S	H	Mg	Ca	S
		fluxes in keq per ha and year							
precipitation deposition PD	$\bar{x}$	0.81	0.15	0.51	1.47	0.81	0.15	0.51	1.47
	s_x	0.14	0.06	0.22	0.12				0.78
throughfall TD	$\bar{x}$	1.40	0.33	1.23	3.24	3.21	0.40	1.65	5.45
	s_x	0.20	0.04	0.19	0.45	0.62	0.08	0.25	0.78
interception deposition in particles (eq.4) ID particulate	$\bar{x}$	0.62	0.13	0.38	1.14	0.95	0.18	0.56	1.72
	s_x	0.20	0.07	0.20	0.37	0.38	0.11	0.22	0.66
interception deposition as gas ID gas	$\bar{x}$	(0.46)	–	–	0.46	(2.18)	–	–	2.18
	s_x	(0.37)			0.37	(0.62)			0.62
Q=TD-PD-ID particulate (eq. 5)	$\bar{x}$	-0.05	0.05	0.37	–	1.4	0.08	0.6	–
	R	-0.6–0.4	(-0.3) -0.18	(-0.3) -0.6		0.5–3.4	(-0.3) -0.2	(-0.2) -0.9	

$Q_H \hat{=}$ unbuffered acidity due to SO_2 absorption

$Q_{Ca,Mg} \hat{=}$ cation exchange in cell walls (H/Ca, Mg)

values (H, Ca, Mg: equation 5), respectively, are plotted in
fig. 1 on the right hand side of the columns representing SO_2
absorption. The numbers on top of the columns represent (upper
row) the fraction of H^+ at the ID of SO_2 (=unbuffered acidity
due to SO_2 absorption), and (lower row) the pH in throughfall
(weighted mean for the period May to October).

The data presented in fig. 1 and table 3 show a great difference
between beech and spruce in dealing with acid deposition. The
beech buffers most or all of the protons produced by SO_2 absorption by release of Ca and Mg (cation exchange). The unbuffered
acidity does'nt exceed 40 % (see fig. 1). If this buffering
becomes impossible, the SO_2 sorption ceases (1971, 1972, 1979).
This is connected with low pH values in throughfall (see fig.1).
It is also probably connected with root damage which followed
the acidification pushes in the warm dry years 1967/69 and
1975/76. The decrease in buffering capability of the trees
showed up with some delay (1971 and 1979, respectively). In 1977
and 1978, the pH in throughfall was low, but there was still
considerable buffer capacity. This may indicate that the trees
possess also a storage of basicity which delays the effects of
root injury, but is exhausted within a few years.

The behaviour of spruce is different. In most years the acidity
produced by SO_2 absorption was buffered only to a small extent.
The acidity was transferred to the throughfall which reached
quite low pH values during the vegetation period (see fig. 1).
As in beech, the sum of H+Mg+Ca balances or exceeds the dry
deposition of SO_2 in most years. There is a marked exception in
1981 (to a lesser degree also in 1973 and 1979). This exception
may indicate that not only Ca and Mg are participating in
buffering, but also other cations like K. The buffering can occur
by transferring $KHCO_3$ into the free space of the stomata. From
there K^+ is leached together with SO_4^{2-}, and H_2CO_3 ($H_2O + CO_2$)
is formed. After the warm dry periods mentioned above (1967/69
and 1975/76), the stand shedded needles in excess of needle
formation between 1970 and 1973 and again from 1977 to 1979.
Another feature showing the limitations in buffer ability of the
canopy are the minimum pH values which have been measured in
February 1969, January 1977, November 1978, and October 1979
(monthly mean values below 2.85). According to equation 6, this
indicates a high fraction of HSO_3^- existing in the solution. It
is not known how the metabolism of the needles can withstand
such high acidities as measured in the throughfall. It may be
that the SO_2 which is not buffered is absorbed not inside the
stomata, but at the outer leaf (needle) surfaces. During autumn
and winter, depending upon the weather conditions, the spruce
canopy can be wet continuously for weeks and months. Thus, the
water films covering needles, and a wet snow cover, may be the
main accumulators of SO_2. The SO_2 buffered inside the stomata

may be equivalent to the basicity consumed (Ca + Mg in fig. 1, plus an unknown fraction of bicarbonates in some years).

LITERATURE

1. Mayer, R. and B. Ulrich (1974): Ecol.Plant. 9, pp. 157-168
2. Ulrich, B., R. Mayer and P.K. Khanna (1979): Z.Pflanzenernähr.
    Bodenk. 142, pp. 601-615
3. Ulrich, B., R. Mayer und P.K. Khanna (1979): Deposition von
    Luftverunreinigungen und ihre Auswirkungen in Waldöko-
    systemen im Solling. Sauerländer Verlag Frankfurt
4. Matzner, E. et al. (1982): Elementflüsse in Waldökosystemen
    im Solling - Datendokumentation. Göttinger Bodenkdl.Ber.
    71, 267 p.
5. Ellenberg, H. (ed) (1971): Integrated Experimental Ecology.
    Ecol. Studies (Springer Verlag) 2, pp. 9-15
6. Granat, L., H. Rodhe and R.O. Hallberg (1976): The global
    sulfur cycle. Ecol.Bull.(Stockholm) 22, pp. 89-134
7. Ulrich, B., U. Steinhardt und A. Müller-Suur (1978):
    Göttinger Bodenkdl.Ber. 29, 133-192

INTERACTION OF FOREST CANOPIES WITH ATMOSPHERIC CONSTITUENTS:
ALUMINUM AND HEAVY METALS

Robert Mayer

Professur für Landschaftsökologie, FB 13 (Stadtplanung/
Landschaftsplanung), Gesamthochschule Kassel,
Postfach 101380, D-3500 Kassel

ABSTRACT

A method for the assessment of wet and dry deposition rates for
aluminum and heavy metals from flux balance measurements is
described. It is shown, that dry deposition may exceed wet
deposition with precipitation to forests in rural areas. Dry
deposition rate clearly depends upon surface type in a given
area. With the aid of atmospheric input rates the gains and
losses of metals in various compartments of the forest ecosystem
(canopy, humus layer, mineral soil) can be calculated. All
metals investigated except Al, Zn and Cd tend to accumulate
strongly in the humus layer while most metals except Pb are
mobilized and become lost from mineral soil. All metals under
investigation were accumulated in the forest canopy, some of
them in large quantities (Pb, Cu, Ni, Cr). The ecological consequences of the accumulation of toxic metals in the ecosystem
are discussed.

Dry deposition of easily soluble atmospheric constituents to
forested surfaces such as NaCl, SO_2, or H_2SO_4, can be measured
in precipitation below the canopy, thus using the tree canopy
itself as "natural" collector instead of an inadequate "artifical"
collector (Ulrich et al. 1979). This method is not applicable in
the case of aluminum and the heavy metals. These elements remain
partly or totally undissolved when coming in as solid particles.
When deposited as gases, or dissolved in aerosol droplets or
rain, they tend to be retained partly by the canopy, on leaf

B. Ulrich and J. Pankrath (eds.), Effects of Accumulation of Air Pollutants in Forest Ecosystems, 47–55.
Copyright © 1983 by D. Reidel Publishing Company.

surfaces and on the bark of stems and branches. These circumstances require a different way of estimating total and dry
deposition.

The main sources for aluminum and heavy metals in the atmosphere are:

1. Soil dust (Al, Mn, Fe, Co, Ni, Cu), in which the elements
   are present in solid form. Particle size is up to 1o μm.
2. Emissions by combustion processes (oil, coal, refuse).
   These emissions contain Cr, Mn, Ni, Cu, Zn, Cd, and Pb.
   Emission is predominantly in the form of very fine particles (around 1 μm), sometimes in gaseous form (Ni, Pb,
   Cd). These fine particles or gases are often adsorbed on
   larger particles which then control deposition.

Being bound in fine particles (aerosols) or gases, the metals
are subject in the atmosphere to transport over large distances
in the order of hundreds of km.

During transport and deposition the metals may be dissolved
partly or totally in liquid droplets. This is mainly true for
Zn and Cd. A corresponding fraction of the metal load of precipitation is, therefore, present in a dissolved form when precipitation reaches the canopy surface.

Dissolution may continue inside the canopy layer, and may
include substances which have previously been deposited in the
form of dry deposition. The process of dissolution is strongly
pH dependent and, therefore, controlled by wet and dry deposition of SO_2 and strong acids. On the other hand, dissolved
and undissolved metals may be adsorbed, bound chemically, or
taken up by leaves and needles and the bark of stems and
branches. It follows that only part of the substances deposited
to the canopy are reaching soil with precipitation. Total deposition is the sum of the element load of precipitation reaching
the ground below the canopy plus the rate of uptake (or
retention) of elements by the canopy. In case an element is taken
up by the roots, translocated to the canopy and subsequently
leached from the leaves or needles by rain, it also adds to the
element load of precipitation below the canopy and has to be
taken into account as an internal turnover to be distinguished
from atmospheric input.

For some elements it is now possible to estimate the rate of
uptake (retention) and of internal turnover, integrated over a
longer period of time, such that total deposition and, from the
flux balance, dry deposition can be calculated for that same
period of time. Since individual metal elements differ considerably
in their chemical behavior and their physiological function, the
following considerations have to treat the elements individually.
Lead is given as an example, the data given and the reasoning
for estimating unknown parameters are valid for beech and spruce
stands on acid soils in the Solling (Mayer 1981).

Measured and calculated fluxes of lead in a beech and a
spruce forest ecosystem of the Solling are given in Table 1.

Measured fluxes (in mg $Pb.m^{-2}.a^{-1}$) are:
- Bulk precipitation (dissolved fraction) measured in the open field or above the canopy layer
- Precipitation measured below the canopy
- Litterfall
- Storage of Pb in the annual increment of wood an bark
- Output with seepage water below the rooting zone of the soil

Table 1: Fluxes of Pb in a beech and a spruce forest ecosystem of the Solling (annual average 1974-1978) from Mayer (1981
(m) measured fluxes

		Beech	Spruce
		\multicolumn{2}{c}{mg $Pb.m^{-2}.a^{-1}$}
	Total deposition from atmosphere	44	73
(m)	Bulk precipitation (dissolved fraction) measured in the open field or above the canopy layer	29	29
	Leaching (internal turnover)	O	O
	Dry deposition plus wet deposition, undissolved fraction	15	44
	Dissolution of solid particles from wet or dry deposition inside the canopy layer	1	18
(m)	Precipitation below the canopy	30	47
(m)	Litterfall	12	26
(m)	Storage in the annual increment of wood and bark	5	8
	Total uptake/retention by the vegetation	17	33
	- uptake by roots	3	7
	- uptake/retention by canopy from atmospheric deposition	14	26
(m)	Output with seepage water below root zone	2	1
	Input-output-balance for total ecosystem (= annual gain)	42	72
	Input-output-balance for soil (= annual gain)	36	64

Total uptake or retention of lead by the vegetation is calculated as the sum of litterfall and storage in the annual increment. This figure of total uptake has to be considered more closely:

When vegetation is taking up Pb via roots, this uptake must be supplied by the solution phase of the soil. Concentration of dissolved Pb in soil solution has been measured as well as the amount of solution taken up by vegetation (transpiration rate). The product of transpiration rate and Pb-concentration in solution gives the amount of Pb taken up by vegetation due to mass flow. The total uptake, as given in Table 1, is exceeding the massflow thus calculated considerably (factor 5). This could only be explained by a selective uptake of Pb which would give rise to a strong concentration gradient between root and free solution and a diffusion transport to the root. Similar conditions are known from plant nutrients which are deficient (P,K) but not from elements that do not have any nutrient function, or are even toxic for plants like Pb. This kind of elements is rather excluded from uptake by roots, i.e. less is taken up than is reaching root surface by mass flow.

For a conservative estimate it is assumed that Pb is taken up proportional to its concentration in soil solution, that it is neither excluded nor taken up selectively by the roots. By this assumption, only about 20 % of total uptake is covered. The remaining 80 % of Pb uptake must be convered by deposition of atmospheric Pb onto the canopy.

The same reasoning may be used to exclude a considerable leaching of Pb from the canopy. When such a large amount of Pb is taken up or retained from atmospheric deposition by the canopy, any leaching seems very unlikely or at least negligible compared to the external fluxes.

Based on these assumptions, the remaining, unknown flow rates of Pb can be calculated: Dissolution of solid particles inside the canopy is equal to the difference between precipitation below canopy and bulk precipitation (dissolved fraction); dry deposition plus wet deposition (undissolved fraction) is equal to the sum uptake/retention by canopy and dissolution of solid particles; total deposition is the sum of bulk precipitation plus dry deposition.

It has been pointed out that the rate of root uptake is probably overestimated in the data presented in Table 1. All experimental evidence shows that Pb is, to a large extend, excluded from root uptake. If this is true, the calculated rate of deposition increases, given no root uptake at all, from 44 to 47 and for 73 to 80 mg $Pb.m^{-2}.a^{-1}$ in the beech and the spruce stand, respectively.

Dry deposition of Pb, as presented in Table 1, amounts to 35 % of total deposition in the case of beech, and to 60 % of total deposition in the case of spruce. This difference is due to the larger filter efficiency of the spruce canopy, especially during the winter months, compared to the deciduous beech stand.

Lindberg et al. (1979) found dry deposition making up 55 % of total deposition of Pb in a deciduous forest in Tennessee.

Dry deposition as well as total deposition of Cr, Ni and Cu can be estimated in a similar way as shown for Pb. In the case of Cd the calculation of the mass flow to the roots indicates a discrimination of the element in root uptake. Here the maximum value for root uptake of Cd can be calculated with the assumption, that no internal Cd is leached from the leaves or needles by rain. There is experimental evidence that this assumption is reasonable. The maximum value then equals the total uptake of Cd (measured flow rate) minus the amount retained in the canopy from precipitation or dry deposition. As minimum value, zero uptake of Cd via roots is assumed. The difference between the minimum and maximum rate of dry deposition calculated in the described way is very small compared to wet deposition. Similar arguments lead to an estimate of dry and total deposition rates of Co.

For estimating dry and total deposition of Zn and Mn, a different approach has to be chosen. Both elements are relatively mobile within the plant, therefore a stronger uptake by roots and leaching can not be excluded. Therefore dry deposition of these elements is calculated from a comparison of deposition rates during the vegetation period and during the dormant season of the forest vegetation (Ulrich et al. 1979). The same method has been used to estimate dry deposition of alkali and earth-alkali metals (see contribution by Ulrich, same volume).

Similar to Zn and Mn, the root uptake of Fe and Al can not be neglected from physiological reasons, also leaching could possibly occur. On the other hand, under the pH regime of precipitation in the canopy layer, formation of metal oxides/hydroxides or assimilation of metals by the leaf surfaces can lead to a retention of Al and Fe. The soluble fraction of these metals in aerosol and precipitation droplets in produced by neutralization of strong acids, mainly H_2SO_4. Therefore, deposition of soluble Al and Fe must be proportional to the deposition of sulfur (as SO_2 or H_2SO_4). Since the ratio between dry and wet deposition of S is known, the same ratio can be used to calculate dry deposition of Al and Fe from their wet deposition rates.

Table 2 gives mean annual rates of dry deposition (figures include wet deposition, undissolved fraction), wet deposition (dissolved fraction) and deposition with precipitation below the canopy for a beech and a spruce forest in the Solling, calculated or measured as described. It can be seen that wet deposition is the prevailing process in the case of Co, Zn, and Cd, while dry deposition is dominating for Cr, Mn, Ni. The third column shows that precipitation below canopy carries large amounts of non-atmospheric Mn under spruce (from leaching), and smaller amounts of non-atmospheric Zn under beech and spruce. On the other hand, large quantities of Cr, Co, Ni, Cu and Pb from atmospheric deposition are retained in the canopy layer and are not reaching

the ground together with precipitation. Most of these quantities
are reaching soil together with litterfall, a small part remains
in the non-cycling part of vegetation (stems and branches).
Concentration measurements of heavy metals in different
fractions of forest vegetation in fact show increased values of
these elements in surface parts (bark) and in long-exposed parts
(old needles) compared to younger needles).

In addition to the deposition rates given in Table 2, the
remaining metal fluxes coupled with litterfall and soil water
flow have been measured in two forest ecosystems in the Solling
(Mayer 1981). Soil water fluxes including ecosystem was collected
by use of suction lysimeters installed below the humus layer of
the forest soil and in the mineral soil below the root zone of
the trees. Litterfall was collected in litter traps. The rate of
change of metal stores in the biomass (equal to total metal
uptake via roots and canopy) was calculated from measured metal
concentrations in biomass compartments and biomass increments.
In Table 3 the flux balances for different compartments of the
forest ecosystems are shown. The accumulation of metals from
atmospheric input in the forest canopy is calculated as difference between total uptake minus uptake via roots. For litterfall,
as for total biomass, the contribution of root uptake and surface
uptake (from atmosphere) can be estimated. Subtraction of the
surface uptake fraction of litterfall from total uptake of
atmospheric deposition by the forest stand yields the accumulation of atmospheric metals in bark and wood. Changes of metal
stores in the humus layer and in mineral soil are calculated by
balancing metal fluxes coupled with litterfall, soil water flow
and root uptake. Negative values indicate a net loss of metal in
the respective compartment, positive values indicate a net gain.

It can be seen from these data that up to 80 % of the metals
coming in from the atmosphere are retained in the canopy and
30 to 50 % of Cr, Co, Ni, and Cu remain in the non-cycling parts
of biomass (bark and wood). Under prevailing conditions a one
hundred years old stand would have accumulated 6 to 8 kg of
Cr, 5 to 6 kg of Ni, 25 kg of Cu, and 4 to 6 kg of Pb.

Another characteristic feature of the element balance of the
forest ecosystems is the accumulation of most metals in the
humus layer, while the mineral soil shows losses of most metals
with the exception of Pb.

The ecological consequences of atmospheric deposition of
metals and their retention in the ecosystems can not be judged
finally based on the present knowledge. It is though obvious,
that the accumulation of highly toxic elements like Cu or Pb
in large amounts in a relatively labile pool must be considered
as hazardous. Any decrease of pH in precipitation water, for
example, will likely be followed by a mobilization of metals
in the canopy and cause damage to the plant itself and to the
organisms living in the phyllosphere or on the forest floor.
Changes in the treatment of the forest stand (as, e.g.,clearcut)

Table 2: Mean annual deposition rates in a beech (B) and a
          spruce (F) forest ecosystem in the Solling (1974-1978)

          Rates are given in $g.ha^{-1}.a^{-1}$ (= 10^{-1} $mg.m^{-2}.a^{-1}$)

Element		Wet deposition (dissolved)	in % of total depos. from atmosph.	Dry deposition (plus wet dep. undissolved)	in % of total depos. from atmosph.	Deposition to soil below forest canopy	in % of total depos. from atmosph.
Al	B	1080	50 %	1080	50 %	1750	81 %
	F		38 %	1760	62 %	3010	106 %
Cr	B	14	9 %	135	91 %	16	11 %
	F		8 %	151	92 %	23	14 %
Mn	B	260	15 %	1490	85 %	3840	219 %
	F		5 %	4890	95 %	5190	101 %
Fe	B	880	50 %	890	50 %	1400	79 %
	F		42 %	1240	58 %	2190	103 %
Co	B	14	90-100 %	0.-2.2	0-10 %	8	50 %
	F		70-100 %	0.-5.8	0-30 %	12	60-80 %
Ni	B	27	22 %	96	78 %	33	27 %
	F		19	113	81 %	39	28 %
Cu	B	236	50 %	234	50 %	162	34 %
	F		36 %	423	64 %	227	34 %
Zn	B	1377	84 %	255	16 %	2169	133 %
	F		80 %	355	20 %	2121	122 %
Cd	B	16	100 %	0.-0.4	0 %	13	79 %
	F		80 %	4.2	20 %	20	101 %
Pb	B	285	65 %	152	35 %	120	27 %
	F		39 %	448	61 %	256	35 %

Table 3: Input, output, and gains and losses of Aluminum and heavy metals in a beech (B) and a spruce (F) forest ecosystem in the Solling (1974–1978). Rates are given in $g.ha^{-1}.a^{-1}$ ($= 10^{-1}$ $mg.m^{-2}.a^{-1}$)

		Al	Cr	Fe	Co	Ni	Cu	Zn	Cd	Pb
Total input by wet and dry deposition	B	2160	149	1770	16	123	470	1630	16	440
	F	2840	165	2120	20	140	660	1730	20	730
Accumulation from atmospheric deposition										
– in forest canopy	B	600	120	450	8	90	310	130	3	135
	F	0	130	10	8	100	430	180	0	270
– in bark and wood	B	300	80	430	5	60	250	100	1	40
	F	0	60	2	4	50	240	140	0	60
Gains and losses in humus layer	B	-2700	40	1200	2	44	60	400	-3	110
	F	-2400	80	8900	6	75	310	-880	2	550
Gains and losses in mineral soil	B	-12000	1	-1800	-56	-30	-8	-3	1	260
	F	-21000	1	-8300	-400	-80	-60	2	-11	100
Output with seepage below roots	B	15500	7	210	64	21	106	1130	17	24
	F	24000	6	160	415	66	110	2360	26	13

may lead to a rapid mineralization of the metal stores bound in the humus layer, thus presenting a hazard for the groundwater. It is obvious that the strain imposed on the ecosystems under consideration as well as on linked ecosystems by deposition of atmospheric pollutants can only be estimated by assessment of a material balance. Conditions of rapid accumulation of toxic metals, as observed in the Solling forests, clearly demand a drastic reduction of emission of this type of pollutants at all possible sources.

LITERATURE

Lindberg, S.E., R.C. Harriss, R.R. Turner, D.S. Shriner and D.D. Huff (1979): Mechanisms and Rates of Atmospheric Deposition of Selected Trace Elements and Sulfate to a Deciduous Forest Watershed. Oak Ridge Ntl. Lab., Environ.Sci.Div.Publ. No. 1299, 514 p.

Mayer, R. (1981): Natürliche und anthropogene Komponenten des Schwermetall-Haushalts von Waldökosystemen. Habil.Schr. Univ.Göttingen, 292 S.

Ulrich, B., R. Mayer und P.K. Khanna (1979): Deposition von Luftverunreinigungen und ihre Auswirkungen in Waldökosystemen des Sollings. Schriften Forstl.Fak.Univ.Göttingen u. Nds. Forstl.Vers.Anst., Bd. 58, 291 S.

INPUT OF ACIDIFIERS AND HEAVY METALS TO A GERMAN FOREST AREA DUE
TO DRY AND WET DEPOSITION

Klaus D. Höfken

Institut für Atmosphärische Chemie, KFA Jülich,
Postfach 1913, D-5170 Jülich

ABSTRACT

Concentrations and deposition of the soluble ions Cl^-, NO_3^-, $SO_4^=$,
NH_4^+ and the heavy metals Mn, Fe, Cd, Pb were measured without
and within a beech and a spruce forest during several months of
1980. The amounts of dry deposition fluxes to the canopies are
compared with wet deposition and flux measurements of other pub-
lications. Filtering of aerosol constituents by the canopies is
evaluated and related to the size of the particles. Deposition
velocities of the aerosol compounds onto the canopies are given.
They characterize forests as a considerable sink for atmospheric
particulate trace constituents.

INTRODUCTION

The uptake of dry and wet deposition by plants and soil is considered to be one of the most critical interactions of atmosphere
and forest ecosystems (KNABE, 1982; SMITH, 1980; OVERREIN et al.,
1981; ULRICH, 1982; ULRICH et al., 1979.

On regard to the deposition of air pollutants onto forest ecosystems, it has been customary to measure accumulations of trace
substances and their effects within the forest ecosystem in order
to calculate the flux of these substances out of the atmosphere
into the forest (ABRAHAMSEN et al., 1976; KNABE, 1982; MAYER &
ULRICH, 1981; STEUBING & KLEE, 1970). This contribution, however,
presents values of fluxes and deposition obtained in a more
direct way. The effectivity of a forest as a sink for a certain
atmospheric trace constituent is seen very clearly, when the trace

57

B. Ulrich and J. Pankrath (eds.), Effects of Accumulation of Air Pollutants in Forest Ecosystems, 57–64.
Copyright © 1983 by D. Reidel Publishing Company.

constituent is measured in the atmosphere. The concentration of
this constituent is influenced by the forest canopy by filtering
(i.e. dry deposition). In the case of some gases, especially SO_2,
some investigations already exist (DOLLARD, 1980; GARLAND, 1976;
GOTAAS, 1977; MARTIN & BARBER, 1971). But information concerning
filtering of particulate heavy metals and acid ions ($SO_4^=$, NO_3^-,
Cl^-,...) is relatively scarce. Many qualitative investigations
and a recent publication (LINDBERG & HARRISS, 1981) show the
importance of deposition of atmospheric aerosol upon forest
canopies.

METHODS

The measurement sites are located in a rural area of the Solling
forest that consists of extensive woodlands approx. 500 m above
sea level. Measurements were carried out in a spruce forest (95
years, 25-30 m high) and a beech forest (130 years, 25-30 m high).
For further details see ELLENBERG (1971), MAYER (1981).

To determine both dry and wet deposition it was necessary to obtain aerosol as well as precipitation samples above and beneath
the forest canopies. During the year 1980 aerosol samples were
taken by air flitration and impactor measurements simultaneously
above, beneath, and outside the forest. Collecting surfaces were
membrane filters of Teflon (Millipore, Fluoropore 0.2 um) for the
soluble ions and cellulosenitrate (Sartorius, SM 0.45 μm) for the
heavy metals respectively. The impactors divided the aerosol into
size fractions of >1.7, 1.7-0.9, 0.9-0.4, 0.4-0.2, <0.2 μm radius.
Precipitation was collected with polyethylene funnels above, outside and at 8 places each in the beech and the spruce forest. The
samples were analyzed for the soluble ions NH_4^+, Cl^-, NO_3^-, $SO_4^=$
(colorimetry, ion chromatography) and the heavy metals Mn, Fe, Cd,
Pb (atomic absorption). A more detailed description of sampling
and analysis procedures is given elsewhere (HÖFKEN et al., 1981).

RESULTS

Table 1 gives the weighted means of ions and heavy metals in precipitation obtained above the forest canopy during the summer
months of 1980. The results are in the lower range of values found
by GEORGII et al. (1981), NÜRNBERG et al. (1981) for the precipitation measurement networks over Germany.

Beneath the canopies of the forests the concentration of each
constituent was higher (tab. 1). Concentrations beneath spruce
trees were found to be even higher than beneath the beech canopy.
The relative enrichment ranges up to a factor of 41 for the case
of Mn. The simplest explanation of this enrichment is that the
precipitation washes off particles that were deposited earlier on
the canopy. Because leaching of ions out of the plants themselves

could also cause concentration enrichments, in situ leaching experiments were also carried out (HÖFKEN et al., 1981). These experiments showed a considerable amount of Mn - and to a lower
extent Cl^-, NH_4^+ - to be leached out of the branches by rain.

Taking these results into account, and estimating the amount of
deposited material absorbed on the branches that is not washed
off (using litter analyses), dry deposition onto the canopies can
be calculated. Table 2 shows the values for the dry deposition
rates that were calculated by a balance of the flux rates:

$$D = T - W + A - L$$

D: dry deposition
T: throughfall
W: wet deposition (tab. 2)
A: absorption
L: leaching

Dry deposition was higher for the spruce trees than for the beech
(tab. 2). In winter, when the beech branches were leafless, the
difference between spruce and beech was even higher. This is probably due to the large difference in canopy structures and surfaces.
Most of the ions and elements show higher dry deposition than wet
deposition values in the case of the coniferous trees (fig. 2).
The ratio of wet deposition to dry deposition varies and is dependent on the season in the case of the deciduous trees (tab. 2).

These results are in good agreement with the findings of
HEINRICHS & MAYER (1980) and ULRICH et al. (1979), who took precipitation samples over a period of several years in the same
forest sites. These are shown in figure 2.

There is a significant amount of aerosol deposition onto the canopies. This can be clearly seen from the aerosol analyses (tab. 3).
The particles are filtered out so that beneath the canopies there
are concentrations of only about 60 % of those measured outside
the forests (tab. 3). Figure 1 shows that the filtering of particles is dependent on their size, with the lowest filtering efficiency for particles of about 1 µm mass-mean-diameter (fig. 1).
This could be the major explanation of the different enrichment
of trace substances in the precipitation throughfall (fig. 2).

Finally, deposition velocities can be calculated for the aerosol
constituents using deposition rates (tab. 2) and above forest
aerosol concentrations, tab. 3 (we have taken into account that
some constituents have a gas-phase). Table 4 gives an example.
These values are relatively high when compared to those measured
for different deposition surfaces like grass and soil. This
characterizes the forest canopy as a very effective sink for par-

	NH_4^+	Cl^-	Pb	Fe	Cd	NO_3^-	$SO_4^=$	M_n
above (µg/l)	1010	970	15	115	0.4	2830	4100	22
beech*	1.7	1.5	1.6	2.5	3.8	1.8	2.9	26.5
spruce*	3.3	3.9	3.9	6.7	6.7	7.4	13.4	41.2

* ratio of the concentration in the throughfall to the above-forest concentration

Tab. 1: Concentrations in precipitation (May-Oct 1980)
        (after Höfken & Gravenhorst, 1981)

components	wet deposition		dry depo sition rate			
			beech		spruce	
	W	S	W	S	W	S
M_n	2	2	<8	<6	<9	<18
Fe	13	10	11	10	26	23
Cd	0.04	0.03	0.09	0.06	0.12	0.08
Pb	1.3	1.2	1.1	2.3	4.3	3.3
NH_4^+	130	84	61	28	120	61
Cl^-	160	80	6	13	90	67
NO_3^-	360	230	97	90	1200	670
$SO_4^{3=}$	660	340	350	440	2600	1400

(W: Feb.-Apr. 1980, S: May-Oct. 1980)

Tab. 2: Deposition rates in mg m^{-2} month^{-1}
        (after Höfken & Gravenhorst, 1981

	concentration above forest ng/m³ (100 %)	concentration beneath the canopy (in %)	
		beech	spruce
Cd	1.3	66	59
$SO_{4}^{=}$	8400	71	68
NH_4^+	3000	73	72
Pb	100	71	67
Mn	24	75	77
Cl^-	130	66	62
Fe	380	63	61
NO_3^-	1600	63	60

(summer (May - Oct.) 1980)

Tab. 3: Aerosol concentrations
        (after Gravenhorst & Höfken, 1981)

Cd	$SO_4^=$	NH_4^+	Pb	M_n	Cl^-	Fe	NO_3^-
1.8	1.1	1.0	0.9	0.7	1.0	1.0	1.3

(summer 1980)

Tab. 4: Deposition veloci ties for beech (cm/s)

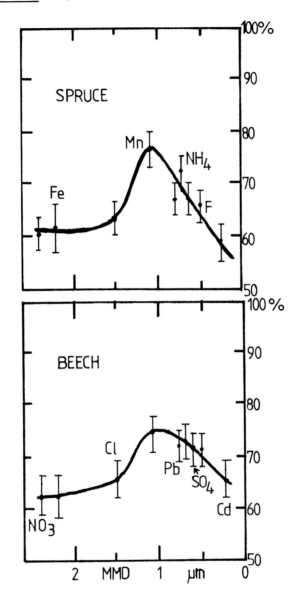

Fig. 1: Relative aerosol concentration within the forests

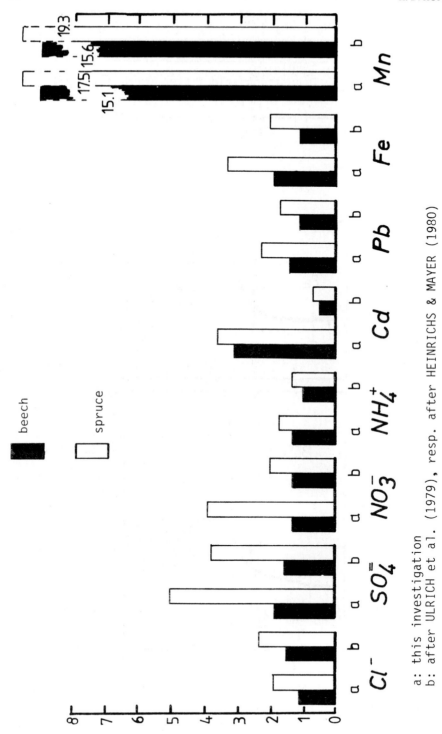

a: this investigation
b: after ULRICH et al. (1979), resp. after HEINRICHS & MAYER (1980)

Fig. 2: Ratio of throughfall flux to wet deposition flux

ticulate atmospheric trace constituents.

ACKNOWLEDGEMENT

This work was mainly carried out in the Institute for Meteorology and Geophysics, Univ. Frankfurt, as a part of a research project financed by the German Federal Environmental Agency, principal investigator Prof. H.W. GEORGII. The author wishes to thank all members of the institute as well as Dr. G. GRAVENHORST (Grenoble), Prof. R. MAYER (Kassel) and Prof. B. ULRICH (Göttingen), Mrs. G. AHEIMER, Mr. H. FRANKEN, Mr. K.P. MÜLLER (Jülich) for their helpful support.

REFERENCES

Abrahamsen, G., Bjor, K., Horntvedt, R., Tveite, B.: Effects of Acid Precipitation on Coniferous Forest, in: Braekke, F.H. (ed.): Impact of Acid Precipitation on Forest and Freshwater ecosystems in Norway, Research Report FRG/76, SNSF, Oslo 1976, pp. 36-63

Dollard, G.J.: Intern Rapport, IR 54/80, SNSF, Oslo 1980

Ellenberg, H.: Integrated Experimental Ecology, Berlin 1979

Garland, J.A.: CONF-740921, ERDA 1976, pp. 212-227

Georgii, H.W., Perseke, C., Rohbock, E.: Feststellung der Deposition von sauren und langzeitwirksamen Luftverunreinigungen aus Belastungebieten, Forschungsbericht, Universitätsinstitut für Meteorologie und Geophysik, Frankfurt 1981

Gotaas, Y.: Intern Report, IR 30/77, SNSF, Oslo 1977

Gravenhorst, G., Höfken, K.D.: Proc. Int. Symposium: Deposition of Atmospheric Pollutants, Oberursel 1981; Georgii, H.W., Pankrath J. (eds.), D. Reidel Publ. Co., p. 187.

Heinrichs, H., Mayer, R.: J. Environ. Qual., Vol. 9, 1980, pp. 111-118

Höfken, K.D., Georgii, H.W., Gravenhorst, G.: Untersuchung über die Deposition atmosphärischer Spurenstoffe an Buchen- und Fichtenwald, Berichte des Instituts für Meteorologie und Geophysik, Nr. 46, Univ. Frankfurt 1981

Höfken, K.D., Gravenhorst, G.: Proc. Int. Symposium: Deposition of Atmospheric Pollutants, Oberursel 1981; Georgii, H.W., Pankrath, J. (eds.), D. Reidel Publ. Co., p. 191.

Knabe, W.: Mitt. Landesanst. Ökologie (Sonderheft) Recklinghausen 1982, pp. 43-57

Martin, A., Barber, F.R.: Atm. Env. 5, 1971, pp. 345-352

Mayer, R.: Natürliche und anthropogene Komponenten des Schwermetallhaushaltes von Waldökosystemen; Göttinger Bodenk. Berichte 70, 1981

Mayer, R., Ulrich, B.: Proc. Int. Symposium: Deposition of Atmospheric Pollutants, Oberursel 1981; Georgii, H.W., Pankrath, J. (eds.), D. Reidel Publ. Co., p. 195.

Nürnberg, H.W., Valenta, P., Nguyen, V.D.: Proc. Int. Symposium: Deposition of Atmospheric Pollutants, Oberursel 1981; Georgii, H.W., Pankrath, J. (eds.), D. Reidel Publ. Co., p. 143.

Lindberg, S.E., Harriss, R.C.: Water, Air & Soil Pollution, Vol. 16, 1981, pp. 13-31

Overrein, L.N., Seip, H.M., Tollan, A.: SNSF, Final Report, Oslo 1981

Smith, W.H.: Air Pollution and Forests, Now York 1980

Steubing, L., Klee, R.: Angew. Botanik (44), 1970, pp. 73-85

Ulrich, B.: Mitt. Landesanst. Ökologie (Sonderheft), Recklinghausen 1982, pp. 9-25

Ulrich, B., Mayer, R., Khanna, P.K.: Deposition von Luftverunreinigungen und ihre Auswirkungen in Waldökosystemen im Solling, Schriften der Forstlichen Fakultät der Universität Göttingen, Band 58, 1979

PROCESSES AND RATES OF DEPOSITION OF AIR POLLUTANTS IN DIFFERENT
ECOSYSTEMS

Thomas, W., Rieß, W. and Herrmann, R.

Lehrstuhl für Hydrologie, Universität Bayreuth

Abstract

This paper describes some studies on the transport mechanisms
(dryfall, wetfall, aerosols) of some air pollutants (trace metals,
polycyclic aromatic hydrocarbons and chlorinated hydrocarbons)
related to their deposition on epiphytic moss. Further, the different transport mechanisms of these pollutants are compared between urban, suburban, industrial, agricultural and forested sites
together with their temporal and seasonal variations. Multivariate statistical modeling is applied in order to establish relationships 1) between traditionally measured air pollutants (e.g.
SO_4) and those rarely observed (e.g. PAH, trace metals), 2) between air pollutants of different transport characteristics and
their deposition on epiphytic moss in forests. Finally, region-
ally different temporal behaviour of air pollution will be explained as dominated by immission or, on the other hand, by air
flow.

1. INTRODUCTION

Within governmental or other air quality monitoring systems generally the pollutants NO_x, SO_2, C_mH_n, in a few cases the trace
metals and more seldom organic micropollutants with temporal and
spatial variance are measured.
The multiform spatial air pollution patterns are results of the
structure of the emission sources, the transport behaviour of the
pollutants and the actual atmospheric transport processes. Therefore, the impact of air pollutants on forest ecosystems depends
on temporally and spatially different combinations of air pollu-

65

B. Ulrich and J. Pankrath (eds.), Effects of Accumulation of Air Pollutants in Forest Ecosystems, 65–82.
Copyright © 1983 by D. Reidel Publishing Company.

tants with different transport mechanisms. In order to estimate the impacts, especially those of synergistic nature, of a spectrum of pollutants on forest ecosystems one has to have a good knowledge of the temporal and spatial air pollution pattern.

2. MATERIAL AND METHODS

2.1. Sampling stations and periods

Between April 1980 and February 1981 seven sampling stations were run in Bavaria (fig. 1). Two stations were installed in forests (Brotjacklriegel (6) and Claffheim (4)), three stations in towns or suburban areas (Nürnberg-Tiergarten (3), Kelheim (5) and München-Effnerplatz (7)), one station in an agricultural area (Hohenberg/Eger (1)) and another one in an industrial area (Arzberg (2)). Integrated samples were taken every three weeks. In addition two more sampling series were carried out at the two forest stations, with an interval of one week at Claffheim (26.10.80 - 20.02.81) and of three days at Brotjacklriegel (28.08.80 - 22.10.80).

2.2. Instrumentation

Each station was equipped with a totalisator (funnel = 885 cm^2) for taking bulk precipitation samples (dry and wet outfall). Besides, atmospheric particulate matter (0.45 µm) was sampled with a low volume sampler (Thomas, 1981a). At the same time the accumulation of the pollutants in the epiphytic moss Hypnum cupressiforme L. ssp. filiforme Brid. was determined.

2.3. Chemical analysis

Thomas (1981a) gives a detailed description.
    Chemical analysis of chlorinated hydrocarbons and PAH in
    epiphytic moss
The organic micropollutants were analyzed according to a modified instruction of 'Methodensammlung zur Rückstandsanalytik von Pflanzenschutzmitteln der DFG (1974)'. Instead of a separation of DDT and PCB by thinlayer chromatography one single purification and separation step was used (Wells and Johnstone, 1977) by means of a repeated column chromatographic elution. Out of the second eluate the PAH 3.4-benzopyrene, fluoranthene and benzo-ghi-perylene were determined. The chlorinated hydrocarbons were determined by gas-liquid chromatography using capillary columns and an ECD, the PAH by high performance thin layer chromatography with fluorescence spectroscopy.

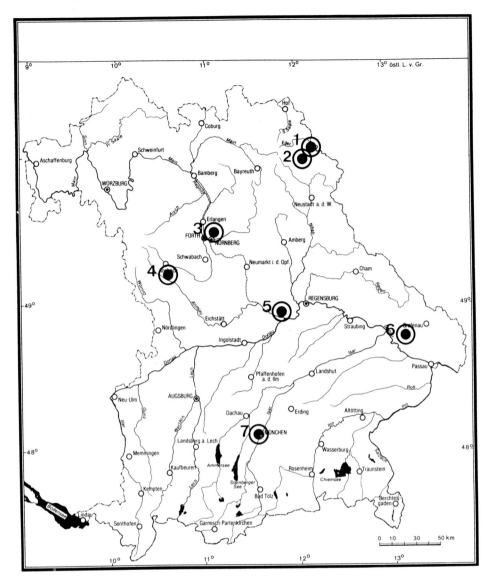

Figure 1. Sampling stations.

Analysis of chlorinated hydrocarbons and PAH in bulk precipitation

The organic micropollutants were obtained out of the bulk precipitation by liquid-liquid extraction. The reduced extract was
analyzed by high performance thinlayer chromatography with fluorescence spectroscopy as to its content of PAH. Subsequently the
extract was cleaned by column chromatography, and the chlorinated
hydrocarbons were determined by capillary gaschromatography with

an ECD.

Analysis of trace metals in bulk precipitation

The samples were acidified and taken to the laboratory. The analysis of the trace metals Zn, Cd, Pb and Cu was carried out by
anodic stripping voltammetry with sodium acetate (pH 4.62). Fe
was analysed by atomic absorption spectroscopy.

Analysis of organic and inorganic micropollutants in atmos-
pheric particulate matter

Atmospheric particulate matter was sampled with low volume samplers using cellulose-nitrate filters (0.45 µm). During each sampling period 100 - 150 m^3 of air were filtered. The filters were
divided into two equal parts. One part was digested with nitric
acid for trace metal analysis (s. above), the other half was extracted using dichloromethane for PAH-analysis (s. above). After
precleaning by column chromatography, the extract was dissolved
in petroleum ether and analysed for chlorinated hydrocarbons by
capillary gaschromatography as above.

Analysis of trace metals in epiphytic moss

Two g of dry moss sample were required for trace metal analysis.
In the laboratory, digestion on the sand bath was made using 20
ml nitric acid and 5 ml perchloric acid mixed with the sample.
Determinations of trace metals were made using atomic absorption
spectroscopy.

3. RESULTS AND DISCUSSION

3.1. Regional distribution of air pollutants

A cluster analysis of the seven sampling stations with mean concentration values was carried out in order to compare the air
pollution structure of the seven sampling stations (s. table 1
and figs. 2 - 5). Grouping of stations follows two criteria
(Ward, 1963): to get minimal distance of values between stations
within a group and to get maximal distance of values between the
groups.
The results of the cluster analysis of the seven sampling stations
are presented in fig. 6. Within the first level the sampling station Hohenberg whose land use is dominated by agriculture is combined with Kelheim, a small town. The low information loss by
combining the two stations is caused by a higher Pb-content in
the bulk precipitation and atmospheric particulate matter and a
higher Fe-content in the bulk precipitation at the sampling station Kelheim. Just as little information loss is caused by adding
to this group the sampling station Claffheim whose land use is
dominated by forests. Claffheim shows higher contents of Fe in
atmospheric particulate matter, lower contents of fluoranthene
and higher contents of γ-BHC in bulk precipitation. On the third
level the highly polluted industrial resp. urban sampling stations Arzberg and München-Effnerplatz are combined. The high

Table 1. Mean values of air pollutants at seven sampling stations in Bavaria.

Pollutant \ Station	Hohen-berg	Arz-berg	Nürn-berg	Claff-heim	Kel-heim	Brot-jackl-riegel	Mün-chen
Zn (S) ng m^{-3}	208	226	230	263	171	245	294
Cd (S) ng m^{-3}	0.3	3.1	0.9	0.7	0.4	1.5	0.7
Pb (S) ng m^{-3}	60	77	101	136	191	55	295
Cu (S) ng m^{-3}	22	49	25	40	27	30	45
Fe (S) ng m^{-3}	231	492	277	329	272	150	527
LF μS cm^{-1}	34	33	33	33	42	27	33
pH	5.6	5.4	4.6	4.3	6.1	5.6	6.3
Zn (N) μg l^{-1}	31	41	42	55	86	140	59
Cd (N) μg l^{-1}	0.3	0.3	0.8	0.4	0.3	0.6	0.4
Pb (N) μg l^{-1}	9	14	17	18	25	8	41
Cu (N) μg l^{-1}	5.9	17.2	8.7	5.9	7.9	7.5	16.6
Fe (N) μg l^{-1}	61	136	104	82	134	47	250
Bper (N) ng l^{-1}	9.5	21.5	14.2	9.8	12.7	8.3	9.3
BaP (N) ng l^{-1}	4.2	6.3	11.4	3.0	5.7	3.5	5.2
Flu (N) ng l^{-1}	35.3	64.4	42.2	17.7	34.2	28.6	51.7
Bper (S) ng m^{-3}	0.08	0.19	0.28	0.24	0.19	0.10	0.17
BaP (S) ng m^{-3}	0.02	0.03	0.30	0.03	0.02	0.03	0.08
Flu (S) ng m^{-3}	0.36	0.80	1.74	0.50	0.57	0.38	1.23
α-BHC (N) ng l^{-1}	6.5	5.9	7.9	6.6	4.6	11.3	5.1
γ-BHC (N) ng l^{-1}	6.9	7.7	20.6	10.1	6.9	12.8	7.7

(N) = bulk precipitation
(S) = atmospheric particulate matter
Bper = Benzoperylene
BaP = 3.4-Benzopyrene
Flu = Fluoranthene
LF   = conductivity

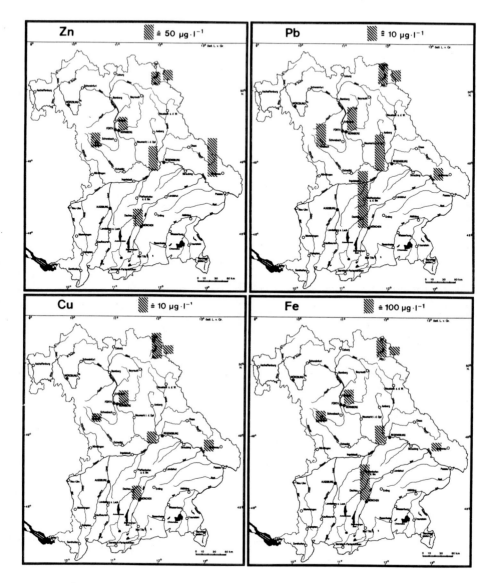

Figure 2. Regional distribution of mean concentration of trace
metals in bulk precipitation.

information loss is caused by the combination of two sampling
stations with different pollution structure: The urban sampling
station is dominated by high contents of Pb in bulk precipitation
and atmospheric particulate matter. In contrast, the PAH-content
in bulk precipitation are higher in the industrial area than in
the urban area. With the PAH-content in atmospheric particulate
matter it is just the other way round. There is the same high

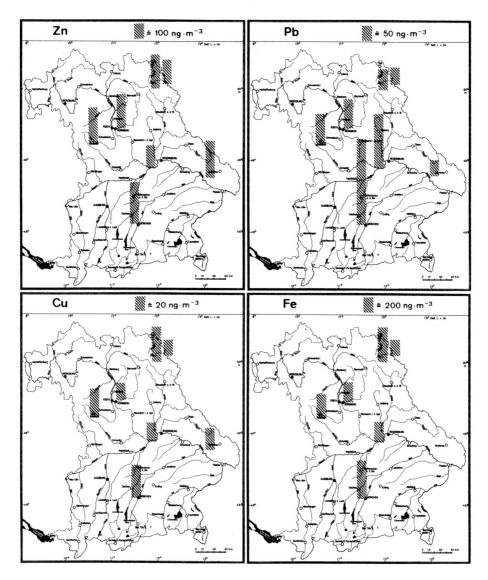

Figure 3. Regional distribution of mean concentration of trace
metals in atmospheric particulate matter.

information loss when the forest station Brotjacklriegel is combined with group no. I, since this station shows higher contents
of pesticides, higher geogenic Zn- and Cd-contents and lower Fe-
and conductivity values than all other stations. At last the station Nürnberg-Tiergarten is added to group no. I. Though of urban
character it has high pesticide contents together with lower Cu-
and Fe-contents than the highly polluted stations of group no. II.

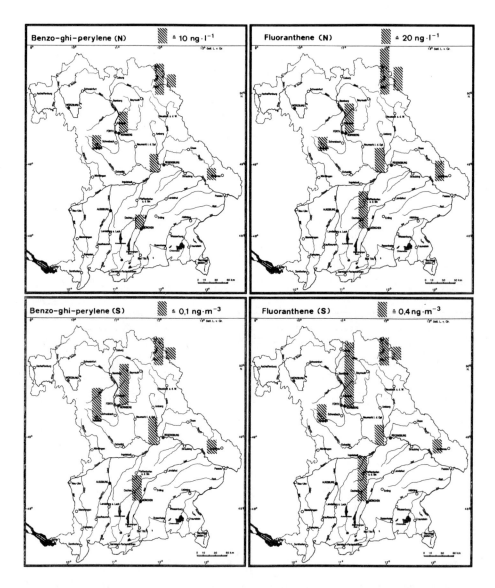

Figure 4. Regional distribution of mean concentration of PAH in
          bulk precipitation (N) and in atmospheric particulate
          matter (S).

The spatial structure of the air pollution is mainly depending on
the neighbouring emission sources. But it can also be dominated
by atmospheric long distance transport processes (e.g. high pesticide contents at Brotjacklriegel, see below).

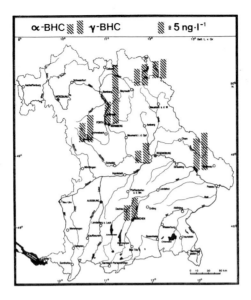

Figure 5. Regional distribution of mean concentrations
of α- and γ-BHC in bulk precipitation.

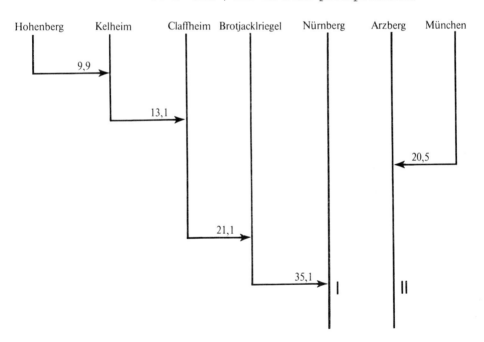

Figure 6. Dendrogramme of hierarchical grouping analysis of the
sampling stations. The figures show the information loss
caused by combining the groups.

The regional correlations of the pollutants were analyzed by
principal component analysis. Principal component analysis points
out the relationship between the variables. Thereby new dimensions, orthogonal to each other, and their correlations with the
original variables are determined. These dimensions are called
principal components, the correlations between those and the measured variables are called principal component loadings (Anderson,
1958).
The results of the principal component analysis confirm (s. table
2) that γ-BHC, PAH and the precipitation are intercorrelated and
that these relations can be explained by washout and fallout. On
the whole there are only weak correlations between meteorologic
variables and pollutants, which points to a more emission induced
rather than a transport induced pollutant structure. As Fe and Pb
at the highly polluted stations (München, Kelheim and Nürnberg)
show the same behaviour in atmospheric particulate matter as well
as in bulk precipitation these variables are found on the same
principal components. High pollution of bulk precipitation by
these variables are accompanied by high conductivity. Besides Cd
(N) and Cu (N) show high loadings on the fourth principal component. It is striking that the PAH in bulk precipitation are
closely related whereas those in atmospheric particulate matter
weakly load different principal components (not shown in table 2).
Whereas Herrmann (1978) and König et al. (1981) show that PAH
have stable relations in well mixed air masses, this investigation confirms that in partially mixed air masses (z. B. Arzberg,
Kelheim, München) a combination of PAH dominates which is depend-
ant on local emission sources (cp. Waibel, 1974).

3.2. The temporal variation of air pollution

In addition to the differences of the mean pollution values there
are temporal variations of air pollution. In winter e.g. the impact of a pollutant on a deciduous forest differs greatly from
that on a coniferous forest.
The single pollutants show different temporal variations at different locations depending on transport mechanisms and emission:
In Hohenberg e.g. a clear fluoranthene maximum is measured in winter, a consequence of the heating period and atmospheric inversions. On the other hand, the irregular temporal variation of
this pollutant in nearby Arzberg may be caused by time variant
local emissions of potteries and power stations (fig. 7). On the
Brotjacklriegel (fig. 8) summer droughts cause a Zn- and Cd-maxi-
mum in atmospheric particulate matter which was also found by
Thomas (1981a) for similar geological forest sites in the Fich-
telgebirge and depends on high Zn- and Cd-contents in the local
rock formations.
There are low concentrations of PAH, Pb and Fe at this station in
atmospheric particulate matter, whereas there develops a distinct

Table 2. Principle component analysis of micropollutants and
meteorological variables, sampling period: 3 weeks.
All loadings < |0.5| were omitted.

Variable	Principal Components 1	2	3	4	5
Variance	16.3	9.8	8.0	6.9	6.7
Conductivity			.51		
pH					
Zn (N)					
Cd (N)					
Pb (N)			.91		
Cu (N)					
Fe (N)			.86		
γ-BHC (N)	-.98				
Benzo-ghi-perylene (N)	-.97				
3.4-Benzopyrene (N)	-.98				
Fluoranthene (N)	-.87				
Zn (S)					
Cd (S)				.96	
Pb (S)					.94
Cu (S)				.89	
Fe (S)					.87
Benzo-ghi-perylene (S)					
3.4-Benzopyrene (S)					
Fluoranthene (S)					
Precipitation	-.81				
Percentage (%) of cyclonal weather		.79			
% days v < 10 kn					
% days 10 < v < 20 kn					
% days P < 1 mm		-.86			
% days 1 < P < 9 mm		.97			
% days P > 9 mm					

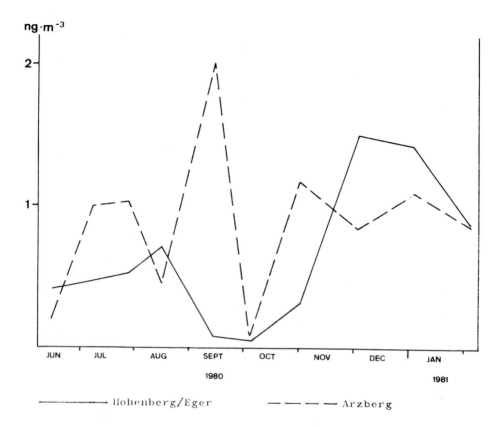

Figure 7. Temporal variation of fluoranthene in atmospheric par-
         ticulate matter, sampling stations Hohenberg/Eger and
         Arzberg.

summer maximum of the pesticides α- and γ-BHC, which are emitted
from the intensive agricultural area of the Dungau and transported within cyclonic winds to the heights of the Bayerischer Wald
(fig. 9).
In the winter months all sampling stations have similar α- and
γ-BHC concentrations. With the use of γ-BHC (Lindane) in summer
the proportion of γ-BHC increases whereas the by-product α-BHC
does not show considerable increase (Thomas, 1981b). This quotient increases towards areas of higher application or favourable
transport conditions (fig. 5).

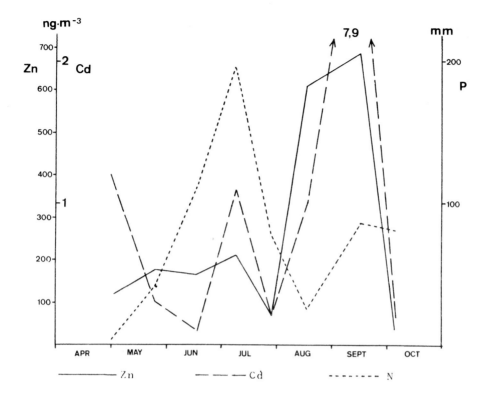

Figure 8. Temporal variation of Cd and Zn in atmospheric particulate matter, sampling station Brotjacklriegel.

3.3. The distribution of the pollutants between atmospheric
particulate matter and bulk precipitation as well as
their accumulation in epiphytic moss

The principal component analysis (table 2) has already shown that
there is no distinct covariance between the pollution in atmospheric particulate matter and that in bulk precipitation. So when
we compare the station at Nürnberg with that at München we find
less pollution in bulk precipitation at Nürnberg, whereas the
pollution in atmospheric particulate matter is equally high at
both stations (s. table 1). Or: Brotjacklriegel has high concentrations of Zn in bulk precipitation in comparison with all other
stations, whereas there is only mean pollution in atmospheric
particulate matter.
The impact of micropollutants on forest ecosystems cannot be directly estimated from the concentrations of these pollutants in
bulk precipitation or atmospheric particulate matter. Moreover,
it is necessary to develop accumulation models of the various
pollutants and their sources for the different compartments of

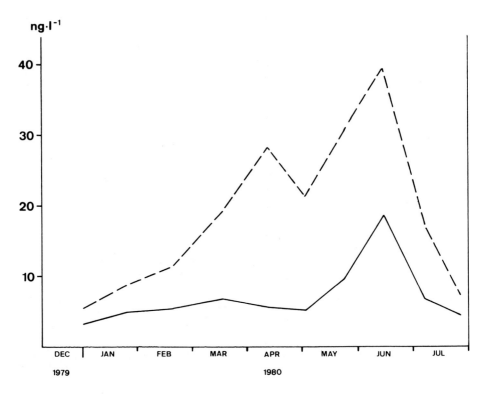

Figure 9. Temporal variation of α- and γ-BHC in bulk precipita-
          tion, sampling station Brotjacklriegel.

the forest ecosystems. The example of the epiphytic moss will demonstrate the extent of concentration and what part of its variance will be explained by the pollutant concentration in bulk precipitation and atmospheric particulate matter and the influence
of precipitation.
We define two accumulation indices:

$$A_N = \frac{\text{concentration in moss}}{\text{concentration in bulk precipitation}} \quad \frac{\text{mg kg}^{-1}}{\text{mg l}^{-1}}$$

$$A_S = \frac{\text{concentration in moss}}{\text{concentration in atmospheric part. matter}} \quad \frac{\text{mg kg}^{-1}}{\text{mg m}^{-3}}$$

Thomas (1981a) reported about transports in the gas phase. These
transports are of no importance for the pollutants treated here
with the exception of HCB.
Fe has the highest index as well as for A_N and A_S (table 3). As
the transport of Fe does not differ basically from that of the
other metals, there may occur a physiological enhanced accumula-

tion. Because of their different physical units and different transport mechanisms the two indices A_N and A_S cannot directly be compared with one another.

The values of A_N for trace metals und PAH are equally high whereas the values of A_S show a lower input of PAH by atmospheric particulate matter than of trace metals. The accumulation of chlorinated hydrocarbons from bulk precipitation shows considerable differences (A_N). Thomas (1981a) explains that HCB is solely transported in gas phase and γ-BHC solely adsorbed on particles.

Table 3. Accumulation indices for trace metals, chlorinated hydrocarbons and PAH in Hypnum cupressiforme.

	A_N	A_S		A_N	A_S
Zn	1336	644	α-BHC	709	n.d.
Cd	1354	606	γ-BHC	897	58
Pb	2811	714	HCB	3156	n.d.
Cu	2639	1146	Benzo-ghi-perylene	1845	141
Fe	7182	3230	3.4-Benzopyrene	2573	180
			Fluoranthene	1325	46

By means of regression analysis we examined the relation between the pollutant contents in epiphytic moss and the contents in bulk precipitation, particulate atmospheric matter and the amount of precipitation. All variables were tested as to their Gaussian distribution (5 % level) by the Kolmogorov-Smirnov-test and transformed if necessary. R^2 was tested against zero by an F-test. The time series were separated in order to test the stability of the regression.

The percentage of the β-values is to indicate the transport mechanisms: (N) ≡ transport by bulk precipitation, (S) ≡ transport by atmospheric particulate matter and P ≡ influence of precipitation. The results of the regression analysis are laid down in table 4. Zn and all PAH show significant and stable relations between their content in epiphytic moss and that in atmospheric particulate matter and bulk precipitation. If the pollutants are mainly transported by atmospheric particulate matter high stability and a high multiple correlation coefficient can be expected. If the pollutants are mostly transported by bulk precipitation, the significance is high but the stability is low (e.g. Cu). Thomas (1981a) gives a detailed discription of this regression analysis.

Table 4. Results of the regression analysis with the pollutant
         content in epiphytic moss as criterion variable.

Pollutant content in Hypnum		(N)	(S)	N	R	p (%)
Zn	1979	29	51	20	0.79	0.1
	1980	23	43	34	0.91	0.0
Cd	1979	7	35	58	0.49	6.1
	1980	21	63	16	0.75	0.1
Pb	1979	69	31	0	0.48	6.9
	1980	56	39	5	0.74	0.2
Cu	1979	20	80	0	0.87	0.0
	1980	82	7	11	0.77	0.1
Fe	1979	40	49	11	0.56	2.9
	1980	7	55	38	0.69	0.5
Benzo-ghi-	1979	20	46	34	0.68	0.6
perylene	1980	20	48	32	0.74	0.2
3.4-Benzo-	1979	24	38	38	0.74	0.2
pyrene	1980	24	56	20	0.66	0.7
Fluoran-	1979	10	82	8	0.90	0.0
thene	1980	19	73	8	0.91	0.0
γ-BHC	1980	79	2	19	0.79	0.1

3.4. Comparison of routine meteorological and pollutant variables
     with trace metal and PAH pollution

A relation between routine meteorological and pollutant variables
with trace metal and PAH pollution can only be found for short
periods (one to a few days). Table 5 gives examples of regression
equations for the sampling stations Claffheim and Brotjacklriegel.
These regression analyses are of great practical importance for
the prediction of micropollution. In Rieß (1982) a detailed list
and description of the prediction equations can be found.

Table 5. Prediction equations for fluoranthene (Claffheim) and
Zn (Brotjacklriegel).

Fluoranthene = 0.007 dust + 0.006 SO_2 - 0.058 S in dust

$\qquad$ + 0.015 N - 0.05 v_G + 0.377

$\qquad$ R = 0.95, p = 0.0 %

Zn $\qquad$ = 2.39 SO_2 + 4.99 T + 1.48 P - 1534

$\qquad$ R = 0.86, p = 0.0 %

---

v_G = windspeed near the ground (kn)
N = precipitation (mm)
dust ($\mu g \ m^{-3}$)
S in dust ($\mu g \ m^{-3}$)
SO_2 ($\mu g \ m^{-3}$)
Zn in dust ($ng \ m^{-3}$)
fluoranthene in dust ($ng \ m^{-3}$)
T = temperature (°C)
P = air pressure at 12 h GMT (mbar)

The regression analysis shows that it is possible to predict the
concentrations of rarely measured micropollutants by means of
routine meteorological and pollution variables. Mostly high contents of atmospheric particulate matter and low wind speed are
accompanied by high contents of micropollutants.

Conclusions

The impact of micropollutants shows considerable regional and temporal variation. Urban types of pollution have a correlation between the PAH and some trace metals. Here the seasonal variation
is determined by the heating period und low atmospheric exchange.

Industrial areas show a micropollution which is dominated by local emission and its temporal variation. Areas with intensive
agricultural land use have higher pesticide pollution, e.g. BHC,
with a temporal variation due to the application.
The quotient of the two BHC-isomeres α and γ shows a distinct
seasonal variation and becomes narrower away from the places of
application.
Areas far away from emission sources have vague temporal and seasonal pollution patterns - with the exception of PAH.
The input of PAH into forest ecosystems - measured by their concentration in epiphytic moss - occurs rather through atmospheric
particulate matter than through bulk precipitation.

Trace metals in epiphytic moss accumulate through atmospheric
particulate matter as well as through bulk precipitation. Therefore a high proportion of the variance of the trace metals is explained by the variance of the precipitation. γ-BHC, the only
chlorinated hydrocarbon with a significant accumulation model, is
enriched first of all through bulk precipitation.
For a few micropollutants it is possible to build linear regression models with routine meteorological and pollution data as independent variables.
The impact of micropollutants on forested ecosystems cannot be
simply estimated by direct measurements of their concentration in
bulk precipitation and atmospheric particulate matter. On the contrary, for each compartment and for many pollutants the accumulation must be determined separately.

Acknowledgments

We are grateful to Mrs E. Misch and Mrs E. Chinta for typing the
paper and drawing the figures. The 'Bayerisches Landesamt für Um-
weltschutz' gave financial support for part of the research. The
Umweltbundesamt helpfully supplied us with macropollutant data
(chapter 3.4.) of the stations Brotjacklriegel and Claffheim.

References

Anderson, T.: 1958, An introduction to multivariate statistical
    analysis. New York: Wiley, pp. 1-374.
Deutsche Forschungsgemeinschaft: 1974, Methodensammlung zur Rück-
    standsanalytik von Pflanzenschutzmitteln. Weinheim: Verlag
    Chemie.
Herrmann, R.: 1978, Catena 5, pp. 167-175.
König, J. et al.: 1981, Staub - Reinhalt. Luft 41, pp. 73-78.
Rieß, W.: 1982, Unveröff. Staatsexamensarbeit, Univ. Bayreuth,
    Lehrstuhl für Hydrologie.
Thomas, W.: 1981a, Bayreuther Geowissenschaftl. Arbeiten Bd. 3,
    pp. 1-143.
Thomas, W.: 1981b, Deut. Gewässerkdl. Mitt. 25, pp. 120-129.
Waibel, M.: 1974, Diss. ETH Zürich.
Ward, J. H.: 1963, American Statistic. Assoc. J. 58, pp. 236-244.
Wells, D. E. and Johnstone, S. J.: 1977, J. Chromatography 140,
    pp. 17-28.

MEASUREMENTS OF SURFACE RESISTANCE DURING DRY DEPOSITION OF SO_2
TO WET AND DRY CONIFEROUS FOREST

L. Granat

Department of Meteorology, University of Stockholm,
Arrhenius Laboratory, S-106 91 Stockholm

ABSTRACT

Direct measurements of dry deposition of SO_2 to pine trees in
central Sweden have so far given a dry deposition velocity of
0.4 cm s^{-1} in "summer" and 0.1 or less in "winter". These values
are tentative and might be subjected to revision after further
measurements. Deposition to a wet canopy is found to be small.
The reason is the acidic nature of the liquid which limits
oxidation and need therefore not be inconsistent with findings
of a more rapid uptake in areas with another composition of the
throughfall. Dry deposition of S in gas and particles is smaller
than wet deposition. This is characteristic for a northern country at some distance from major anthropogenic SO_2 emission areas.

INTRODUCTION

There is a well recognized need to obtain good quantitative
estimates of the dry deposition of trace gases of anthropogenic
origin especially for sulfur dioxide. Of particular concern in
Sweden is the deposition to a forest. This paper will explain
the reasons for the experimental approach taken by our group,
present some results from measurements and extend them to a
tentative estimate of dry deposition of SO_2 in central Sweden
(approximately the area between Stockholm and Gothenburg). This
is then compared with measured wet deposition. A few concluding
remarks will also be given on how these findings apply in areas
closer to major source regions.

Dry deposition of several trace gases - we shall here mainly

B. Ulrich and J. Pankrath (eds.), Effects of Accumulation of Air Pollutants in Forest Ecosystems, 83–89.
Copyright © 1983 by D. Reidel Publishing Company.

deal with SO_2 ⁻ is to a large degree controlled by the living
vegetation (stomatal and other resistances). The resistance
changes considerably with time both in a more regular way (diurnal, seasonal) and more randomly (due to dry periods etc.). We
must also consider processes outside the needles which might
affect the rate of deposition for instance when they are wetted
by rain or dew. Large spatial variations in the resistance must
be assumed due to variations in species, growing conditions and
canopy density, to mention a few factors. These variations may
or may not coincide with changes in trace gas concentration and
varying transfer resistance in the air which is controlled mainly
by wind speed and stability. It is therefore quite obvious that
the dry deposition is likely to vary substantially both in time
and space and this must be kept in mind when selecting suitable
measuring techniques and estimating procedures.

A large number of experimental methods to measure and estimate
dry deposition has been suggested and the advantages and
disadvantages of the method have also been discussed (see for
instance Hicks et al., 1980).

Several of the established methods may fail to give the required
information with regard to a forest in Sweden - especially when
the large temporal and spatial variability of the deposition
velocity is considered. Micro meteorological methods (for instance the flux gradient method) which have been used with lower
vegetation is not very suitable for a tall forest primarely due
to the rather high surface resistance which leads to a very small
concentration gradient above the forest where measurements should
be performed. Practical problems when measuring at several different locations also arise once the gradients can be measured and
interpreted as a flux. The eddy correlation method is claimed
to be useful even in the case with the small gradient over a
forest (i.e. Hicks et al., 1980) but we observe that even today
there is no sufficiently fast and sensitive sensor available
(median SO_2 value is 1-2 ppb in central Sweden and even lower in
the northern part). Radioactive tracer technique has several
merits but also drawbacks, especially when estimating temporal
and spatial variation which requires a large number of measurements. Measurements of throughfall and mass balance calculations
for certain small water sheds might provide further support in
calculations of dry deposition of SO_2.

Our preferred method has been to enclose parts of branches into
chambers and determine deposition velocity or surface resistance
from the concentration difference of trace gases (NO_2, NO, O_3,
SO_2) as well as of CO_2 and water vapour in an airstream passing
through the chamber. Measurements are made on living trees in the
field and by allowing a short residence time, adequate temperatu-

re control and near ambient SO$_2$ concentrations in the chamber, the disturbance of the branch is probably small. Deposition to wet canopy (wetted by rain or dew) can be measured with the same technique either directly on the branches or on collected liquid alone.

Special attention will be given to find means to verify the areally averaged estimate obtained with this technique.

EXPERIMENTAL

Measurements on "dry canopy (no liquid water outside the needles)

Parts of branches were enclosed in well thermostated 3 ℓ chambers transparent for photosynthetically active light. A stream of ambient air was passed through the chamber (at about 5 ℓ/min) and the concentration of trace gases, water vapour and carbon dioxide was measured before and after the chamber.

Measurements were first made at Jädraås, 200 km north-west of Stockholm in the experimental site for the project "Ecology of coniferous forest". The rate of uptake of sulfur dioxide together with stomatal and mesophyll conductance, light intensity and temperature was here measured continuously for more than three months. (See further in Hällgren et al., 1982). Later on the equipment for analysis of trace gases (SO$_2$, NO, NO$_x$ and O$_3$), CO$_2$ and water vapour was installed in a mobile van which also contains necessary valves for automatic control of zero and span of the instruments and for their connection to air entering and leaving the chamber. This equipment is suitable for investigating the spatial (as well as temporal) variation in the dry deposition velocity in that branches at many different places can be investigated, although this is a time-consuming work. In the early experiments, SO$_2$ was added but the equipment has eventually been improved and is now capable of measuring small changes at near ambient levels (a few ppb). Measurements have so far been made at three different locations during part of the growing season and at two sites during winter.

Wet canopy

The uptake on wet vegetation was primarily investigated in the laboratory by measuring the decrease in SO$_2$ concentration in an air stream after contact with either collected throughfall or water solution with similar pH and catalyst concentrations (Mn, Fe) as in throughfall.

RESULTS

The results are first discussed in terms of deposition velocity relative to projected needle area (v_s) which is the quantity obtained from the field measurements. In the next paragraph these findings are used to estimate deposition to a "typical" forest.

The deposition velocity (v_s) was found to vary from day to day. A diurnal variation which is quite well related to the stomatal conductance can, however, easily be observed. Day and night time values can thus be a factor of two higher and lower, respectively, than the average value. The average value from measurements made so far during the growing season and weighted to account for variations between day and night is $v_s = 0.13$ cm/s. Large seasonal variations were also confirmed when deposition velocity during winter was found to be very small ($v_s = 0.03$ cm/s or less). During these winter measurements no transpiration or carbon dioxide uptake or emission was found.

In the Jädraås experiment, the deposition velocity was on the average about half of the calculated stomatal conductance for SO_2 (see Hällgren et al., 1981). If this difference is evaluated in terms of an internal resistance, data suggest that it has a minimum value during early morning hours when compared both to afternoon and night values. It was also found that part of the apparent internal resistance was due to an emission of a reduced sulfur compound. Although interesting in itself, this emission appears to be of little quantitative importance at ambient SO_2 levels. Later measurements have shown cases where the deposition velocity is higher than the stomatal conductance for SO_2 which suggests that an active uptake outside the needles or on the twigs takes place in addition to the uptake through stomata. Further measurements are needed to clarify these matters but they are mentioned here as an indication of the complexity of the deposition process and that present quantitative estimates of the deposition are uncertain.

Deposition velocity in the case of a wet surface is initially high until an equilibrium is obtained between SO_2 in air and dissolved in the liquid (if an equilibrium is not reached already in the falling raindrops) and is thereafter controlled by the (usually slow) oxidation in the liquid. Laboratory studies of the rate of oxidation in collected throughfall and in synthetic solutions indicate that a high acidity (pH usually in the range 3.5 to 4.5) and a quite high Mn concentration (about 10^{-5} M) are the two factors which are most crucial for the observed rate of oxidation.

Deposition of SO_2 on snow was also measured to provide an estimate with regard to snow-covered vegetation in the winter. It
was found that the deposition velocity to cold snow (temperatures
ranging from -3°C to -14°C were investigated) was about 0.1 cm/s.
The results were obtained with the chamber technique where the
air is well stirred and the value given might be an overestimate
due to penetration of air in deeper snow layers. The surface
resistance over melting snow was found to be very small at least
during the initial phase when the snow melt was not saturated
with sulfur dioxide. From measurements of SO_2 deposition in
synthetic and real rain water it is likely that deposition of
SO_2 on wet snow will decrease to very low values as soon as the
water is saturated with SO_2.

TOWARDS AN ESTIMATE OF ANNUAL DRY DEPOSITION OF SULFUR DIOXIDE TO A SWEDISH CONIFEROUS FOREST

As a tentative estimate of v_S for dry canopy during the vegetation period we take the average value for three different sites
(v_S = 0.13 cm/s). During winter v_S for the canopy is 0.03 cm/s
or less. We assume that these values represent a typical forest
for which we put average leaf area index (projected needle area
per ground area) to 3 as an average for pine and spruce and
obtain corresponding v_g values (deposition velocity relative to
ground area) of 0.4 cm/s (summer) and 0.09 cm/s (winter). For
a canopy wetted by rain, the deposition velocity is estimated
from the laboratory measurements and we estimate that a typical
forest might hold about 2 mm of rain or less and this persists
for 20% of the time. The time weighted contribution to "summer"
deposition velocity would then amount to 0.1 cm/s (as v_S) but
at the same time stomata are likely to be blocked by the water.
Deposition to snow on or below the canopy is difficult to
estimate but is assumed to be quite small on the average, and a
value of 0.06 cm/s is adopted as a time weighted average
although larger values have been suggested (Sehmel, 1980). We
adapt v_g = 0.15 cm/s (0.09 + 0.06) during winter, which is about
three months in central Sweden, and v_g = 0.40 for the remaining
period. (Two decimals are given here and in Table 1 merely for
the purpose of internal comparisons - it goes without saying that
the accuracy is far less).

In central Sweden (the area between Gothenburg and Stockholm)
average SO_2 concentration is about 4 µg SO_2-S/m³ during winter
(three months) and 1.8 µg SO_2-S/m³ during the remaining period.
The measurements are obtained from sites located in clearings in
forests and can therefore approximately represent the atmosphere
in contact with the SO_2 absorbing needles.

We can now estimate the deposition as the product of v_g and the

SO_2 concentration. The result is given in Table 1 and compared with deposition of sulfate particles which can be assumed to have relatively small seasonal variation (from Lannefors et al., 1981) and wet deposition.

DISCUSSION

The tentative estimates of dry deposition of sulfur dioxide to a coniferous forest in Sweden indicate that this process is less important than wet deposition provided that the few trees which measurements so far have been made on are representative for a forest in central Sweden. Comparison with the few pre- viously reported deposition velocities to coniferous forest supports these findings (Garland and Branson, 1977; Galbally et al., 1979). The quantities obtained for dry deposition (table 1) are valid for a northern country somewhat outside the major an- thropogenic emission area in Europe. Closer to the source region the dry deposition is very likely to be much more important as pointed out by Garland and Branson (1977). There are several reasons for this. As can be seen from the previous calculations the deposition velocity is very small in winter when the tempera- ture is below zero (and when we actually have the highest sulfur dioxide concentration levels). Higher deposition velocity in summer is counteracted by lower sulfur dioxide concentration. Closer to large anthropogenic SO_2 source regions in Europe, the sulfur dioxide concentration is higher and the winter period shorter. At the same time the sulfate concentration in rain- water is only somewhat higher when compared to central Sweden.

From our measurements in Sweden, dry deposition to a wet coni- ferous forest was estimated to be insignificant (when compared to a dry surface) essentially due to the very acid throughfall. In other areas, especially close to agricultural regions, where alcaline dust or ammonia is deposited on the vegetation, a higher pH might be encountered which would promote rapid oxidation of sulfur dioxide and hence give a high deposition velocity during periods when the canopy is wet.

The relation between wet and dry deposition of sulfur compounds thus depends on a number of factors including climate, the kind and density of the forest and proximity to large source areas for SO_2 and neutralizing compounds.

ACKNOWLEDGEMENT

Financial support for work reported here has been obtained from the Swedish National Protection Board, contract No. 641-3082-81Fp.

REFERENCES

GALBALLY, I.E., GARLAND, J.A. and WILSON, J.G., 1979: Sulfur
    uptake from the atmosphere by forest and farmland. Nature
    280, 49-50

GARLAND; J.A. and BRANSON, J.R., 1977: The deposition of sulfur
    dioxide to a pine forest assessed by a radioactive tracer
    method. Tellus 29, 445-454.

HICKS, B.B., WESELY, M.L., DURHAM, J.L., 1980: Critique of
    methods to measure dry deposition. Workshop summary. U.S.
    Department of Commerce. National Technical Information Service,
    Springfield, U.S.A.

HÄLLGREN, J-E, LINDER, S., RICHTER, A., TROENG, E. and GRANAT, L.,
    1982: Uptake of SO$_2$ in shoots of Scots pine: field measure-
    ments of net flux of sulphur in relation to stomatal con-
    ductance. Plant, Cell and Environment 5, 75-83.

LANNEFORS, H., HANSSON, H-C and GRANAT, L., 1982: Background
    aerosol composition in southern Sweden - 14 micro and macro
    constituents measured in seven particle size intervals at
    one site during one year. Accepted for publication in Atmo-
    spheric Environment.

SEHMEL, G.A., 1980: Particle and gas dry deposition: A review.
    Atmospheric Environment 14, No. 9, 983-1011.

TABLE 1 Estimates of dry deposition in a forest in central
         Sweden compared to wet deposition. LAI = 3 is used.
         Values in g S/m^2, year.

		Summer	Winter	Average
SO$_2$-S	dry dep.	0.23	0.19	0.22
SO$_4^{2-}$	dry dep.	0.10	0.19	0.10
SO$_4$-S	wet dep.	0.98	0.66	0.90
Total		1.31	0.95	1.22

Topic 2:

Processes and Rates of Proton Production by Discoupling
of the Ion Cycle, and of Proton Consumption
by Silicate Weathering

THE TURNOVER OF PROTONS BY MINERALIZATION AND ION UPTAKE IN A
BEECH (FAGUS SILVATICA) AND A NORWAY SPRUCE ECOSYSTEM

E. MATZNER and B. ULRICH

Institut für Bodenkunde und Waldernährung, Universität Göttingen, Büsgenweg 2, D 34 Göttingen

ABSTRACT

On the base of the flux balance of two forested ecosystems, the annual rates of mineralization within the humus layer and ion uptake are discussed in respect to H^+ turnover. In both ecosystems (beech and spruce), ion uptake in mineral soil is connected with a strong H^+ production in most years between 1971 and 1980. The rate of H^+ production resulting from ion uptake is regulated by the NH_4/NO_3 ratio and is higher under spruce than under beech in accordance with this ratio. Assuming steady state conditions, the H^+ production as a result of ion uptake is followed by an equivalent H^+ consumption during mineralization. Nevertheless, the rate of H^+ consumption during mineralization is, due to spatial and temporal discouplings of the ion cycle as well as due to the production of organic acids, less than the rate of H^+ production resulting from ion uptake. Thus the sum of both processes leads to ecosystem internal H^+ net production. The influence of the NH_4/NO_3 ratio during mineralization and ion uptake on the rate of ecosystem internal proton production is discussed.

1. INTRODUCTION

In previous papers (7,13) the annual rates of the total ecosystem internal H^+ net production of two forests were calculated from the balance of the annual element fluxes from 1969 to 1975. The cause of the calculated H^+ production was not discussed in detail and no relation to special processes acting within the

B. Ulrich and J. Pankrath (eds.), Effects of Accumulation of Air Pollutants in Forest Ecosystems, 93–103.
Copyright © 1983 by D. Reidel Publishing Company.

ecosystem was looked for. In principle, the H^+ turnover in
ecosystems can only be estimated by indirect calculations. Such
calculations are made by Likens et al. (6) and by Andersson et
al. (1), but these authors neglect either the fluxes of anions
(6) or underestimate dry deposition (interception deposition)(1).
The hydrogen ion budget given by Sollins et al. (10) does not
include N turnover and the connected H^+ fluxes during mineralization and ion uptake.
However, mineralization and ion uptake represent the central
processes inside the ion cycle of forest ecosystems (14). In this
paper these both processes are regarded seperate from each other
to evaluate their influence on the H^+ budget of the ecosystem.
The calculations include the data from 1971 to 1980. Further the
relation of the H^+ fluxes resulting from mineralization and ion
uptake to the total ecosystem internal H^+ net production will be
discussed in order to evaluate their relative importance.

2. METHODS

The data used for this paper are obtained from the element fluxes
within a beech stand (135 years old) and a spruce stand (100
years old) located in the Solling region (W.Germany). A detailed
description of the site is given by Ellenberg (5). The way of
sampling procedure, data evaluation and the model used are reported by Matzner and Ulrich (7) together with the results of the
flux balance from 1969 to 1975. The annual element fluxes from
which these balances were calculated can be found in Matzner et
al. (8).

Matzner and Ulrich (7) concluded that an important part of the
total H^+ load of the forest ecosystems investigated results from
ecosystem internal sources. The annual rates of H^+ input from
ecosystem internal sources were calculated from the flux "input
to the mineral soil" (humus lysimeters including uptake of
elements by living roots within the humus layer) and from the
cation/anion balance of the changes of element storage within the
mineral soil. These calculations include ion uptake and mineralization as well as all other processes occuring within the mineral
soil. As mentioned above, ion uptake and mineralization within
the humus layer will now be regarded separately because of their
linkage with H^+ turnover.
In both ecosystems, the decomposition of above ground litter is
restricted to the humus layer. Thus the annual rate of mineralization can be calculated from the difference between the element
fluxes "canopy drip" and "input to the mineral soil" (using in
this case data from funnel lysimeters without uptake of elements
by living roots within the humus layer) according to equation 1.
(see compartment model in figure 1).

$$F32 = F24 - F12$$ (1)

F32 = net mineralization, F24 = canopy drip, F12 = input to the
mineral soil.

Fig.1: Compartment model for the fluxes of elements within a
       forest ecosystem

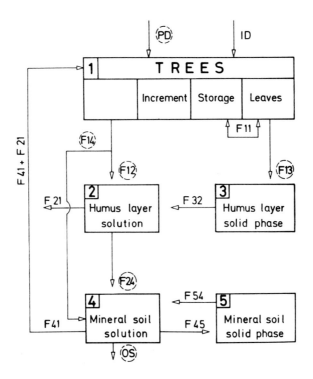

PD = precipitation-deposition; ID = interception-depositon;
OS = output with seepage water; ecosysteminternal fluxes: F11 =
translocation of elements inside the plant; F12 = canopy drip;
F13 = stem-flow; F13 = litterfall; F24 = input to the mineral
soil; F21 = uptake by the stand out of the humus layer; F32 = net
mineralization; F54 = release from the soil by desorption,
dissolution, weathering; F45 = fixation inside the soil by adsorption, exchange and precipitation; F41 = uptake by the stand
out of the mineral soil.
The element fluxes marked with dotted circles were measured
directly, the others were calculated.

The application of equation 1 to protons gives the H^+ net production during mineralization. The origin of the H^+ turnover
connected with mineralization can be calculated by comparing the
rates of mineralized cations (formation of bases) with the rates
of mineralized anions (formation of acids (14)). Furtheron the

rate of formation of organic anions has to be considered. The
difference between the fluxes of cations and of anions is taken
as a measure of the flux of organic anions.
Since the rate of ion uptake can not be measured directly, it has
to be calculated indirectly using the flux balance principle. For
Na, K, Ca, Mg, Mn, P and S the rate of ion uptake is assumed to
be the sum of the rates of storage within the increment, of
litterfall and of plant leaching. The separation of plant
leaching from dry deposition was discussed in detail by Ulrich
et al. (13). Uptake of Al and Fe results from the sum of increment and the amount of elements stored in the green leaves
(needles). The uptake of Cl was neglected because of its discrimination during ion uptake (9). As mentioned above N plays in two
ways a significant role during mineralization and ion uptake. On
the one hand N is the quantitatively most important element and
on the other hand it is taken up to unknown fractions in the form
of NH_4 and NO_3. The rates of N uptake were determined assuming
that the rate of NH_4 and NO_3 measured in the flux "input to the
mineral soil" (data from funnel lysimeters), deducting NH_4 and
NO_3 output with seepage water are taken up in this ratio. The
difference between the rate of uptake of all cations (including
NH_4) and the rate of uptake of all anions (including NO_3) yields
the rate of H^+ production connected with ion uptake.
All calculations were done for each year from 1971 - 1980. This
approach of evaluation includes the following assumptions and
neglections. Because of the neglection of the fine root dynamic,
rates of ion uptake and mineralization within the mineral soil
are underestimated. Assuming that this "little ion cycle" shows
no discoupling, the effect on the H^+ budget of the ecosystem will
be neglegtible. Furtheron, changes of N store in the mineral soil
are neglected. Nevertheless, changes of fine root and organic
matter store within the mineral soil are to be expected and if
included may modifie the rates of H^+ turnover calculated.

3. RESULTS

The annual rates of cations (without H^+) and anions mineralized
within the humus layer are given for the beech and the spruce
stand in tab. 1. In addition the changes of H^+ fluxes during
mineralization are presented.
Since only total N values (NH_4+NO_3) have been analytically determined in 1969 and 1970, this calculation can not be made for this
both years.
The mineralization of cations leads to the formation of base (H^+
consumption) while the mineralization of anions is followed by
the formation of acids (H^+ production (14)). The difference of
both these processes (row 3 tab. 1) amounts to the theoretical
H^+ turnover during mineralization. Negative values in row 3
correspond to a net H^+ consumption, positive values indicate a net

Table 1 : Cation/anion balance of mineralization within the humus layer
$(keq.ha^{-1}.a^{-1})$

BEECH

		1971	1972	1973	1974	1975	1976	1977	1978	1979	1980
1	$\sum$ cations mineralized	0.9	0.8	2.0	3.5	1.9	1.3	2.7	2.8	3.3	2.3
2	$\sum$ anions mineralized	0.9	-0.2	1.4	3.3	2.5	2.0	2.7	2.1	2.7	2.8
3	2 - 1	0.0	-1.0	-0.6	-0.2	0.6	0.7	0.0	-0.7	-0.6	0.5
4	H^+ production measured (eq.1)	0.9	0.6	0.6	0.9	0.5	0.7	1.5	1.2	0.9	1.7
5	4 - 3	0.9	1.6	1.2	1.1	-0.1	0.0	1.5	1.9	1.5	1.2

SPRUCE

		1971	1972	1973	1974	1975	1976	1977	1978	1979	1980
11	$\sum$ cations mineralized	1.8	2.5	2.6	3.0	1.5	1.5	0.8	2.1	2.3	1.9
12	$\sum$ anions mineralized	0.5	1.7	1.7	0.9	0.5	1.5	-1.4	1.6	1.0	1.7
13	12 - 11	-1.3	-0.8	-0.9	-2.1	-1.0	0.0	-2.2	-0.5	-1.3	-0.2
14	H^+ production measured (eq.1)	-0.8	0.0	0.1	-0.6	-0.5	-0.1	-1.9	-0.6	-0.7	0.1
15	14 - 13	0.5	0.8	1.0	1.5	0.5	-0.1	0.3	-0.1	0.6	0.3

H^+ production. If the flux of anions with "canopy drip" (F12) exceeds the "input to the mineral soil (F24), negative values for the flux of mineralized anions result (Beech 1972, spruce 1977).

Under beech, the balance of mineralized cations and anions indicates an equilibrium in the years 1971 and 1977, that is no H^+ turnover by mineralization should occure in these years. In 1975, 1976 and 1980 a net H^+ production is obvious due to high nitrification rates in these years. For all other years investigated, theoretical the effect of mineralization should be a net H^+ consumption. Comparing the theoretical calculated H^+ turnover with the fluxes of H^+ measured in the field, a satisfying agreement is only observed in 1975 and 1976. For the other years the net H^+ production measured is much higher than should be expected from the mineralization of anions. The difference between calculated and measured H^+ turnover (row 5 tab. 1) is attributed to a net efflux of dissociated organic acids from the organic top layer. The rate of H^+ production by the formation of dissolved organic acids reaches rather high values and shows a great variation over the period of investigation. The low rates in 1975 and 1976 indicate better conditions for mineralization and a high microbial activity as a result of temperatures somewhat closer to the optimum during the vegetation period of these years.

This results first in an increased rate of ammonification. As
long as nitrification does not follow immediately, the accumulation of NH_4 is connected with the consumption of one proton
per N atom, that is with a pH increase in the microcompartments
where this process is occuring. As a consequence, the solubility
of Al and heavy metals will decrease and the oxidation of water
soluble phenolic compounds will increase, leading to the formation of higher polymerized water insoluble organic substances.
For the spruce stand, H^+ consumption during mineralization is
evident from the data measured in the field except 1973 and
1980 (row 14). The rates of H^+ turnover calculated from the cation/
anion balance and the rates of H^+ turnover measured do not agree
for most of the years investigated. Only in 1976 and 1978 the
charge balance is close to zero (row 15 tab. 1). The difference
between H^+ turnover calculated and H^+ turnover measured is again
attributed to a net efflux of dissociated organic acids from the
humus layer. The average rate of the production of dissolved
and dissociated organic acids within the humus layer of the
spruce stand is about 50 % of that for the beech stand. The
reason for the lower output rates of dissolved organic acids
from the humus layer under spruce may be the lower pH values of
the percolating water. The average pH value from 1969 to 1979
in the water collected below the humus layer was 3.43 under
spruce and 3.65 under beech. This lower pH can result in a
decrease of the dissociated fraction of organic groups.
The cation/anion balances for ion uptake are given for both
stands in table 2.
Because of missing data for the element flux "output with seepage water" (OS), the calculation of N uptake for the spruce
stand was only possible from 1973 onword.
The rate of ion uptake for the beech stand is relatively low in
1971 and 1972 as a result of limited N mineralization in these
years. Calculating N uptake, this leads to a rate of N uptake
which is much lower than the N requirement of the stand. Balancing annual element fluxes for the ecosystems (7), this difference was attributed to a taking out from the N storage of the
trees.
With the exception of 1977, ion uptake of the beech stand is
connected with a considerable H^+ production, due to a surplus of
cation uptake rate over anion uptake rate. This surplus of
cations results mostly from the unfavourable NH_4/NO_3 ratio during
ion uptake.
The ion uptake of the spruce stand yields much higher rates of
H^+ production than in the beech stand. The NH_4/NO_3 ratio during
ion uptake is higher under spruce, hence the sum of cation
uptake is much higher than the rate of anion uptake.
Assuming steady state conditions, the H^+ production following
ion uptake should be connected with an equivalent H^+ consumption
during mineralization. Because of temporal and spatial discouplings of the ion cycle, steady state conditions are not

Table 2 : Cation/anion balance of ion uptake (keq.ha^{-1}.a^{-1})

B E E C H

		1971	1972	1973	1974	1975	1976	1977	1978	1979	1980
1	$\sum$ cations taken up	3.4	3.4	5.5	5.6	4.5	3.8	3.7	4.9	5.3	5.1
2	$\sum$ anions taken up	2.0	1.1	2.4	4.2	3.3	3.0	3.6	3.7	3.6	4.2
	2 - 1 (H$^+$production)	1.4	2.3	3.1	1.4	1.2	0.8	0.1	1.2	1.7	0.9

S P R U C E

				1973	1974	1975	1976	1977	1978	1979	1980
11	$\sum$ cations taken up			7.3	7.8	6.0	4.4	4.9	6.1	6.6	6.4
12	$\sum$ anions taken up			2.4	1.5	1.7	1.6	1.1	1.7	2.8	2.7
	12 - 11 (H$^+$production)			4.9	6.3	4.3	2.8	3.8	4.4	3.8	3.7

reached and ecosystem internal net production of protons results.
The accumulation of biomass either in form of increment or in
form of humus can be defined as temporal discoupling. In the
case of the beech and the spruce stand, mineralization of the
accumulated biomass will be followed by H$^+$ consumption in most
of the years. In the case of accumulation of biomass, proton
production by ion uptake exceeds proton consumption by mineralization.
The cation exchange reactions occuring in the crown with the
effect of H$^+$ consumption during canopy passage (7) can be considered as a spatial discoupling of the ion cycle. Partly, the
H$^+$ consumption connected with mineralization takes place in the
crown by ion leaching. But this buffering action of the stand
is further transferred into the mineral soil via the cation/
anion balance of ion uptake and the connected H$^+$ production.

Starting from the question, which processes within the ion cycle
of the investigated stands are responsible for the observed
ecosystem internal H$^+$ net production, mineralization and ion
uptake were regarded separated from each other. The comparison
of the H$^+$ turnover from these processes with the rates of ecosystem internal H$^+$ net production calculated from the flux
balance of the ecosystems (5) is given in tab. 3. The calculation of the ecosystem internal H$^+$ net production was done from

the cation anion balance of the changes of element storage
within the soil and is not totally independent from the other
calculations, but includes mineralization processes within the
mineral soil. The rates of total ecosystem internal H^+ net
production now include the changes of storage of organic anions
inside the mineral soil which were not considered in the previous paper (7).

Table 3: Comparison of the total ecosystem internal H^+ net production
with the rates of H^+ turnover from mineralization and ion uptake
$(keq.ha^{-1}.a^{-1})$

	BEECH	1971	1972	1973	1974	1975	1976	1977	1978	1979	1980	x̄ 1971-1980
1	total ecosystem internal H^+ net production	3.2	2.3	3.1	1.5	2.6	1.6	1.2	1.8	2.4	2.7	2.2
2	H^+ production during mineralization	0.9	0.6	0.6	0.9	0.5	0.7	1.5	1.2	0.9	1.7	1.0
3	H^+ production during ion uptake	1.4	2.3	3.1	1.4	1.2	0.8	0.1	1.2	1.7	0.9	1.4
4	2 + 3	2.3	2.9	3.7	2.3	1.7	1.5	1.6	2.4	2.6	2.6	2.4
	SPRUCE											x̄ 1973-1980
11	total ecosystem internal H^+ net production			4.1	5.6	4.1	2.9	1.2	4.5	2.7	4.6	3.7
12	H^+ production during mineralization			0.1	-0.6	-0.5	-0.1	-1.9	-0.6	-0.7	0.1	-0.5
13	H^+ production during ion uptake			4.9	6.3	4.3	2.8	3.8	4.4	3.8	3.7	4.2
14	12 + 13			5.0	5.7	3.7	2.7	1.9	3.8	3.1	3.8	3.7

The comparison of the rates of ecosystem internal H^+ net production (row 1) with the rates of H^+ turnover following mineralization within the humus layer and ion uptake indicate that
the sum of both processes is of the same order of magnitude. The
total ecosystem internal H^+ net production can be seen as the
result of these two processes. The data for the spruce stand
show better agreement than those for the beech stand. Since
there are plenty of assumptions and neglections for these calculations, best agreement is not expected.

4. DISCUSSION

In spite of the restrictions mentioned above, mineralization
and ion uptake when considered separately give an important
inference concerning the H^+ turnover   in   the

ecosystem. For the beech stand mainly the input of dissolved
organic anions in the mineral soil was responsible for H^+ net
production inside the humus layer. In the absence of dissolved
organic acids mineralization should be followed by a H^+ consumption. The formation of water soluble organic acids appears
to have resulted from the detrimental conditions for decomposition which are, under our climatic conditions, due to toxic
substances present in the soil solution (Al, heavy metals,
phenols).

The deciding factor controlling the rates of H^+ turnover by ion
uptake is the ratio of NH_4/NO_3. This was also emphasised by
Ulrich (13,14) and Andersson (1). Only under the condition that
N is mainly taken up in the form of NO_3 (NH_4/NO_3 ca. 0.2 for
beech) the cation anion balance for ion uptake is equalised
and no H^+ turnover results.

The NH_4/NO_3 ratio of the annual element fluxes "input to the
mineral soil" varies between 0.2 and 1.1 under beech, under
spruce between 0.2 and 11. According to this ratio, the H^+
production resulting from ion uptake reach higher values under
spruce. The equivalent H^+ turnover during mineralization,
which should be a H^+ consumption in the case of H^+ production
during ion uptake, is prevented by spatial (H^+ buffering in the
crown) and temporal (accumulation of biomass) discoupling of
the ion cycle and by the formation of organic acids and net
production of protons results.

The ratio of NH_4/NO_3 has a significant effect on the cation/
anion balance and the connected proton turnover of mineralization
and ion uptake when regarded separately. Looking at a system
without significant input and output of N, rates of N uptake
and mineralization are equal even in the case of biomass
accumulation. N is transformed from the soil organic N store
with a charge of O by mineralization and ion uptake into the
biomass N store also having the charge O. Equal rates of N
mineralization and N uptake means that the N cycle shows no
discoupling and no net production of protons will result from
N cycling. Hence the ratio of NH_4/NO_3 has no influence on the
rate of ecosystem internal proton production. Net production
of protons will only occur if the rate of N mineralization
exceeds the rate of N uptake caused by climatic conditions or
by "humus disintegration" (Ulrich this volume). Providing a
closed N cycle protons are produced in the soil by the surplus
of"cation excess" uptake rate over "cation excess" mineralization rate, a condition that occurs during accumulation of
biomass. (The term "cation excess" stands for all cations
except NH_4 deducting all anions except NO_3). The accumulated
cation excess is taken from the exchangeable pool of the soil
and protons must be released to keep up electroneutrality.
Summarizing these considerations, one can conclude that
ecosystem internal proton production will occur in every
ecosystem in the case of biomass accumulation independently

from the dominant form of N in the soil solution. The rates of
proton production given by Nilsson (this volume) who used
defined ratios of NH_4/NO_3 calculating the rate of cation
accumulation therefore may be overestimated in this respect.

Acid precipitation was shown to cause soil acidification
(4,15) and deterioration of decomposition (2,11,12). Deterioration of decomposition will lead to accumulation of biomass
therefore increasing rates of ecosystem internal H^+ net production are the consequences.

5. LITERATURE

1: Andersson, F.T., T. Fagerström and S.I. Nilsson (1980):
      Forest ecosystems responses to acid deposition -
      hydrogen ion budget and nitrogen/growth model approach.
      In: T.C. Hudchinson and M.Havas (ed): Effect of acid
      precipitation on terrestrial ecosystems. Plenum press
      New York, 319-334.
2: Baath, E., B. Berg, U. Lohm, B. Lundgren, H. Lundkvist,
      T. Rosswall, B. Söderström and A. Wiren (1980): Soil
      organisms and litter decomposition in a Scots Pine forest
      - effects of experimental acidification. In: Effect of
      acid precipitation on terrestrial ecosystems. Ed. T.C.
      Hutchinson and M. Havas, Plenum, New York, 375-381.
3: Baum, U. (1975): Stickstoff-Mineralisation und Stickstoff-
      Fraktionen von Humusformen unterschiedlicher Waldökosy-
      steme. Göttinger Bodenkundliche Berichte 38, 1-96.
4: Butzke, H. (1981): Versauern unsere Wälder? Erste Ergebnisse
      der Überprüfung 20 Jahre alter pH-Wert Messungen in Wald-
      böden Nordrhein-Westphalens. Forst und Holzwirt 36, 542-
      548.
5: Ellenberg, H. (ed) (1971): Integrated experimental ecology,
      methods and results of ecosystem research in the German
      Solling project. Ecological studies 2.
6: Likens, G.E., F.H. Borman, R.S. Pierce, J.S. Eaton and N.M.
      Johnson (1977): Biogeochemistry of a forested ecosystem.
      Springer Verlag, 1-146.
7: Matzner, E. und B. Ulrich (1981): Bilanzierung jährlicher
      Elementflüsse in Waldökosystemen im Solling. Z.Pflanzen-
      ernaehr.Bodenkd. 144, 660-681.
8: Matzner, E., P.K. Khanna, K.J. Meiwes, M. Lindheim, J.Prenzel
      und B. Ulrich (1982): Elementflüsse in Waldökosystemen im
      Solling - Datendokumentation - Göttinger Bodenkundliche
      Berichte 71, 1-266.
9: Prenzel, J. (1979): Mass flow to the root system and mineral
      uptake of a beech stand calculated from 3-year field data.
      Plant and Soil 51, 39-49.

10: Sollins, P., C.C. Grier, F. M. McCarison, K. Cormack,Jr. and R. Foyel (1980): The internal element cycles of an old growth Douglas fir ecosystem in Western Oregon. Ecological Monographs 50, 261-285.

11: Strayer, R.F. and M. Alexander (1981): Effects of simulated acid rain on glucose mineralization and some physicochemical properties of forest soils. J.Environ.Qual. 10, 460-465.

12: Strayer, R.F., C.J. Lin and M. Alexander (1981): Effects of simulated acid rain on nitrification and nitrogen mineralization in forest soils. J.Environ.Qual. 10, 547-551.

13: Ulrich, B., R.Mayer und P.K. Khanna (1979): Die Deposition von Luftverunreinigungen und ihre Auswirkungen in Wald-ökosystemen im Solling. Schriften Forstl.Fak.Univ.Göttingen, Bd.58, 1-279, Sauerländers Verlag Frankfurt.

14: Ulrich, B. (1981): Theoretische Betrachtungen des Ionenkreislaufs in Waldökosystemen. Z.Pflanzenernaehr.Bodenkd. 144, 647-659.

15: Ulrich, B., R.Mayer and P.K. Khanna (1981): Chemical changes due to acid precipitation in a loess derived soil in Central Europe. Soil Science 130, 193-199.

EFFECTS ON SOIL CHEMISTRY AS A CONSEQUENCE OF PROTON INPUT

S. Ingvar Nilsson

Department of Ecology and Environmental Research
Swedish University of Agricultural Sciences
S-750 07 Uppsala, Sweden

ABSTRACT

    Documented increases of soil acidity in Swedish forest soils
are discussed. It is concluded that to date no unequivocal evidence exists, that points to a soil acidification mainly caused
by atmospheric deposition. Tree species replacement and ion accumulation in plant biomass and humus seem to be the most important
causes. Deposition of SO_4^{2-} and SO_2 is important for the lake
acidification process through interactions with protons and alu-
miniumhydroxyions in the soil, although the abovementioned ion
accumulation in the terrestrial system is also likely to be a
contributing factor.

INTRODUCTION

    In a steady-state ecosystem proton fluxes induced by unequal
fluxes of cations and anions in the ion uptake by plants are balanced by reverse fluxes during the complete mineralization of
organic matter (1, 2, 3). This claim is essentially correct if a
sufficiently large area is taken into account and the processes
are regarded in the time-scale of several hundred years. If a
disturbance, natural or man-made, is imposed on the system, one
of the consequences is a discoupling of the ion cycle sensu
Ulrich (4). This results in a net acidification or alcalinisation
depending on the direction of the processes. The discoupling
manifests itself by a relative retardation of either ion uptake
or mineralization. A retarded mineralization shows up as an accumulation of soil organic matter ($\approx$humus) and can be induced for
instance by tree species replacement. If a deciduous species such

105

B. Ulrich and J. Pankrath (eds.), Effects of Accumulation of Air Pollutants in Forest Ecosystems, 105–111.
Copyright © 1983 by D. Reidel Publishing Company.

as European beech is replaced with e.g. Norway spruce a build-up
of a mor-layer can take place during the early stages of the
spruce stand development. Harvesting of biomass is another way
of discoupling the ion cycle as ions previously taken up by the
trees, are exported from the system, and part of the reverse flow
of protons during mineralization is prevented.

The relative importance of biological proton transfers connected with ion uptake and mineralization versus the proton input from atmospheric deposition as determining factor for soil
acidity and proton transport in the soil solution has been a
matter for discussion for quite a long time (3, 5, 6, 7). It is
widely accepted that atmospheric input is an important factor
for the acidification of softwater lakes in Scandinavia, northeastern U.S. and eastern Canada, although it is far from self-
evident, that it is the only factor (6, 7, 8). As far as soils
are concerned the matter is far from being settled.

The most straightforward way to detect changes in soil chemical properties such as acidity or pH, is to make a number of
samplings and chemical analyses within appropiate time intervals.
This approach does not allow any farreaching conclusions concerning the actual cause or causes behind observed changes but
some inferences are permissible, if the similar type of data is
available for a number of sites with sufficiently similar soil-
and vegetation properties. The ideal is of course to combine the
soil inventories with actual measurements of ion fluxes as has
been done within the Solling project (2, 9).

In this paper an evaluation is given of some reported pH and
acidity changes in Swedish forest soils. Furthermore, a theoretical calculation is presented which estimates the proton fluxes
connected with different combinations of tree thinning and final
tree harvest. The influence from different tree species will also
be shown. Finally the occurrence of basic aluminium sulphates and
their importance as buffer substances are discussed.

RESULTS

Soil pH changes due to tree species replacement were documented by Nihlgård (10). He studied 23 sites with adjacent stands
of European beech and Norway spruce, the latter being planted on
former beech forest soil. A pH decrease of 0.5 units was recorded
in the topsoil. This was mainly due to an increased accumulation
of mor humus. A decrease could be traced down to the B/C-horizon
(appr. 0.2 units). The pH in the beech forest soils was in the

order of 4.0 and 4.5 for topsoils and B/C-horizons respectively.
At three of the studied sites a thorough soil investigation was
undertaken, which showed a marked increase of the acidity in the
topsoil of the spruce stands. Base saturation went down from
10-15 units to 5-8 units (10).

The importance of the dominating tree species as a soil pH
determinator has also been shown by Troedsson (11). pH in the
humus layer was 4.1, 4.0, 4.3 and 4.5 in stands of Norway spruce,
Scots pine, European beech and birch respectively. The results
were based on a total of 3538 plots.

The importance of biomass harvest as a means of discoupling
the ion cycle was discussed by Ulrich (4). Different forest
practises and their possible influence on soil fertility and soil
acidity have also been a topic for discussion in Sweden during
recent years. Theoretical calculations indicate that whole tree
harvesting induces a proton flux equal to or larger than the proton input in atmospheric deposition, even when quite high deposition levels are considered. The actual figures based on the ion
capital in the harvested biomass are of course strongly dependent
on site quality and the growth rate, which in turn determine the
ion accumulation rates. (3, 12). The following calculations which
refer to a Scottish stand of Corsican pine planted on sand dunes
(13) are cited from Nilsson, Miller and Miller (3). In this
special case the ratio between NH_4^+ and NO_3^- is supposed to be 3 :1.
The ratio is important provided that the atmospheric input of
nitrogen is quantitatively significant compared to the mineralization rate, or if there is any substantial leaching of NO_3^-.

The various permutations shown in Table 1 indicate a steadily increasing proton flux with increasing biomass export. This is
partly confirmed from data by Bringmark (14). He made a lysimeter
study at a clearfelled site, which was divided into a number of
plots where either the slash was left of removed. After two years
there was a marked decrease in the H^+ concentration in soil water
collected under plots with remaining slash compared to soilwater
collected under plots where the slash had been removed (53 µM-or-
ganic topsoil; 40 µM 20 cm down in the mineral soil). The former
tree stand was 120-year-old Scots pine. The soil was a nutrient
poor sandy iron podzol. The lowest H^+ concentrations were connected with the highest concentration of organic carbon, which
could be taken as an indication of a higher mineralization rate
in the plots with slash than in the plots without slash. (15).

Harvesting policy	Growth rate	Acidity on harvesting
A	Slow	0.10
	Fast	0.15
B	Slow	0.16
	Fast	0.48
C	Slow	0.25
	Fast	0.80
D	Slow	0.35
	Fast	1.16

Table 1. Acidity associated with removal of harvested produce
calculated for Corsican pine. A - stems removed only at final
clearfelling; B - stems removed at final felling and in intermediate thinnings at approximately five year intervals; C - stems
removed at thinnings, but whole trees removed at final felling;
D - whole trees harvested at both thinnings and final felling.
Figures are shown for both a slow-growing stand (unfertilised)
and a fast-growing stand (fertilised) and are expressed as
keq $H^+ \cdot ha^{-1} \cdot yr^{-1}$.

So far no soil chemical changes have been found in Swedish
forest soils, which can be unequivocally attributed to influences
from acid atmospheric deposition. Troedsson (16), compared pH
values in the organic topsoil from a large number of coniferous
forest stands, which were sampled in 1961-1963, and resampled in
1971-1973. The values were lower 1971-1973 than ten years earlier, but the reason for this is unclear. For practical reasons
the measurements were performed on airdried samples which means
that the results must be treated with much caution. One thing was
evident however, namely decreasing pH values with increasing
stand age. This tendency was seen on both sampling occasions and
could mainly be attributed to increasing humus accumulation with
increasing stand age. Troedsson also showed a regional pH gradient with the lowest values in southeast Sweden and the highest
values in the central and northeastern parts of the country. This
gradient was mainly attributed to differences in geology and
climate. The soil pH gradient largely coincides with a similar
gradient in lake water pH.

It is a crucial question, to what extent lake acidification
can be attributed to the atmospheric deposition of H^+ and SO_4^{2-}.
A hypothesis stating an increased anion flux in the soil solution
due to an increased atmospheric input of SO_4^{2-} and SO_2 has been
put forward by Seip and coworkers (17, 18). This increased flux
would in turn increase the fluxes of protons and positively
charged aluminium complexes as these two cation categories are
the dominating ones in strongly acid soils. An alternative hypo-

thesis says that sulphate ions originating either from atmospheric input or from internal release processes such as mineralization and oxidation could be involved in an actual equilibrium
with some basic aluminium sulphate (2, 19, 20). It has recently
been shown that equilibria involving either jurbanite or basaluminate to a large extent regulate the concentrations of H$^+$, aluminium complexes and sulphate in the B horizon of a number of
Swedish forest soils with a fairly wide pH spectrum (21, 22).
Figure 1 shows an equilibrium diagram with the sulphuric acid
potential as abscissa and the aluminium potential as ordinate.
These types of equilibria are not necessarily unique for the last
thirty years or so which has been claimed by some authors. They
rather seem to constitute buffer systems which become of increasing importance at increasing sulphate levels, and probably
constitute one important group of factors which contribute to the
high aluminium concentrations observed in some shallow groundwater aquifers (23) and acid lakes (6, 7, 8).

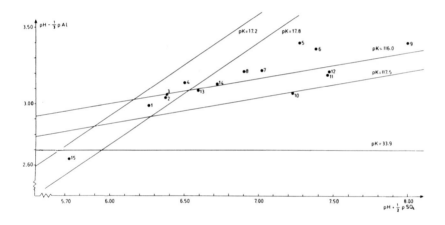

Figure 1. Equilibrium diagram for a number of Swedish forest
soils (B-horizons).
Average pH values in the soil solution:
1.-4.4; 2.-4.4; 3.-4.5; 4.-4.6; 5.-5.3; 6.-5.3; 7.-5.0; 8.-4.7;
9.-6.0; 10.-5.1; 11.-5.3; 12.-5.3; 13.-4.7; 14.-4.8; 15.-4.0.
1-4 are replicates of the same soil. 6. is the same soil as 5.
but treated with sulphuric acid of pH 3.4. 8. is the same soil
as 7, and 10 is the same soil as 9. 8 and 10 are treated with
nitric acid. 11-14 represent the same soil. 14 and 15 represent
treatment with sulphuric acid, 12 and 14 and 13 and 15 two different sampling occasions. 1-14 are lysimeter leachates. 15. represents the average composition of an equilibrium soil solution
(ESS). The pk values are all taken from reference 20 and papers
cited therein: 33.9 - gibbsite; 17.2 - Al(OH)SO$_4$ according to
van Breemen. 17.8 - jurbanite; 116.0 - amorphous basaluminite;

117.5 - crystalline basaluminite.

CONCLUSIONS

A manyfold of factors contribute to soil- and lake acidity. In Sweden no unequivocal evidence exists that points to a soil acidification mainly caused by atmospheric deposition. Documented increases of soil acidity seem mostly to be attributed to a dis-coupling of the ion cycle, caused by change of tree species, accumulation processes during the stand development and export of biomass. Changes in lakewater acidity are extensively documented (6, 7, 8) and in this case atmospheric input of mainly SO_4^{2-} and SO_2 is likely to be one important factor, although other factors such as acidification of surrounding soils due to ion accumulation in plant biomass and in the humus also could contribute. (5, 7, 24).

ACKNOWLEDGEMENT

This paper is based on work carried out within an integrated project on the effects of acid deposition on terrestrial and aquatic ecosystems, financed by the Swedish Environmental Protection Board. L. Bringmark and B. Popović are thanked for providing unpublished data and information.

REFERENCES

1. Ulrich, B.: 1980, Effects of Acid Precipitation on Terrestrial Ecosystems pp. 255-282. Eds. Hutchinson, T.C. and Havas, M. Plenum Publishing Corporation, New York.
2. Ulrich, B., Mayer, R. und Khanna, P.K.: 1979, Deposition von Luftverunreinigungen und ihre Auswirkungen in Waldöko-systemen im Solling. J.D. Sauerländers Verlag Frankfurt am Main, 291 pp.
3. Nilsson, S.I., Miller, H.G. and Miller, J.D.: 1982, Oikos in press.
4. Ulrich, B.: 1981, Z. Pflanzenernaehr. Bodenk. 144, pp. 647--659.
5. Rosenquist, I. Th.: 1977, Sur jord - surt vann. Ingenioerforlaget A/S, Oslo.
6. Overrein, L.N., Seip, H.M. and Tollan, A.: 1980, Acid Precipitation - Effects on Forest and Fish. Final Report of the SNSF Project 1972-1980. Research Report 19 175 pp.
7. Chester, P.F. (ed.): 1981, Effects of SO_2 and its Derivatives on Natural Ecosystems, Agriculture, Forestry and Fisheries. Report of an IERE Working Group. Central Electricity Research Laboratories, Leatherhead, England.

8. Drabløs, D. and Tollan, A. (eds.): 1980, Proc. Int. Conf. Ecol. Impact Acid Precip. Norway 1980, SNSF Project.
9. Matzner, E. und Ulrich, B.: 1981, Z. Pflanzenernaehr. Bodenk. 144, pp. 660-681.
10. Nihlgård, B.: 1971, Oikos 22, pp. 302-314.
11. Troedsson, T.: 1977, Sveriges Skogsvårdsförbunds Tidskrift 2-3 (special issue).
12. Nilsson, I. and Nilsson, J.: 1981, Olika källor till markförsurning SNV PM 1411, Swedish Environmental Protection Board 36 pp.
13. Miller, H.G., Cooper, J.M., Miller, J.D. and Pauline, O.J.L.: 1979, Can. J. For. Res. 9, pp. 19-26.
14. Bringmark, L.: 1981, Swedish Coniferous Forest Project. Internal Report 103, pp. 100-103.
15. Bringmark, L. Pers. comm.
16. Troedsson, T.: 1981, Monitor 1981, pp. 67-74. Ed. Bernes, C. Swedish Environmental Protection Board.
17. Seip, H.M.: 1980, Proc. Int. Conf. Ecol. Impact Acid Precip. Norway 1980, pp. 358-366.
18. Christophersen, N., Seip, H.M. and Wright, R.F. Water Resour. Res. in press.
19. Eriksson, E.: 1981, Nordic Hydrology 12, pp. 43-50.
20. Nordstrom, D.K.: 1982, Geochim. Cosmochim. Acta 46, pp. 681--692.
21. Nilsson, S.I. and Bergkvist, B., Water Air and Soil Poll. Submitted.
22. Nilsson, S.I. Nature. Submitted.
23. Hultberg, H. and Johansson, S.: 1981, Nordic Hydrology 12, pp. 51-64.
24. Harriman, R. and Morrison, B.: 1980, Proc. Int. Conf. Ecol. Impact Acid Precip. Norway 1980, pp. 312-313.

HOLOCENE VERSUS ACCELERATED ACTUAL PROTON CONSUMPTION IN GERMAN FOREST SOILS

Mazzarino, M.J., Heinrichs, H. and Fölster, H.

Institut für Bodenkunde und Waldernährung
Mineralogische Anstalten, Geochemisches Institut
Universität Göttingen

ABSTRACT

Weathering balance studies of 3 Holocene soils and one Riss-Wuerm-interglacial soil were carried out in order to obtain information on base level proton consumption during undisturbed forest conditions. The balance calculations are based on chemical analysis and chequed by calculation of mineral assemblages with the epinorm-method. Average annual proton consumption per ha amounts to 0.6 kmol in the fossil soil and 0.9 to 1.9 kmol in the Holocene soils, both on carbonate-free base.

INTRODUCTION

Industrialisation and the growing consumption of fossil fuel has led to an accelerating input of protons into forest ecosystems. ULRICH (22) furthermore assumes that internal proton production started to rise already since man began to interfer with the forests. In the beech and spruce stands of the Solling project in Central Germany, actual input and internal production of protons run up to an estimated 4 resp. 8 $keq.ha^{-1}.yr^{-1}$ (8) of which 2 resp. 4 $keq.ha^{-1}.yr^{-1}$ are due to H^+ input. This acid load can be considered more or less equal to proton consumption.
The question arises to what extent these annual rates surpass the base level of proton input and production respectively consumption during Holocene soil formation under forest previous to man's interference. Disregarding percolation losses of H^+, proton consumption should find expression in the weathering balance of the soil column. For this reason, we selected 3 Holocene forest soils on loess for such a study. Assuming a soil formation period

113

B. Ulrich and J. Pankrath (eds.), Effects of Accumulation of Air Pollutants in Forest Ecosystems, 113–123.
Copyright © 1983 by D. Reidel Publishing Company.

of approximately 10ooo years, the annual consumption rates can be
calculated from the total of silicates weathered, or the cations
lost, although these figures would average the cumulative effects
of long term base level rates and more recent accelerated
rates.
In order to obtain an idea of proton consumption in undisturbed
forest soils, we included a fossilized, Riss-Wuerm-interglacial
soil on loess, of an assumed formation time of 40 000 years.

PROBLEMS AND LIMITATIONS OF WEATHERING BALANCE STUDIES

Weathering balance studies are harrassed by a number of conceptual
and methodological uncertainties. Fundamentally, the approach is
to compare silicate contents, or the chemical composition, of the
weathered part of the soil column with an unweathered C-horizon.
However, such unweathered reference horizons still containing
$CaCO_3$ are rare in the study area (Northern Hesse, Lower Saxony)
and, when encountered, may rather form part of truncated profiles
from which the total Holocene weathering impact cannot be derived.

Uncertainties about intra-Holocene disturbances including those
caused by man's earlier activities do exist and they are not the
only reason for not concentrating the analytical effort to one
selected profile. The other reason is that strict homogeneity of
the parent material is a rare feature even in loess. Combined or
independent variation of texture and mineral composition is
common, and of these, the texturally independent variation of
mineral composition poses the greatest problem as it becomes
apparent only after completion of the rather laborious mineralogical and/or chemical analysis. Thus, from a preselected group of
40 profiles we finally selected 5 for balancing but had to drop
one after complete analysis, while in the remaining 4 profiles we
had to operate with certain assumptions and balance alternatives
in order to arrive at the most consistent solution.
Weathering balance studies can be based on data of total mineralogical or chemical analysis. We chose the second approach as
mineralogical analysis by phase contrast microscopy and
X-ray becomes rather semi-quantitative in the mineral fractions
with diameters below 20 μm, because the knowledge of how to transform grain count % or reflection intensities (X-ray) into weight
percentage figures is still very unsatisfactory, and because the
optical identification of mineral species does not indicate the
degree of imperfection due to weathering, esp. the degree of
cation replacement connected with proton consumption.
On the other hand, there is no doubt that mineralogical data provide a more satisfactory basis to control homogeneity, to discover
inconsistencies due to inhomogeneities in mineral composition,
and to obtain more specific data on clay formation and dissolution.
Still more important is the quartz content which represents the

relatively best reference base for calculating the gains and
losses of the soil horizons (1,3,7,10).
An approach that permits the combined use of chemical data without
indulging into the rather laborious direct mineralogical analysis,
has been the application of geochemical methods of calculating
mineral composition from chemical data (4,13) as proposed and
adapted by VAN der PLAS and SCHUYLENBORGH (21) and BOUMA and VAN
der PLAS (2). This method has certain problems and limitations to
be discussed below (for details see 9) but in general seems applicable for the intended purpose.
There is no doubt, however, that because of the uncertainties and
limitations involved in both the methods and the objects, the
balance results will not be able to yield more than approximative
data.

SOIL MATERIAL

The soils selected for this study can be characterized as follows:

DAS (Daseburg): Fossilized Riss-Wuerm-interglacial Hapludalf
(Parabraunerde) in 3 m thick Riss-glacial loess, with 75 cm A-
and a strongly developed argillic horizon (225 cm). Due to coverage
with carbonaceous Wuerm loess, the soil is today highly saturated
with exchangeable Ca^{2+} and near neutral (pH 6.7).

SPAN (Spanbeck, 650 mm): Aquic Hapludalf (Parabraunerde-Pseudo-
gley) in a 190 cm thick loess deposit, plateau position, under
low deciduous forest. The A-horizon is 40 cm (SA1), the argillic
horizon (SBt) 150 cm with clay contents decreasing from 27 to
22 %. The pH($CaCl_2$) rises from 3.5 to 4.3, the Al-saturation
drops from 80 to 10 %.

HOF (Hofgeismar, 725 mm): (aquic) Hapludalf (Pseudogley-Parabraun-
erde) in 175 cm loess deposit. The 30-50 cm thick A-horizon
represents a colluvium originating from earlier cultivation
practices. The material of this layer (10-18% clay) is translocated A-horizon material and mineralogically very similar to the
underlying B(clay contents similar to SPAN). The pH increases
downward from 3.1 to 4.2, the Al-saturation decreases from 88 to
16 %. The present vegetation is a mature stand of Norway spruce.

WEST (Westerhof, 720 mm): Alfic Dystrochrept
developed in 180 cm thick colluvial loess, of which the lowest
horizon represents a parautochthonous remnant of the lower B of
the original Holocene Hapludalf. The clay content (between 15
and 22 %) shows a minor bulge between 50 and 100 cm indicating
some clay movement. The pH rises from 3.1 to 6.3.

METHODS

The balance of element oxides was based on the chemical analysis
of the total soil (effectuated by X-ray fluorescence specto-
graphy) in the Institute für Geochemie, Mineralogische Anstalten,
Universität Göttingen.

For the purpose of mineral balance, the soil was separated into
fine clay (< 0.2 µm), coarse silt (20-60 µm), and coarse clay
and fine silt (0.2-20 µm). The former two fractions were analyzed
separately, the chemical composition of the latter calculated by
subtracting the fine clay and coarse silt oxides from those of
the total soil. Fine sand (normally below 5 %) was assumed to be
similar to coarse silt in composition except in SPAN which was
richer in fine sand and a separate analysis was considered
necessary.

Calculation of mineral composition was bases on the chemical composition of the 3(4) fractions after subtraction of the free Si-
and Al-oxides which were determined independently by means of a
NaOH-extraction (adapted after FOSTER (6), see 9). For the
coarser Fractions (> 20 µm) we followed BOUMA and VAN der Plas
(2) applying the Standard-Epinorm-procedure of BURRI as simplified by the authors.

In the 0.2 to 20 µm fraction, the program had to be modified as
the Standard Epinorm - like the mineralogical analysis - does not
consider the here already strongly reduced K-concentrations in
the micaceous minerals. Depending on the respective degree of
weathering, we had to assume K concentrations of 8.6 and 6.8 % K.
This assumption was counterchequed in parts by selected microscopic
counts, and in parts by alternative calculations (see 9). Also,
we had to assume an unchanging kaolinite content (4 %, in DAS 6%),
as this mineral is not considered in the original Standard
Epinorm.

On the base of the so calculated mineral composition of the soil
> 0.6 µm we arrived at the quartz concentrations needed for the
quartz factors, as well as obtained information on mineralogical
inhomogeneities. No epinorm procedure was applied to the fine
clay fraction. The Goethite-Norm proposed by VAN der PLAS and
SCHUYLENBORGH (21) and apparently successful in tropical Oxisols,
proved inapplicable in loess-derived soils rich in 2:1 layer
silicates (details see 9).

The soil column had been separated into 7-8 horizons, of which
only 3 were analysed in the described manner: the basal reference
horizon, the central part of the argillic horizon, and the upper
part of the A-horizon. The chemical composition of the latter
horizons were taken as representative for the total A- and B-
horizons respectively. The mass of the soil and grain-size
fractions, however, represent cumulative figures of all
subhorizons.

An examples for balance sheets of minerals and oxides is provided
in Tables 1 and 2 (SPAN). By means of the quartz factor

Table 1:   Weathering balance sheet, profile SPAN
(kg/m^2 or % of soil material expected)(Q: quartz,
Alb: albite, Or: orthoclase, M: mica, Kaol: kaolinite,
HM: heavy mineral).

A-horizon (f = 0.998)

	Q	Alb	Or	M	Kaol	HM
60	21.8	3.2	3.2	3.7	-	0.9
20-60	167.9	25.9	26.8	13.1	-	7.5
0.2-20	87.5	16.3	10.9	24.3	6.2	5.8
found	277.2	45.4	40.9	41.1	6.2	14.2
expected		52.4	55.2	78.3	6.6	31.6
gain/loss						
kg/m^2		-7.0	-14.3	-37.2	-0.4	-17.4
%		-1.5	- 3.0	-7.8	-0.1	-3.7

B-horizon (f = 2.724)

	Q	Alb	Or	M	Kaol	HM
60	109.5	16.0	16.0	18.7	-	4.7
20-60	378.6	68.8	89.1	36.5	-	22.1
0.2-20	268.4	67.8	38.7	85.5	21.5	32.3
found	756.5	152.6	143.8	140.7	21.5	59.1
expected		143.0	150.7	213.6	18.0	86.4
gain/loss						
kg/m^2		+9.6	-6.9	-72.9	+3.5	-27.3
%		+0.6	-0.4	-4.6	+0.2	-1.7

C-horizon

	Q	Alb	Or	M	Kaol	HM
60	81.1	11.9	11.9	13.8	-	3.5
20-60	137.3	29.7	25.8	20.9	-	10.9
0.2-20	59.3	10.9	17.6	43.7	6.6	17.3
	277.7	52.5	55.3	78.4	6.6	31.7

Clay balance (< 0.6 μm)

	expected	found	gain/loss
A-horizon	74	46	- 28 kg/m^2
B-horizon	201	288	+ 87
Gain of fine clay			+ 59
Loss of mica			-110
Dissolution of fine clay			51

Table 2: Balance calculation of element oxides, profile SPAN (in $kg.m^{-2}$), separate for free oxides and elements from silicates (found: analysis, exp(ected): calculated with quartz factor f).

		SiO_2	Al_2O_3	Fe_2O_3	FeO	MgO	CaO	Na_2O	K_2O	TiO_2	P_2O_5
A-horizon	found	385.7	39.9	5.4	3.3	1.7	1.9	5.7	12.2	3.7	—
f = 0.998	exp.	424.9	63.6	19.1	5.3	7.0	2.8	6.1	22.0	4.0	—
		−39.2	−23.7	−13.7	−2.0	−5.3	−0.9	−0.4	−9.8	−0.3	—
free oxide	found	11.6	2.9								
	exp.	8.9	3.3								
		+2.7	−0.4								
B-horizon	found	1170.7	167.0	54.1	13.4	12.5	6.3	18.4	48.7	12.7	1.0
f = 2.724	exp.	1159.7	173.5	52.0	14.4	19.1	7.6	16.6	59.9	10.9	2.2
		+11.0	−6.5	+2.1	−1.0	−6.6	−1.3	+1.8	−11.2	+1.8	−1.2
free oxide	found	52.1	30.6								
	exp.	24.3	9.0								
		+27.8	+21.6								
C-horizon	found	425.7	63.7	19.1	5.3	7.0	2.8	6.1	22.0	4.0	0.8
free oxide		8.9	3.3								
total oxides											
− dissolved		28.2	30.2	11.6	3.0	11.9	2.2	−1.4	21.0	−1.5	1.2
− lixivated		8.9	9.0								

$(\varrho_A/\varrho_C$ resp. $\varrho_B/\varrho_C)$ the amount of minerals respectively oxides
to be expected in B- and A-horizon has been calculated and
balanced against the amounts actually found. The fine clay
balance (Table 1) shows a nominal gain to be explained by the
strong tendency of silt size mica towards mechanical disminution
(5, 14,15,16,20). Because of this strong inclination we can imply
that the difference between mica lost from the soil > 0.6 μm and
fine clay gained represents the amount of fine clay dissolved.

In general, the balance sheets of the profiles DAS, WEST and SPAN
appear to be consistent and satisfactory while that of HOF suffers
from an inconsistently high quartz percentage in the silt of the
reference horizon. In this case we used an alternative balance
assuming an average quartz percent in the reference horizon. The
results should of course be considered with certain reservation.

RESULTS: PROTON CONSUMPTION DURING SOIL FORMATION

The cumulative effect of weathering can be demonstrated as change
in the mineral stock (Table 3) or as balance of losses of element
oxides (Table 4). We shall first consider the former as it provides a better qualitative understanding of the pathways of
weathering.
Of the total loss of minerals in the soil > 0.6 μm, felspars and
heavy minerals on the one hand and mica on the other contribute
with near equal rates. The oberved differences (37-65 %) can be
attributed first of all to primary differences in mica content.

Both groups of minerals show a different pattern of weathering.
While in the former, dissolution prevails over disminution, mica
tends to disminute into clay size dimension before being dissolved. K-concentration, though, may already decrease strongly in
the fine silt mica. The chemical composition of the fine clay
fraction does not show very strong changes from C- to A-horizon
as dissolution of clay is constantly compensated by addition of
disminuted mica. The percentage of the original mica thus lost
to the fine clay fraction is surprisingly uniform (32 - 40%), a
fact that might be explained by the much lower resistance of
biotite to weathering (3,11).
While there is qualitative and quantitative agreement with other
authors on this process of clay formation from silt size mica,
results differ in respect of dissolution of felspars and fine
clay. Equally strong or even more advanced weathering of felspar
was found by KUNDLER (7), SCHLICHTING, BLUME (17), BOSSE (1),
SCHROEDER (18) and BRONGER et al (3) in soils from loess and
other parent material. On the other hand, FÖLSTER et al. (5),
SCHEFFER et al. (16) and ROHDENBURG and MEYER (14) registered no
loss of felspar.
The same authors are generally less inclined to accept the

Table 3:   Absolute and relative losses during soil formation:
           results of mineral balance calculations of 4 profiles.

	SPAN	WEST	HOF	DAS
Loss of minerals in soil >0.6 μm				
kg/m^2	170	112	155	296
% of expected soil >0.6 μm	20	18	29	27
Fine clay formation from mica				
kg/m^2	110	55	58	141
% of expected mica	38	32	40	35
Fine clay dissolution				
kg/m^2	51	51	82	176
% of expected soil total	2.4	3.1	5.7	6.2
Total loss				
kg/m^2	111	108	179	331
% of expected soil total	5.2	6.6	12.4	11.7
Fine clay dissolution % of total loss	46	47	46	53

possibility of fine clay dissolution. Those working with soils
still containing carbonaceous C-horizons deny this process while
others accept it as possible. Such divergence of results or
concepts is not necessarily due to differences in methodology
applied, but may be caused by differences in the parent material
(original content of $CaCO_3$) or the conditions of soil formation
(TRIBUTH, 20); erosion of the A-horizon and subsequent lowering
of the argillic horizon may also simulate a higher gain of clay
from mica disminution, or even neoformation of clay, and thus
conceal the real extent of clay dissolution.
Comparing the four soils (Table 3), it becomes obvious that the
fossil soil DAS has suffered a higher percentage loss of minerals >0.6 μm and fine clay than SPAN and WEST (about 100 %
greater loss of total soil) while the less certain balance of
HOF shows similarity with DAS. The balance of the colluvial
profile WEST, originally selected to provide information on
weathering after human interference, had to be interpreted as
representing the total Holocene weathering impact because the
basal layer was not part of the colluvial deposit but rather the
reference horizon of the Holocene soil. Even if the HOF balance
may give an overestimation, the degree of weathering in DAS is
relatively less advanced. If the assumption of a 4 times longer
period of soil formation for DAS as compared to the Holocene
soils is correct, one might conclude that the actual state of

weathering in the latter soils has been substantially influenced
by accelerated proton consumption during the last 1000 years of
human interference. The conclusion can be supported not only by
the fact that the encreasing figures for clay dissolution in the
order SPAN, WEST, HOF agree with the concordantly greater intensity of the human influence (conifers in WEST and HOF, visible
signs of previous cultivation in HOF). We also found in the group
of 40 preselected forest soils a good correlation between the
free Al-oxide concentration (Al/clay) and these interference
indicators (9).
If we accept these results of the weathering balances as at
least approaching the real losses, one can calculate the proton
consumption involved from the amount of cations lost (Table 4).
Problems arise in the case of aluminum because the consumption
of protons replacing Al in the crystall lattice is partly or
totally compensated by proton production during the formation of
hydroxy cations or $Al(OH)_3$. However, great uncertainties exist
as to the Al-forms prevailing during the different phases of
soil formation. Considerable amounts of Al (like Si) have been
translocated vertically within the profiles and even beyond the
soil column. One can estimate that 29 to 78 % of the dissolved Al
has been leached. We don't know whether this Al was able to move
in the highly hydroxilized forms, whether less hydroxylized,
more mobile forms moved because they changed too slowly to
equilibrium forms (12) or because of intermittently more acid
phases, or whether - and what type of - organic compounds were
involved. Contrasting the balanced gain of B- and C-horizons of
2000 to 4000 kmol of free $Al_2O_3.ha^{-1}$ with the actual percolation
loss at 50 cm depth of 0.5 to 1.0 $kmol.ha^{-1}.yr^{-1}$ or an actual
H^+ input of 2-4 $kmol.ha^{-1}.yr^{-1}$ shows that dissolution and
movement of Al cannot be explained only by more recent acidification processes. The assumption of an average negative charge of
0.5 per Al as applied in Table 4 may be considered a conservative
estimate.
Total cumulative proton consumption calculated on this assumption
and disregarding the original outfit of exchangeable cations,
runs up to 0.9 - 1.9 $kmol.m^{-2}$ (or $kmol.ha^{-1}.yr^{-1}$) for the Holocene
soils and 2.5 $kmol.m^{-2}$ (or 0.6 $kmol.ha^{-1}.yr^{-1}$) for the fossil soil
DAS. This estimate neglects a possible primary admixture of
carbonates. 2 respectively 5 % $CaCO_3$ in the original loess would
increase the estimate by 0.8 to 1.3 $kmol.ha^{-1}.yr^{-1}$ for the Holocene soils and by 0.2 - 0.5 $kmol.ha^{-1}.yr^{-1}$ for the older soil.
These consumption rates compare well with those that can be
infered from the balanced cation losses (except Al) published by
STAHR (19) for South German soils which, in the majority of
profiles, vary between 0.8 and 1.3 $kmol.m^{-2}$ as against 0.6 -
1.1 $kmol.m^{-2}$ in the soils of our study.
Following the original approach, we tend to interpret the difference in the annual consumption rates between the Holocene soils
and DAS as being caused by accelerated proton input and

Tab. 4:   Total loss of cations and H^+ consumption (in $kmol.m^{-2}$) during soil formation.

	SPAN		WEST		HOF		DAS	
	x^{n+}	H^+	x^{n+}	H^+	x^{n+}	H^+	x^{n+}	H^+
Al^{3+}	0.59	0.30	0.57	0.29	1.26	0.63	1.56	0.78
Fe^{2+}	0.04	0.08	-	-	0.07	0.14	-	-
Mg^{2+}	0.30	0.59	0.11	0.21	0.13	0.26	0.30	0.60
Ca^{2+}	0.04	0.08	0.06	0.11	0.07	0.13	0.08	0.16
Na^+	-	-	0.10	0.10	0.34	0.34	0.21	0.21
K^+	0.45	0.45	0.16	0.16	0.36	0.36	0.75	0.75
total	1.42	1.50	1.00	0.87	2.23	1.86	2.90	2.50
$kmol.ha^{-1}.yr^{-1}$		1.50		0.87		1.86		0.63

production during the last 1000 years. The existing uncertainties about the real age of the Riss-Wuerm interglacial soil as well as the methodological problems of the weathering balances should, however, favour a cautious attitude.
Still, a base level consumption of 0.5 - 1 $kmol.ha^{-1}.yr^{-1}$ does not appear to be an overestimate. This figure is 5 times higher that that originally assumed by ULRICH (22) and forces us to accept the possibility that even under undisturbed forest conditions, soil formation has always experienced temporary phases of stronger acidification through internal proton production caused by fluctuations of weather conditions and resulting changes in accumulation-oxidation-tendencies of the organic matter. The cumulative effect of such temporary acidifications may explain the high rates of aluminum dissolution and displacement in the soils.
On the other hand does the presently possible rate of acidification exceed this base level by a factor of 5 to 10, an assumed cumulative average of Holocene proton production of 1.0 - 1.5 $kmol.ha^{-1}.yr^{-1}$ still by a factor of 3 to 6. There seems to be little scope at present to arrive at a closer estimate.

REFERENCES

1   Bosse, I. 1964. Verwitterungsbilanzen von charakteristischen Bodentypen aus Flugsanden der nordwestdeutschen Geest (Mittelweser-Gebiet). Diss.Univ.Göttingen.
2   Bouma, J. and van der Plas, L. 1971. J. Soil Sci. 22, pp 81-93.
3   Bronger, A.; Kalk, E. und Schroeder, D. 1976. Geoderma 16, pp 21-54.
4   Burri, C. 1959. Petrochemische Berechnungsmethoden auf äquivalenter Grundlage. Birkhäuser Verlag. Mineralogisch-

Geotechnische Reihe. Band VII.

5 Fölster, H.; Meyer, B. und Kalk, E. 1963. Z.Pflanzenernähr.,
   Bodenkd. 100, pp 1-11.

6 Foster, M. 1953. Geochim. et Cosmochim. 3, pp 143-154.

7 Kundler, P. 1959. Z.Pflanzenernähr., Bodenkd. 86, pp 215-222.

8 Matzner, E. und Ulrich, B. 1981. Pflanzenernähr., Bodenkd.
   144, pp 660-681.

9 Mazzarino, M.J. 1981. Holozäne Silikatverwitterung in mittel-
   deutschen Waldböden aus Löss. Diss.Univ. Göttingen.

10 Meyer, B., Kalk, E. und Fölster, H. 1962. Z.Pflanzenernähr.,
   Bodenkd. 99, pp 37-54.

11 Meyer, B. und Kalk, E. 1964. in: Soil Micromorphology, 109-
   129. Elsevier Publ.Co., Amsterdam.

12 Nair, V.D. 1978. Göttinger Bodenkd.Ber. 52, pp 1-122.

13 Niggli, P. 1936. Über Molekularnormen zur Gesteinsberechnung.
   Schweiz. Mineral.Petrog.Mitt. 16, pp 295-817.

14 Rohdenburg, H. und Meyer, B. 1966. Zur Feinstratigraphie und
   Paläopedologie des Jungpleistozäns nach Untersuchungen
   an südniedersächsischen und nordhessischen Lößprofilen.
   Mitt.Dtsch.Bodenkd.Ges. 5, pp 1-137.

15 Scheffer, F.; Meyer, B. und Kalk, E. 1958. Mineraluntersu-
   chungen am Würm-Löß südniedersächsischer Lößfluren als
   Voraussetzung für die Mineralanalyse verschiedener Löß-
   bodentypen. Chemie der Erde, Band 19, pp 338-360.

16 Scheffer, F.; Meyer, B. und Gebhardt, H. 1966. Pedochemische
   und kryoklastische Verlehmung (Tonbildung) in Böden aus
   kalkreichen Lockersedimenten (Beispiel Löß). Z.Pflanzen-
   ernähr., Bodenkd. 114, pp 77-89.

17 Schlichting, E. und Blume, P. 1961. Art und Ausmaß der Verän-
   derungen des Tonmineralbestandes typischer Böden aus
   jungpleistozänem Geschiebemergel und ihrer Horizonte.
   Z.Pflanzenernähr., Bodenkd. 95, pp 227-239.

18 Schroeder, D. 1955. Mineralogische Untersuchungen an Lößprofilen. Heidelberger Beit.z.Mineral. u. Petr. 4,443-463.

19 Stahr, K. 1979. Die Bedeutung periglazialer Deckschichten für
   Bodenbildung und Standorteigenschaften im Südschwarzwald.
   Freiburg.Bodenkund.Abhand. Heft 9.

20 Tributh, H. 1976. Die Umwandlung der glimmerartigen Schicht-
   silikate zu aufweitbaren Dreischicht-Tonmineralen.
   Z.Pflanzenernähr., Bodenkd. 139, pp 7-25.

21 Van der Plas, L. and van Schuylenborgh, J. 1970. Petrochemical
   calculations applied to soils-with special reference to
   soil formation. Geoderma 4, pp 357-385.

22 Ulrich, B.; Mayer, R. und Khanna, P. 1979. Deposition von
   Luftverunreinigungen und ihre Auswirkungen in Waldökosystem im Solling. Schriften aus der Forst.Fak.Göttingen
   u. der Niedersäch.Forst.Versuchsanst.,58.

Topic 3:

Effects on Chemical Soil State

SOIL ACIDITY AND ITS RELATIONS TO ACID DEPOSITION

B. Ulrich

Institut für Bodenkunde und Waldernährung der
Universität, Büsgenweg 2, D-3400 Göttingen

ABSTRACT

The nature of soil acidity as well as measures of the capacity
and the intensity terms are discussed. According to the proton
buffer reactions occuring in soils, buffer ranges are distinguished.
They are defined by pH values. Forest soils on limestones which
should be in the calcium carbonate buffer range, acidify under
the influence of acid deposition as soon as the fine earth is
free of calcium carbonate. The same may be true for soils
staying in the silicate buffer range if the rate of acid load
exceeds the rate of acid buffering by base cation release
during silicate weathering. From existing data on the rate of
acid deposition in Central Europe, it is concluded that soils
staying in the cation exchange buffer range should have lost
considerable amounts of exchangeable Ca due to acid deposition
since beginning of industrialization. The resilience of the
ecosystem becomes very limited if the soil stays with all major
horizons in the aluminium or even in the iron buffer range.
The iron buffer range is characterized by podzolization.

NATURE OF SOIL ACIDITY

Soil acidity has a quantitative (capacity) and a qualitative
(intensity) aspect. The capacity can be defined as the equivalent
sum of acids which can be neutralized by addition of a strong
base to pH 7 or 8. The intensity is expressed as the thermodynamic activity of the protons and is measured as pH value,
it is determined by the strength of the acids controlling the
proton activity in the soil solution.

B. Ulrich and J. Pankrath (eds.), Effects of Accumulation of Air Pollutants in Forest Ecosystems. 127–146.

In soils, two different kinds of Brønsted acids can be
distinguished, R-OH and M^{x+}, with the following proton transfer
reactions:

$$R-OH + H_2O \rightleftharpoons H_3O^+ + R-O^-$$
$$\text{acid} \quad \text{base} \qquad \text{acid} \quad \text{base}$$

$$M^{x+} + 2\ H_2O \rightleftharpoons H_3O^+ + MOH^{(x-1)+}$$
$$\text{acid} \quad \text{base} \qquad \text{acid} \quad \text{base}$$

Strong mineral acids like H_2SO_4, HNO_3 and HCl play an increasing
role at pH $<$ 4 and a dominant role at pH $<$ 3.
It is a very important feature of soil chemistry that $MOH^{(x-1)+}$
is not only a base but can react also as acid:

$$MOH^{(x-1)+} + 2\ H_2O \rightleftharpoons H_3O^+ + M(OH)_2^{(x-2)+}$$
$$\text{acid} \qquad\qquad \text{base} \qquad \text{acid} \quad \text{base}$$

R symbolyses carbon (mainly phenolic hydroxyl groups and carboxy
groups), Si (pH dependent charge of clay minerals, dissolved
silicic acid at pH above 7) and CO (carbonic acid). M symbolyzes
cation acids, that is NH_4^+ (van BREEMEN, this volume) and metal
cations like Al (PRENZEL, NILSSON, this volume), Mn, Fe, and
heavy metals including their anionic complexes. The accumulation
of positively charged ions of these metals in soil corresponds
to the accumulation of acids.

MEASURES OF SOIL ACIDITY

The amount of acid (capacity term) can be determined by the
reaction of a base in solution with the acids in solution and
accumulated in the solid phase. This can be done by a continuous
or discontinuous titration procedure with a strong base. The
amount of base consumed to pH 8 corresponds to the base
neutralization capacity (BNC). A differentiation according to
acid strength can be made by recording the buffer curve
(amount of base consumed versus pH).
The acids reacting include ionic species of Al, Mn, Fe and
heavy metals (which may exist in different binding forms at
outer and inner particle surfaces), soluble metal salts like
AlOHSO$_4$ (which releases cation acids by dissolution),
"exchangeable" protons at particle surfaces (including soil
organic matter), and NH$_4$ ions (at pH above 7). The binding forms
of cation acids and "exchangeable" protons are not fully understood. There is no doubt that the reactions occuring after
addition of a base to an acid soil sample are complex and
kinetically limited. This means that the buffer curve is a
function of reaction time and of the change of pH achieved after
the addition of a base increment. The reaction time is limited
not only for practical reasons, but also by unwanted microbial

and chemical reactions which release acids or bases and may
thus adulterate the buffer curve (e.g. ammonification, nitrification, denitrification, sulfate reduction). The buffer curve
and BNC are therefore conventional measures. To approach
reality best it is suggested to limit the reaction time to
24 hrs., to exceed never pH 7 to 8, and to use closed bottles
(to avoid NH_3 losses if pH exceeds intermittently the value of
7).
Another way is to determine separately the different acids and
acid forming substances present. This may be done by determining
total and effective cation exchange capacity and the different
cations balancing the negative charge of the soil exchangers
(including H, NH_4, Al, Mn, Fe). Difficulties exist especially
with the determination of acid forming metal salts like
$AlOHSO_4$.
The buffer curve and BNC are recommended as standard procedure,
with data about CEC exchangeable cations as additional information about the kind of acids.
Also the pH value as intensity parameter is subjected to
systematic errors. Only in very rare cases the soil solution
can be sampled directly and used for pH measurement. In most
cases water or a solution has to be added to the soil in order
to get a solution or suspension in which the measurement can be
made. Strictly spoken, the soil itself has no pH, only the soil
solution, and their pH depends upon the kind how the solution
(suspension) has been prepared.
The best approach to reality are the pH of the equilibrium soil
solution (pH(ESS)) and the pH of a soil/water suspension
(pH(H_2O)). These values can be expected to represent the
activity (mean free energy) of protons in the natural soil
solution at the time of measurement. Since the acids existing
in the solid phase and determining pH need not to be of constant
composition, these pH values can vary as a function of time. The
variation between seasons may exceed one pH unit (19). These
variations reflect real changes in soil chemistry and may be of
ecological significance.
To avoid the variation characteristic to pH(H_2O), pH is often
measured after addition of salt solutions. Due to the reactions
taking place between the added cations and cation acids present
in the solid phase, the pH of such a soil solution depends upon
the kind of the cation (K or Ca) and the salt concentration
used. In soils with illitic clay minerals, K^+ has a specific
strong exchange power for cation acids and decreases pH therefore
more than Ca^{2+}. Such differences can be used to estimate the
kind and binding form of cation acids present in soil, on the
basis of existing knowledge and experimental correlations.
Since pH(salt) is determined by more strongly bound acids than
pH(H_2O), it is changing less in time. Changes in acid composition
of the solid phase during acidification pushes and deacidification phases may also be reflected in pH(salt).

In a number of investigations it has been shown (1,2,3,4) that
the pH(salt) values of forest soils in West Germany have
decreased during the last 5 to 30 years. A decrease in the mean
pH(salt) of a group of soils indicates that an accumulation of
stronger acids in soil has occured. For soils being in a steady
state, that means not subjected to acidification, the natural
variation in pH(salt) should not result in a decrease of the
mean value.
Whereas pH(H_2O) is an approximation to the activity of protons
in the true soil solution at the time of measurement, pH(salt)
is an approximation to the activity of protons in the soil
solution in case of acid or salt load. Both values together give
in fact a good insight into the actual and potential ecochemical
conditions in the soil environment of microorganisms and plant
roots. In the following buffer ranges are defined as a framework
for interpretation of pH values.

BUFFER RANGES IN SOILS

Considering the stability of minerals and oxides, and the possible buffering reactions involving protons, the following
buffer ranges can be distinguished (5):
-   calcium carbonate buffer range (pH > 8 to 6.2)
-   silicate buffer range (silicates being the only buffer
    between pH 6.2 and 5.0)
-   cation exchange buffer range (pH 5.0 to 4.2)
-   aluminium buffer range (pH 4.2 to 2.8)
-   iron buffer range (pH 3.8 to 2.4)
In the following, buffer capacities are calculated with a bulk
density of 1.5 and for a soil layer of 1 dm thickness and 1 ha
area.

CALCIUM CARBONATE BUFFER RANGE

Only soil horizons containing $CaCO_3$ in their fine earth fraction
stay exclusively in this buffer range. In these horizons the
dissolution rate of $CaCO_3$ is high enough to keep the system
close to the equilibrium described in equation 4:

$$CaCO_3 + H_2O + CO_{2\,(g)} \rightleftharpoons Ca^{2+} + 2\ HCO_3^-$$

The dominating acid is carbonic acid which is produced in large
quantities by root and decomposer respiration. The buffer
capacity amounts to 150 kmole H^+ per 1 % $CaCO_3$. The buffer rate
is high, as long as it is determined by the rate of dissolution
of $CaCO_3$. If $CaCO_3$ is distributed unevenly and diffusion becomes
the rate limiting step, the buffer rate can be considerably
lowered. Depending upon the rate of CO_2 production and of pH,

the concentration of Ca^{2+} and HCO_3^- in the soil solution may
be very high. Due to this high concentration the soil fabric is
very stable. Ca is the dominating cation in the soil solution
and at the exchanger surface. The main form of inorganic phosphates are calcium phosphates. Toxins like Al ions and water
soluble phenols are missing. The organic acids formed in soil
are highly polymerized, water insoluble, and possess weak acidic
phenolic groups (humic acids).
There is therefore no soil-borne limitation due to toxicity to
bacterial activity and growth of plant roots. The humusform is
mull, indicating a high activity of soil burrowing animals like
earth worms and a rapid bacterial degradation of root and leaf
litter if humidity is high enough. The high bacterial activity
is a precondition for the formation of stable humic substances.
The soil organic matter formed has a high N content and a low
C/N ratio (around 10). The level of soil organic matter accumulated is increasing with increasing rate of litter production
and decreasing mean annual temperature.
Limiting factors for phytomass production may be:
- unfavourable cation ratios due to the dominance of Ca^{2+} (e.g.
  K/Ca, relative K deficiency)
- unbalanced NH_4^+/NO_3^- ratio in ion uptake. There may be a
  large nitrate surplus in ion uptake due to strong nitrification and uptake of nitrate
- limited solubility of metal trace elements may result in Mn,
  Fe, Cu, Zn deficiency
The role of these limitations is the smaller, the larger the
internal turnover in soil. The turnover within the mineral soil
can be regarded as being proportional to the rate of decomposition of phytomass. Forest ecosystems which maintain a closed
herb layer including legumes, provide a source of food for
decomposers at high quantity with high quality. Higher turnover
rates of soil organic matter may narrow the NH_4^+/NO_3^- ratio in
ion uptake by allowing the plants to take up some N as NH_4, and
can also improve the uptake of metal trace elements as organic
chelates.
Forests growing on calcareous soils buffer the acidity of the
incoming rain during the vegetation period very efficiently at
the leaf surface. The pH in the canopy drip can be higher than
in the incoming rain (e.g. 4.82 compared to 4.39, from the input
amounting to 0.6 kmol H^+ $ha^{-1}yr^{-1}$ 0.24 kmol H^+ have been
buffered (7)). This buffering is finally taken over by the soil
close to the root surface (ULRICH , this volume). The most
common tree species on calcareous soils in Central Europe is the
beech (Fagus silvatica). For this species the stem flow may
amount to 15 % of the precipitation. The stem flow shows no
buffering, it may be more acid than the incoming rain, indicating
interception deposition. Also the stem flow can be assumed to
reach preferably the soil close to the root surface. From this
it follows that in beech ecosystems on calcareous soils the

larger fraction of the deposited acidity acidifies not the soil
surface, but the soil close to the root surface even in greater
soil depth. Thus the most susceptible soil part in respect to
tree roots is directly affected. Only the acidity carried by
the throughfall in the leafless period reaches the soil surface
and acidifies the top soil.
In fig. 1 two examples are given for the chemical conditions of
the soil below the stem base of beech trees of age about 110
years growing on soils developed on limestone (unterer Muschelkalk) in the Göttinger Wald. Tree No. 3 grows in a Terra fusca
(texture: silty clay), tree No. 6 in a loess layer covering
weathered limestone. The values given represent pH(H_2O), in
brackets pH(KCl). The data indicate great variation in chemical
soil state. No soil volume has been found staying in the carbonate
buffer range, even close to limestones. According to the pH(H_2O)
values, the actual chemical soil state varies between the
silicate and the iron buffer range, the most acid conditions
prevailing close to large roots where the soil is influenced
by the acid stem flow. The pH(KCl) lies in any case below 4.2,
in most cases below 3.8. This means that with an acid input of
pH below 3.8, which is quite common for stem flow in the leafless
period, the soil will shift to the aluminum or iron buffer range.
Soil acidification comparable to podzols can thus be found below
the stem base of beech trees on limestone soils. This acidification is due to acid deposition. This soil acidification may
play a role in the development of bark necrosis which is often
found at the stems of the trees.
But also the densely rooted upper part of rendzina soils is
acidified. Table 1 gives some data on chemical soil state and
Ca/Al ratio in fine roots of ground vegetation in a beech forest
on rendzina, developed on Unterer Muschelkalk in the Göttinger
Wald. The samples are taken from the Ah horizon (0-10 cm) in
June 1980, the soil showed mainly a polyedric fabric.

In rendzina soils being continuously in the calcium carbonate
buffer range there should be no exchangeable Al and no polymeric
Al masking the clay mineral surfaces available. In any of the
samples polymeric Al is present in considerable amounts. This
becomes clear by looking at the ratio CEC_e/CEC_t (effective to
total cation exchange capacity) and to the reduction of total
CEC_t (see last column) with increasing acidification.
The most acid conditions exist close to the stem base of beech
trees.
Typical plants of the soil flora of a rendzina soil under beech
do already reflect the acidification by poorer growth and closer
Ca/Al-ratios in the roots.
Acid rain influences also rendzina soils and their vegetation.
As soon as the fine earth is free of $CaCO_3$, acid rain in combination with internal proton production by nitrification may
shift the buffer system acting to the Al buffer or even the iron

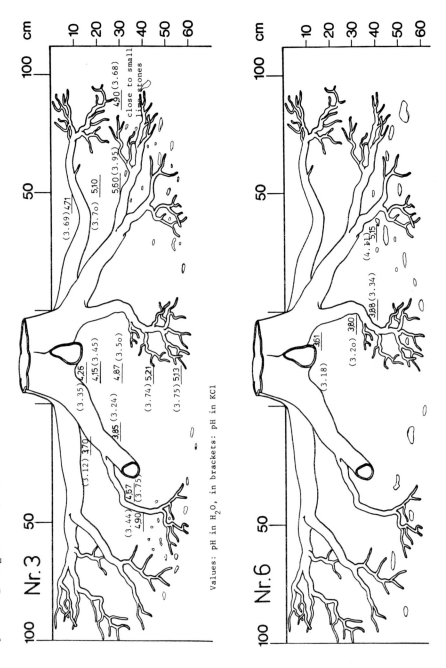

Fig. 1: pH(H₂O) and pH(KCl) (in brackets) below the stem of beech trees on lime stone soil

Values: pH in H₂O, in brackets: pH in KCl

Table 1: Chemical soil state of rendzina soils and plant vitality

	pH GBL	pH CaCL$_2$	CEC$_e$ ueq/g	$\frac{CEC_e}{CEC_t}$	Ca	Fe	Mn	Al	Ca	Al	N$_{org}$	mol Ca/mol Al in ESS	mol Ca/mol Al in roots	CEC$_t$ µeq/g
					-% on exchanger--				-mg/l in ESS--					
Ah-horizon 0-10 cm, polyedric fabric	6.00	5.10	221	.41	82	0	2	11	16	1.5	1.3	7.0	3.1	541
0-10 cm, close to a beech stem	3.88	3.00	173	.47	7	1	1	72	2	.8	.7	1.6	no roots	370
0-10 cm, close to an esh stem	6.82	5.49	353	.57	84	0	.5	.8	22	.9	1.9	16	1.8	616
0-10 cm, adhering to roots of a vital Asarum plant	7.2	6.05	410	.66	95	0	0	0	47	.2	1.3	212	15	619
0-10 cm, adhering to roots of a less vital Asarum	7.10	4.40	210	.48	81	0	1.5	10	20	1.8	2.1	7.3	2.6	438
0-10 cm, adhering to roots of a vital Mercurialis perennis	7.70	6.07	454	.72	95	0	0	0	57	.1	1.5	320	13	634
0-10 cm, adhering to roots of a Mercurialis less vital	7.48	4.49	314	73	92	0	1	.3	22	1.4	1.6	11	9.4	427

buffer range at least for short time periods. As long as small
limestone particles are well distributed within the A-horizon,
the system may swing back to the carbonate buffer range, but the
polymeric Al formed during the acidification push remains. Thus
the soil has a memory for previous acidification pushes which
have not been buffered in the carbonate buffer range.
From these data it must be concluded that even forest ecosystems
developed on limestone are threatened by acid deposition through
soil acidification. Only the homogeneous incorporation of lime
($CaCO_3$) into the fine earth of the whole rooting zone would be
a countermeasure. This countermeasure can only be used for
agricultural crops with short rotation periods on arable land,
but not for deep rooting forests with rotation periods exceeding
100 years.

SILICATE BUFFER RANGE

pH(H_2O) values in soils approaching 6 indicate that $CaCO_3$ does
not play any role in buffering, and is absent at least from the
fine earth fraction of these soils. As long as carbonic acid is
the only acid being produced in the soil, mass action considerations show that pH(H_2O) will stay at values > 5 (cf.6).
In this range the only buffer acting in soils is through the
weathering of silicates. The process involved may be described
as the release of alcali and earth alcali cations from the
silicate lattice under proton consumption. Equation 5 may serve
as an example of the principle involved, describing the weathering
of Ca feldspar to kaolinite (for other examples see 20).

$$CaAl_2Si_2O_8 + 2\ H_2CO_3 + H_2O \longrightarrow Ca^{2+} + 2\ HCO_3^- + Al_2Si_2O_5(OH)_4$$

In reality this reaction is not a straightforward process, but
passes through many stages. The cations liberated during silicate
weathering can be bound as exchangeable cations in the clay
minerals formed from the weathered silicate lattices. The
protons consumed during this process are converted to undissocia-
ted silicic acid which is finally transformed to SiO_2 and H_2O.
Any leaching loss of silicic acid below pH 7 is therefore not
accompanied by cations. For ecosystems in the quasi-steady state
there is by definition almost no nitrate loss by leaching in
these soils, because the rate of nitrification does in the
temporal mean not exceed the rate of nitrate uptake (closed
nitrogen cycle). There is some movement of NaCl through the
system, which has its origin in the sea spray being deposited
from air. The only soluble anion being formed in the soil is
HCO_3^-; its concentration in the soil solution reaches a minimum
as pH approaches 5.0. Under these conditions the ecosystem
exhibits minimal leaching losses from the soil, irrespective of
the amount of precipitation: chemically the soil is tight due to

the low solubility of carbonic acid. It is often said that rate
of natural soil leaching and acidification increases with
increasing precipitation. The fact that forest soils in medium
altitudes are acidified cannot however be explained this way.

The buffer capacity as well as the buffer rate depend upon the
silicate content and the kind of silicate present. The content
of base cations in silicates varies between 3 (muscovite) and
20 (olivine) meq/g silicate. For sedimentary rocks a mean value
of 5 may be used. This gives a buffer capacity of 75 kmol H^+
per 1 % silicate in a soil layer of 10 cm thickness, bulk
density 1.5 and an area of 1 ha. The buffer rate due to base
cation release with proton transfer to silicic acid may vary
between 0.2 and 2 kmol $ha^{-1}yr^{-1}$ in the rooted soil layer of 1 m
depth (5,8,20). For soils on sedimentary rocks values below 1
can be assumed (5, MAZZARINO, this volume). Data based on cation
output from watersheds (8,20) may overestimate the actual rate
of silicate weathering in the rooted soil zone due to three
reasons. Usually the cation input from air is underestimated
(neglection of interception deposition or of total deposition).
A further assumption inherent in this concept is the constancy
of exchangeable base cation storage in the soil. Soils subjected
to acid deposition and/or internal proton production may loose
exchangeable base cations to the seepage water and accumulate
cation acids. These base cations do'nt originate from actual
silicate weathering and lead thus to its overestimation. In
addition, a spatial discoupling within the unsaturated zone has
to be taken into account. From the viewpoint of the forest ecosystem one is interested to judge the processes in the root and
decomposer environment. Acids can be leached from the top soil
and neutralized by silicate weathering during their passage
through the subsoil and underlying (soft) rock. In such a case
the watershed balance method may indicate silicate weathering
rates of 2 keq ha^{-1} yr^{-1}, whereas the silicate weathering rate
in the rooted soil zone may be much less, and the soil acidifies
at the same time.
The rate of protonization of silicates should increase with
increasing proton activity (decreasing pH). Since the weathering
of silicates is not a simple dissolution process but includes
reactions on internal mineral surfaces, the rate limiting step
may be the counterdiffusion of the ions involved. This can
result in only limited increasing rates of protonization with
decreasing pH. The variability due to pH lies probably within a
factor of two, but does'nt exceed substantially 2 keq ha^{-1} yr^{-1}.

The possible buffer rate of 0.2 to 2 keq H^+ ha^{-1} m^{-1} yr^{-1} can be
directly compared with the acid load. In Central Europe, the
acid deposition may vary between 0.8 keq (in sheltered positions)
and more than 6 keq (in positions exposed to high interception
deposition, especially by capture of fog and cloud droplets). The

internal proton production may vary between zero and 4 keq ha^{-1} yr^{-1}. The total acid load may thus vary between 0.8 and more than 10 keq. This comparison shows that, with the possible exception of sheltered positions, almost all forest soils will acidify under the present load, that is pass over into the buffer ranges where cation acids are produced.

If the rate of proton load exceeds the rate of silicate protonization, the chemical soil state passes over into the cation exchange buffer range. This is reversible, as soon as the rate of proton load becomes less than the rate of silicate weathering. From this it follows that the ratio of both rates determines the chemical soil state.

As with calcareous soils, soils in the silicate buffer range and not subjected to acid deposition exhibit no gradients in their chemical properties with depth. Chemical composition of the soil is characterized by the absence of polymeric hydroxo Al in the interlayers of the clay minerals, and by the absence of exchangeable Al ions. Their presence indicates that the chemical soil state has temporarily been in the cation exchange buffer range, i.e. that the soil has passed through one or more periods of higher proton load.

As in calcareous soils, soil-borne toxins are missing. In virgin forest ecosystems producing easily decomposable phytomass at a high rate, earthworms or other soil burrowing animals work over the whole rooting zone of 1-2 m depth, resulting in a crumb-like structure and a organic matter content of low C/N ratio in the rooting zone. The amount of organic matter accumulated in the soil depends upon climatic conditions (temperature and humidity). The humusform is mull which may in cool climates carry an organic top layer comparable to moder. This organic top layer may show lower pH values: the biological gradient creates a chemical gradient.

Soils remaining continuously in the silicate buffer range guarantee a high nutrient storage of cations as well as N, S and P in the deep rooting zone and a high turnover of nutrients in the soil. They thus guarantee a well balanced and always sufficient nutrient supply to plants, provided that the parent material shows no extreme mineralogical composition.

CATION EXCHANGE BUFFER RANGE

The exchangeable base cations, mainly Ca^{2+}, play a deciding role in buffering acidification pushes which are caused by the temporal discoupling of the ion cycle (ULRICH ,this volume). If a net production of a strong acid (HNO$_3$) occurs in the soil, the pH is buffered by proton exchange with Ca^{2+} at exchange sites in a reversible and rapid reaction:

$$\boxed{\begin{array}{c}\text{clay mineral} \\ \hline \text{soil organic matter}\end{array}} \begin{array}{c} - \\ - \end{array} Ca + 2H^+ \rightleftharpoons \boxed{\begin{array}{c}\text{clay mineral} \\ \hline \text{soil organic matter}\end{array}} \begin{array}{c} - H \\ - H \end{array} + Ca^{2+}$$

If the pH drops below around 5, the solubility of the oxides of
Mn and many heavy metals reaches a level where concentrations
of the metal ions in the soil solution become of ecological
significance (toxicity). All these metal ions including Al are
cation acids, they can exist in different ion species. The kind
of the ion species plays a deciding role in its physiological
effect.

The release of Al ions is bound to the lasting presence of
exchangeable protons at the clay mineral surface. If within
days or weeks a net consumption of protons occurs within the ion
cycle, the reaction can go from left to right and the soil
returns into its initial state. As long as the percentage of
exchangeable Ca is above 5 to 10 % of the CEC, an acid load in
soil will mainly be buffered by this reaction. As long as this
is true, there will be no lasting physiological strain to
microorganisms and plant roots by an acid load.

If, however, protons stay long enough (weeks or months) on the
surface of clay minerals, it is known that they are consumed
with the release of Al ions (9). In the pH range between 5.0 and
4.2 most of these Al ions form polymeric hydroxo cations with a
charge of around +0.5 per Al atom. These polymeric hydroxo Al
cations accumulate in the interlayers of the swelling clay
minerals. Their exchangeability is very limited, even at higher
pH values. In soil horizons swinging back and forth between
silicate and exchange buffer ranges, the Al hydroxo polymers
remain more or less unchanged during the period with low proton
load. They are like a memory, indicating that the soil has
passed through periods of high proton load. The more the
exchange sites are occupied by Al, the less are available for
K, Mg and Ca. This is a consequence of their leaching. The
anion accompaning the leached cations is no longer HCO_3^-. It may
be nitrate (indicating humus disintegration and discoupling of
the nitrogen cycle), or organic anions (indicating podzolization),
or sulfate (indicating acid precipitation).

The buffer capacity is equal to the cation exchange capacity
(CEC), in respect to clay a mean value may be 7 kmol H^+ per
% clay. In addition, the CEC of soil organic matter has to be
taken into account. The buffer rate is large enough to prevent
soils from swinging into the aluminium buffer range (pH $<$ 4.2)
as long as the exchangeable Ca saturation is not too low
($>$ 10 %).

As already stated, the acid load of forest soils in Central
Europe may in many cases be around 7 keq acid ha^{-1} yr^{-1}. In
these cases, the annual loss of exchangeable Ca may approach
the CEC due to 1 % clay in a soil layer of 10 cm depth. Within
50 years, the exchangeable Ca may be completely leached from
a soil containing 10 % clay to a depth of 50 cm. This indicates

that a considerable soil acidification must have occured due to
acid deposition during the last century and especially since
1950 in Central Europe. This is in agreement with observed pH
decreases in forest soils (1,2,3,4).
With Al appearing at the clay surface in ionic form, Al
toxicity comes into play, thereby limiting the growth and
regeneration of most intolerant (calcicole) species. In addition
bacteria suffer from Al toxicity. These conditions and the
reduced activity of earthworms initiate the accumulation of
litter on the forest floor. Absence of faunal activities can
easily be recognised by examining the soil structure which is no
more crumb-like but polyedric or coherent. The accumulation of
an organic top layer is usually connected with a substantial
proton production in the mineral soil (ULRICH, this volume).

Many acid forest soils stay in this buffer range. Usually the
soil exhibits chemical gradients with depth. In the case of
humus disintegration (ULRICH, this volume) the lower horizons
may be more acid than the upper ones. Much more common is the
reverse: soil acidity is decreasing with increasing depth, and
the lowest horizon may still be in the silicate buffer range.
Such pH gradients in the soil profile indicate missing bioturbation. They indicate further that the soil processes operate
relatively far from chemical equilibrium and that the ecosystem
may be in a transition state. As a consequence of the leaching
of cations and the opening of the nitrogen cycle, growth is
often limited by deficiency of nutrients, mostly N, sometimes
Mg, K or P in addition. Phosphates exist in binding to Fe and Al
with a lower solubility compared to silicate buffer range. As
long as the percentage of exchangeable Ca is not too low
(exceeding 1o to 15 %), fertilizer salts can be applied to the
soils without doing harm to the plants or decomposers.

ALUMINIUM BUFFER RANGE

At pH 4.2, Al^{3+} is reaching a concentration in the soil solution
which makes it the most important factor in determining plant
growth. Tolerance of Al toxicity becomes the deciding factor in
plant competition, which in turn determines the composition of
the vegetation and the rate of phytomass production. The
chemical status of soils is characterized by the dominance of Al
ions among the exchangeable cations. The sum of exchangeable H,
Mn, Al and Fe, which represents potential acidity, will exceed
9o % of the effective cation exchange capacity of the soil.
Exchangeable Ca and Mg are reduced to values below 5 to 1o %.
Bacterial and soil animal activity as well as growth of roots
can be restricted by Al toxicity in the mineral soil.

The Al stems from the weathering products of the primary

140   B. ULRICH

silicates, mainly from clay minerals. The release of Al may be
described by the following equation

$$AlOOH + 3 H^+ \longrightarrow Al^{3+} + 2 H_2O$$

For calculation of the buffer capacity one can start with the
fact that the total clay content of the soil is available for
buffering, the buffer capacity will be exhausted only after
total clay destruction. Assuming a mean composition of clay
minerals, the buffer capacity amounts to 100 to 150 kmol H^+ per
% clay. The process goes not directly according to the equation
given, in between new solid phases of Al compounds are formed
with a positive charge between zero and 3. Examples are the
polymeric Al hydroxo cations and Al hydroxo sulfates (PRENZEL,
this volume). The buffer rate depends upon the dissolution rate
of the solid phases, being high (several kmol H^+ per ha and
year) for most of the intermediates, and being low for Al bound
in silicate lattices (some tenth of kmol H^+). As a consequence,
a soil horizon stays at the aluminium buffer range as long as
Al hydroxo compounds of high dissolution rate are available. If
they are used up, the buffer rate drops down and the horizon
may switch over to the iron buffer range if the proton load is
high enough. Bleached horizons have therefore lost most of
their interlayer aluminium and do also not contain Al hydroxo
sulfates (10).
Al toxicity is a well known phenomon since long (11,12,13) and
was demonstrated for pine seedlings (14) and beech seedlings
(15). For being toxic, cation acids like Al must come in contact
with cell membranes of roots or mycorrhiza. This depends upon:
- the nature of the ion species present in solution. The
  monomeric trivalent species Al^{3+} possesses the highest
  potential toxicity, polymeric Al species and organic
  chelated Al the lowest one. At pH > 4.2, the ratio of
  monomeric to polymeric Al species is shifted in favour of
  the polymeric species; at pH < 4.2 in favour of monomeric
  species (16).
- the ability of the root or mycorrhiza to change the ionic
  species of Al in the free space before reaching cell
  membranes. This can be achieved by change in pH (see above),
  or by transforming Al^{3+} into phosphat, silicate or organic
  complexes. The formation of Al phosphate complexes may lead
  to phosphorus deficiency.
There are therefore soil conditions (pH value and presence of
organic ligands in the soil solution) as well as plant conditions
(buffering and complexing of cation acids in the free space of
the root) deciding upon the toxicity of Al. The role of
formation of organic complexes in the soil solution is demonstrated by the data in table 2. The data are taken from a
lysimeter study performed in a mixed coniferous stand dominated
by spruce on the Swedish westcoast (NILSSON, personal communication).

Table 2: Ca/Al-ratio in soil solution at Gårdsjön
          (NILSSON, personal communication)

Soil depth	Ca/Al	Ca + Mg	Organic aluminium complexes as a percentage of total aluminium [1]
0 - 5 cm	8.6	16.5	94
0 - 15 cm	0.7	1.5	90
0 - 35 cm	0.3	0.9	36
0 - 55 cm	0.4	1.2	14

1) determination according to DRISCOLL (17)

The data (average values for 12 months) show that in the top soil
90 % or more of the observed Al is tied to organic compounds.
Even in the mineral soil a substantial part of the Al can occur
in organic complexes. Since soil acidification is a natural
process, species growing on acid soils (e.g. Picea abies, Pinus
silvestris, Fagus silvatica) have developed tolerance mechanisms
against toxic cation acids during evolution.
The formation of a mycorrhiza could be one important protection
mechanism for the roots. It has been shown that mycorrhizal
fungi produce substantial amounts of oxalic acid which could
(by complexation) to a large extent detoxify any inorganic
species present. If the mycorrhiza formation is hampered, root
damages caused by aluminium might become more important. One of
the factors that is of importance for the mycorrhiza is the
supply of carbohydrates from the roots which in turn are
translocated from the leaves. If direct leaf or needle damage
(caused by any outside physical or chemical factor) results in
a decrease of photosynthesis, this may in turn result in a lower
production of carbohydrates and a decreased carbohydrate transport to the roots. Thereby the nutrient supply to the
mycorrhizal fungi may be decreased and their development
hampered. This in turn results in a lower production of oxalic
acid and an increased possibility for inorganic aluminium forms
to penetrate the roots and cause damage.
There are no reasons to assume that mycorrhiza fungi are not
themselves subjected to Al toxicity, even if they have developed
more efficient avoidance mechanisms as the roots they are
protecting. It seems that the mycorrhiza does'nt prevent the
entrance of cation acids into the free space of roots (which
would be a bad strategy since some cation acids are needed as
micronutrients). This is indicated by the data in table 3. In
this table the Ca/Al ratios in the equilibrium soil solution as
well as in the fine roots of trees are given for soils of
different exchangeable Ca percentage, varying between $>$ 50 % and
$<$ 5 %. Values are also given for roots developing only in the
organic top layer of moder and raw humus (last line). With
increasing acidity (decreasing Ca saturation) the Ca content of

Table 3: Content of Ca and Al in fine roots in soils
          of different Ca saturation x_{Ca}^S

| n | x_{Ca}^S | $\dfrac{mol\ Ca}{mol\ Al}$ in ESS | fine roots | | | | $\dfrac{mol\ Ca}{mol\ Al}$ |
			ash %	SiO$_2$ %	Ca mg/g dm	Al mg/g dm	
19	> 0,5	75	6,71	3,00	8,80	3,06	2,30
		+49	+2,75	+2,46	+2,41	+1,56	+1,47
9	0,5 to 0,1	15	7,25	4,24	5,24	2,25	1,56
		+14	+2,25	+2,38	+2,95	+0,56	+0,89
18	0,1 to 0,05	3,5	8,16	5,25	2,07	4,06	0,41
		+6,2	+3,54	+3,53	+1,40	+1,83	+0,32
46	< 0,05	0,96	8,83	5,79	1,60	4,82	0,28
		+0,52	+4,90	+3,98	+0,81	+2,15	+0,19
11	OH-horizons	-	3,46	1,04	5,77	2,12	1,75'
			+0,45	+0,54	+3,99	+0,62	+0,97

the roots is lowered (in the mean to 1/6), whereas the Al content
increases only little. The effect is best seen at the Ca/Al mol
ratio in the root, which changes from 2.3 at Ca saturation
> 50 % to 0.3 at Ca saturation < 5 %.
BAUCH (this volume) shows that in diseased trees Ca and Mg is
lost from the cell wall of root cortex cells of fir, whereas
both elements are present in the roots of healthy trees. From a
physico-chemical point of view, a low Ca content in the cell
wall indicates a pH in the water phase being 1 to 2 units below
the pK_{diss} value of the acidic group. Therefore the low Ca
content indicates the acidification of the free space. The data
of table 3 show that this acidification process goes parallel
with soil acidification, reaching critical values at Ca saturation degrees below 10 and especially below 5 % of the effective
CEC. It is not known how this acidification causes cell injury,
whether by inducing Ca and Mg loss from membranes and cells, or
by allowing cation acids to come in contact with membranes and
entering cells. Since it seems common that roots have now
accumulated a wide variety of heavy metals (18), cell injury may
be due to heavy metals more easily soluble than Al polymers.

In the organic top layer with its low primary Al content and
high Al complexing capacity the Ca/Al ratio lies in the mean on
the safe side. Field data show, however, that the OH horizons
are accumulating Al and Fe as part of the podzolization process;
both ions seem to be transported in organic compounds from the
Ae to the OH. Thus the equivalent ratio of Ca to Ca+Al+Fe shows
a decreasing tendency with continuing acid deposition. Field
observations indicate that at Ca/(Ca+Al+Fe) ratios below 0.05

and especially below O.O2 in the OH horizon, spruce stands show
heavy damage not only in the root system, but also in the crown.
It seems thus a very general feature that with very low
basicity in the root environment, the free space of the root
becomes also acidified which finally ends in root injury.

Taking into account this field experience, the following soil
chemical data indicate a high potential for cation acid toxicity
to acid tolerant tree species:
- in mineral soil: exchangeable Ca saturation below 5 %
- in the rooting zone of organic top layers: ratio of Ca to the
  equivalent sum of Ca+Al+Fe below O.O5 (very critical below
  O.O2)
- Ca/Al molar ratio in the soil solution below 1 (very critical
  below O.2)
- Ca/Al molar ratio in roots below 1 (very critical below O.2)
As discussed elsewhere (ULRICH, this volume), root injury
does'nt necessarily lead to growth reduction or crown damage.
However, it may often be the precursor of other kinds of damage.
The parameters mentioned should therefore be taken as indicators
of a very low resilience of the forest ecosystem, they need not
be connected with actual visible damage in trees.

The most direct measure should be the Ca/Al ratio in the soil
solution. The problem in applying this measure lies in the
analytical difficulties to differentiate between monomeric and
chelated Al. In horizons containing dissolved or soluble organic
matter,most of the analytically determined Al can be present as
chelate (see table 2)and thus being less toxic or even untoxic,
depending upon complex stability. In soils staying in the
aluminium buffer range, decomposer activity and root growth are
therefore restricted to the top organic layer and the uppermost
mineral soil horizon which is within the influence of infiltrating dissolved organic matter. Soils being in the phase of
humus disintegration can show dispersable organic matter in
deeper soil horizons, so throughout the whole rooting zone cation
acids may be masked by chelat formation.
Soil horizons in the aluminium buffer range have lost most of
their nutrient cations, they possess only low stores of organic
matter and nitrogen in the mineral soil, and the solubility of
the phosphate bound to Fe and Al is low. On the other hand, the
solubility of heavy metals including trace elements is high and
may even have lead to the leaching of the mobilizable pools also
of the micronutrients including boron. The vegetation on these
soils is therefore subjected to many kinds of growth limitations
and growth disturbances.
Unfortunately, fertilizers consisting of soluble salts (nitrates,
chlorides, sulphates) should not be applied on soils of this
chemical state. The consequence of adding soluble salts is well
known to all soil scientists from the difference observed

between pH measured in soil/water suspension and in $CaCl_2$ or KCl
suspension: on addition of soluble salts to acid soils the pH
drops considerably. In most cases this is due to the exchange of
Al ions by Ca and K ions; the Al ions hydrolyze in the solution
to produce protons. The application of water soluble fertilizers
increases Al toxicity under these circumstances, and may thus
have a detrimental effect on the vegetation. Detrimental effects
of fertilization have been found in fertilizer trials, but they
usually have not been traced back to this cause.

IRON BUFFER RANGE

At pH values below 3.2, in horizons influenced by infiltrating
organic matter already below 3.8, the solubility of iron oxides
becomes high enough to reach Fe concentrations in the soil
solution of ecological significance. The transport of iron leads
to visible (colour) symptoms in the soil profile, which is not
the case for aluminium. Due to an increase in the number of
toxic elements and their concentrations in the soil solution,
the effects of toxicity and nutrient deficiency become stronger
at such low pH values. They are usually bound to a considerable
proton production within the soil. Such phases are of only
limited duration, whereas the visible symptoms, e.g. the
bleached eluvial horizon, remain the same. The existence of a
eluvial horizon does therefore not tell whether podzolization,
the process characteristic for the iron buffer range, does
actually take place.
The actual podzolization which is now widespread in beech and
older spruce forests in Central Europe can only be explained
as triggered by acid deposition (ULRICH, this volume).

RECOGNITION OF SOIL ACIDIFICATION BY ACID DEPOSITION

The fact that acid deposition occurs calls for the hypothesis
that thereby soil acidity is increased. This hypothesis has to
be falsified. There are three approaches which can be used for
this purpose.
From the capacity point of view, an increase in acidity can be
measured as the difference in inventories of the sum of soluble
and exchangeable bases and acids in soil. Acidification is
indicated by a decrease in the sum of bases and/or by an increase
in the sum of acids with time. The use of this approach is very
limited since there exist no data on base and acid storage of
soils in the past.
From the intensity point of view, an increase in acidity may
show up as a decrease in pH with time. Since pH is related to
base and acid capacities only indirectly, this approach is much
more limited. If pH(salt) is taken, a loss of bases should

always show up in a pH decrease. If, however, the acid accumulates as a solid like $AlOHSO_4$, its accumulation may not be connected with a change in pH. In Central Europe, this approach failed to falsify the above mentioned hypothesis (1,2,3,4).

With both approaches, the depth gradient of soil chemical state has to be taken into account. If the top horizons are already staying in the Al or Fe buffer range, an increase in soil acidification may only show up in changes of the chemical state of deeper soil horizons (loss of bases, accumulation of cation acids, decrease in pH). In these cases, measurements limited to the top soil horizons are meaningless.
From the point of view of rates, the rates of deposition and of buffering of acids can be compared. This is the only approach which does'nt need baseline values from the past. The most simple approach is to compare both rates directly, assuming that acid deposition enters the soil at its surface. Calculations of this kind assume that chemical equilibrium is achieved between the acids transported downward with the soil solution and the soil matrix. In this case, an acid input by addition to the soil surface results in the development of strong depth gradients of the chemical soil state: the upper soil layer acidifies before the acid breaks through to the lower layer. Soils then behave like chromatographic columns. Depending upon the rate of water movement, the breakthrough of the acid at the bottom (pH decrease) can be retarded till the buffer capacity of the total soil column (soil depth) is exhausted, or till the rate of acid buffering becomes less than the rate of acid input. Depending upon acid load and acid buffering, the release of acid into drainage waters (in case of column experiments: the breakthrough of acid at the soil bottom) may occur only after years or even decades of continuous acid input. That the leachate of a soil is not acid does therefore not exclude ongoing soil acidification.
The higher the rate of water input (e.g. heavy rain, snow melt), and the coarser the soil texture (e.g. sandy soils with high gravel or stone content), the less favourable are the conditions to reach chemical equilibrium between the solutes and the soil matrix. In such cases substantial fractions of the soil solution may be rapidly transported through macropores and leached from the rooted soil layer already after days. In such cases acidity may be leached with drainage waters already at exchangeable Ca percentages exceeding 5 to 10 %. But in principle, there exists no exception from the physico-chemical rule that a cation exchanger, saturated with base cations, will loose base cations and accumulate acids if an acid solution percolates through. Therefore it need not to be proofed, it is proofed that the store of exchangeable base cations in a soil is diminished by an acid input. If the deposition rate of acids is known, the open problem is only at which place (soil horizon, layer) in the

unsaturated zone (soil and seepage conductor) the loss of bases
and the accumulation of acids occurs. This can be estimated if
the chemical state of the horizons and layers is known
(including buffering rates and buffering capacities), and if the
water flow characteristic in the soil is known. The objective
of further soil chemical investigations in this field should be
directed towards the development of a mathematical model
describing solute transport and buffer reactions.

REFERENCES

1. Blume, H.-P. (1981): Berliner Naturschutzbl. pp.713-715
2. Butzke, H. (1981): Forst- u. Holzwirt 36, pp. 542-548
3. v. Zezschwitz, E. (1982): Forst- u. Holzwirt 37, pp.275-276
4. Wittmann, O. (in press): Untersuchungen zur Ermittlung der
        aktuellen Bodenversauerung in Bayern. Bayer.Geol.Jahrb.
5. Ulrich, B. (1981): Z.Pflanzenernähr.Bodenk. 144, pp. 289-305
6. van Breemen, N. and W.G. Wielemaker (1974): Soil Sci.Soc.
        Amer.Proc. 38, pp. 55-66
7. Ulrich, B., U. Steinhardt und A. Müller-Suur (1973):
        Göttinger Bodenkdl.Ber. 29, pp. 133-192
8. Johnson, N.M., Ch.T. Driscoll, J.S. Eaton, G.E. Likens and
        W.H.McDowell (1981): Geochimica et Cosmochimica Acta 45,
        pp. 1421-1437
9. Schwertmann, U. and M.L. Jackson (1964): Soil Sci.Soc.Amer.
        Proc. 28, pp. 179-183
10. Meiwes, K.J. (1979): Göttinger Bodenkdl.Ber.60, pp. 1-108
11. Mevius, W.: Reaktion des Bodens und Pflanzenwachstum.
        Freising und München, Datterer
12. Ellenberg, H. (1958): In W. Ruhland (ed.): Handbuch der
        Pflanzenphysiologie IV, pp. 638-708, Springer, Berlin
13. Foy, C.D., R.L. Chaney and M.C. White (1978): Ann.Rev.Plant
        Physiol. 29, pp. 511-566
14. Süchting, H. (1948): Z.Pflanzenernähr., Düng.,Bodenk. 87,
        pp. 193-218
15. Süchting, H. (1943): Allgem.Forst- u. Jagdztng. 119, p.34
16. Nair, V.D. and J. Prenzel (1978): Z.Pflanzenernähr.Bodenk.
        141, pp. 741-751
17. Driscoll, Ch.T. (1980): Chemical characterization of some
        dilute acidified lakes and streams in the Adirondack
        region of New York State. Ph.D.thesis Cornell University
18. Mayer, R. und H. Heinrichs (1981): Z.Pflanzenernähr.Bodenk.
        144, pp. 637-646
19. Ellenberg, H. (1939): Mitt.Florist.Soziol.Arb.gem. Nieder-
        sachsen 5, pp. 3-135
20. Bache, B.W. (1982): The implications of rock weathering for
        acid neutralization. In "Ecological Effects of Acid
        Deposition", series PM, Swedish National Environment
        Protection Board

BALANCES OF ELEMENT FLUXES WITHIN DIFFERENT ECOSYSTEMS
IMPACTED BY ACID RAIN

E. MATZNER

Institut für Bodenkunde und Walderernährung, Universität Göttingen, Büsgenweg 2, D-34 Göttingen

ABSTRACT

In order to determine the rate of proton and sulfur load resulting from atmospheric deposition and the interaction of these pollutants in different types of ecosystems, measurements of element fluxes were carried out in stands of pine, oak, beech, spruce and heath. Highest rates of deposition were found for the spruce stand and lowest for the heath stand. The effect of proton consumption in the soil depends on the chemical soil state. For the stands investigated the chemical soil state corresponds to the Exchange- or the Al-buffer range. Therefore the consumption of protons leads either to the liberation and leaching of Ca and Mg or to the leaching of significant amounts of Al. Accumulation of N and other elements within the humus layer of the beech and the spruce stand seems to be due to far reaching changes of decomposition conditions caused by acid rain.
Rates of ecosystem internal proton production are about equal to the rate of proton load from deposition for the beech and spruce stand.

1. SITE DESCRIPTION

The fluxes of chemical elements were measured in 4 different ecosystems in North Germany. The ecosystems selected are stands of beech (Fagus silvatica), spruce (Picea abies), oak (Quercus robur), Scotch pine (Pinus silvestris) and heath (Calluna vulgaris) located in the Solling region (beech, spruce) and in the Lüneburger Heide. The Solling sites are described in more detail

147

B. Ulrich and J. Pankrath (eds.), Effects of Accumulation of Air Pollutants in Forest Ecosystems, 147–155.
Copyright © 1983 by D. Reidel Publishing Company.

by Ellenberg (7) while further information about the heath
ecosystem is given by Matzner (10). The observed oak stand is
about 100 years old, the soil is represented by a sandy podsolized brown earth. The pine stand has an age of about 80
years and the soil type again corresponds to a sandy podsolized
brown earth. Selected chemical soil data of the oak and pine
stand are documented in Tab. 1. Chemical soil data for other
stands are available from Ulrich et al. (22) and Matzner (10).

Tab. 1   Soil chemical properties of the oak and pine stand
          (sampling date: May 1981)

		pH ($CaCl_2$)	%Ca+Mg of CEC	pH ESS	mol Ca/ mol Al
oak	0 - 10 cm	3.4	15	4.35	0.72
	10 - 30 cm	3.6	10	4.48	0.52
	30 - 40 cm	4.0	6	4.53	1.15
	40 - 90 cm	4.2	3	4.43	1.47
	90 -130 cm	3.9	8	4.24	1.60
pine	0 - 10 cm	3.1	8	3.75	0.42
	10 - 15 cm	3.7	3	4.0	0.34
	15 - 40 cm	4.1	2	4.17	0.53
	40 - 90 cm	3.8	2	4.09	0.50
	90 -130 cm	3.8	2	4.09	0.58

The more acidified soil is found under the pine stand indicated
by lower pH values in the equilibrium soil solution (ESS) and
lower degree of Ca and Mg saturation of the effective cation
exchange capacity.

2. METHODS

The following fluxes of elements were measured in the Solling
stands: Precipitation-deposition, canopy drip, stemflow, input
to the mineral soil=output from the humus layer (either including living roots or without living roots in the humus layer),
output with seepage water, litterfall, increment. The period of
investigation evaluated so far ranges from 1969 - 1980. To
characterize element transfer through the heath ecosystem,
precipitation-deposition, canopy drip, litterfall, increment,
output from the humus layer (without living roots) and output
with seepage water was measured over one year (9/78 - 8/79).
Flux measurements in the oak and pine stand included precipitation-deposition, canopy drip, stemflow, output with seepage
water. For these stands the element storage by increment was
estimated. Results are available so far only for 1980.
The soil solution was extracted by tension lysimeters (ceramic
plates or cups) at a depth of 1.5 m (oak, pine), 1 m (beech,
spruce) and 1.2 m (heath). The following elements were analysed:
H, Na, K, Ca, Mg, Fe, Mn, Al, SO_4, PO_4, Cl, NO_3, NH_4 and

total N.
Further information about the field measurements are given by
Matzner (10), Ulrich et al. (22) and Seibt (19) while the element concentrations and element fluxes from which the element
balances for the Solling stands were calculated are published
by Matzner et al. (14). The model used to describe the element
fluxes and the balance equations employed are described for the
Solling stands by Matzner and Ulrich (13), for the heath stand
by Matzner and Ulrich (11).

3. RESULTS AND DISCUSSION

Detrimental effects of deposition and accumulation of air pollutants in forest ecosystems arise from the deposition of sulfur
and protons as well as from organic compounds and especially
from heavy metals. This paper only deals with the deposition of
sulfur and protons and their turnover in ecosystems. In order to
set off the most important interactions of this impact, the
results of the flux balances will be presented only for a few
elements, element balances including all elements are given by
Matzner and Ulrich (13).
Evaluating the ecological effects of proton and sulfur input,
the first information required is the annual rate of total deposition of these elements, which are given together with the
input of Ca, Mg and Al in Tab. 2.

Tab. 2   Annual rates of total deposition in different
          ecosystems (kg . ha^{-1} . a^{-1})

	beech x̄ 1969-1980	spruce x̄ 1969-1980	oak 1980	pine 1980	heath 1979
H	1.69	3.72	1.53	1.64	0.63
Ca	19.9	25.5	7.7	8.8	4.5
Mg	3.8	4.1	2.2	2.5	1.1
Al	2.4	2.8	0.47	0.53	0.29
S	51.7	87.6	33.1	35.3	18.8

Under spruce the rate of incoming protons resulting from deposition of air pollutants exceeds the rate of the neighboured
beech stand more than twice, while the load of acidity for the
oak and pine stand is about equal and corresponds to the mean
proton load of the beech stand. Equal rates of atmospheric
sulfur and proton load under coniferous and deciduous stands
may be characteristic for stands located in remote areas with
low fog frequency. The occurence of fog seems to be a dominant
factor rising the rate of deposition under coniferous stands
because it occures mostly during winter and autumn when hard-
woods are leafless. From 1977 to 1981, the mean annual number
of days where fog was registered was 71 in the Solling region

and 28 in the flat plains (6). High fog frequency should be the
reason for the observed high rates of deposition in the spruce
ecosystem the Solling. The mechanism can be direct filtering of
acid aerosols as well as wet needle surfaces over long periods
allowing adsorption of SO_2.

In relation to the rates of proton deposition, the rates of sulfur deposition under oak and pine are small compared to the
Solling stands. This indicates less buffering by soil dust and
NH_3 before or during deposition in the Lüneburger Heide, a conclusion that is confirmed by comparing the rates of deposition
of Ca and Mg.

Smallest rates of deposition are observed for the heath ecosystem because of the modest roughness of its canopy surface
and because of its protected location between adjacent pine
stands.

The question that arises now is, how do these incoming pollutants
react in the ecosystems. A relatively small part of the proton
load from deposition is consumed during canopy passage by the
exchange of Ca and Mg from the plant tissues. This buffering
capacity of the stands is very limited and amounts to about
0.3 keq.ha^{-1}.a^{-1} for the beech stand and 0.46 keq.ha^{-1}.a^{-1} for
the spruce stand (average values from 1969-1979). In the oak
stand 0.63 keq.ha^{-1}.a^{-1} and in the pine stand 0.42 keq.ha^{-1}.a^{-1}
are consumed during canopy passage. But, this buffering action
of the stands is transferred into the mineral soil via the
cation/anion balance of ion uptake and the connected proton
turnover. Hence the soil has finally to carry the total proton
load.

Before discussing the effects of incoming protons on the soil
buffer system and chemical soil state, another possible effect
of accumulation of air pollutants should be stressed, that
arises from the element balance of the humus layer.

The annual changes of N-storage within the humus layer calculated from the flux balance are presented for the Solling stands
in Tab. 3. Positive values indicate an increase of storage,
negative values a decrease. This also holds for Tab. 4 and
Tab. 5.

Tab. 3   Annual changes of N-storage within the humus layer
          (kg . ha^{-1} . a^{-1})

	1969	1970	1971	1972	1973	1974	1975	1976	1977	1978	1979	1980
beech	+4.0	+39.5	+35.3	+49.8	+8.4	-11.4	+6.1	+27.6	0.0	-3.8	-11.3	-8.0
spruce	+14.0	+4.5	+17.7	+17.5	-3.4	-11.2	+16.4	+11.7	+17.3	-9.3	-21.0	-16.3

The annual changes of N storage of the humus layer especially
under beech indicate a strong N accumulation between 1969 and
1973 that amounts to about 135 kg.ha^{-1}. This rate of N accumulation is confirmed by comparison of direct measurements of N
and C storage by inventory studies (23). Depression of N

mineralization to such an extend could only be explained by
deterioration of decomposition conditions which may be caused
by increasing soil acidity (Al-toxicity) and heavy metal accumulation (16).
Deterioration of mineralization by acid rain was also observed
by other authors (2,20,21).
Coming back to the effects of proton and sulfur input on the
soil buffer system and the chemical soil state, the flux
balance of the soil gives further information. For the Solling
stands the compartment soil has been devided into the compartments "humus layer" and "mineral soil", the following data
therefore only concern the mineral soil. The flux balance for
the mineral soil includes the element fluxes: Input to the
mineral soil = output from the humus layer, stemflow, uptake by
the stand, output with seepage water. For the other stands,
only the sum of humus layer and mineral soil has been regarded,
the flux balance of the so defined "soil" includes: Total atmospheric input, element storage by increment, and output with
seepage water. The results of the flux balances are given in
Tab. 4.

Tab. 4   Changes of element storage within the soil
          (keq . ha^{-1} . a^{-1})

	beech $\bar{x}$ 1969-1980	spruce $\bar{x}$ 1973-1980	oak 1980	pine 1980	heath 1979
H	+1.17	+2.31	+1.35	+1.21	+0.40
Ca	-0.31	-0.72	-1.21	-0.79	-0.28
Mg	-0.10	-0.22	-0.57	-0.70	-0.15
Al	-0.98	-4.35	-0.04	-3.30	-0.35
S	+0.41	+0.21	-0.15	-3.61	-0.14

The main part of the protons resulting from deposition processes
is consumed within the soil of all stands investigated. To keep
up electroneutrality in the soil, the accumulation of protons
must be balanced by the liberation of other cations or by a
connected accumulation of anions. The kind of released cations
gives indications on the buffer system acting. Significant
liberation of Al occures under the beech, spruce and pine stand,
an observation that corresponds to the pH of the extracted soil
solution and to the buffer range of the soil (aluminum-buffer
range (22)).
The buffering action of the soil under oak mostly results in the
release of Ca and Mg. This effect is mainly caused by the displacement of Ca and Mg from the exchange sites by Al ions formed
during the buffer reaction. The soil of the oak stand shows the
highest degree of Ca and Mg saturation of the CEC of all stands
and the pH of the soil solution corresponds to the exchange
buffer range, whose characteristic feature is the change in the
composition of the exchangeable cations as stated above.

Al ions can be defined as Brønsted-acids. Therefore the accumulation of Al ions in exchangeable form is equal to the accumulation of acidity in the soil. The period of accumulation of
acidity in form of exchangeable Al now occures in the oak stand
but has finished in the beech, spruce and pine stand. For the
last named stands the load of protons is mainly converted to a
slighly weaker acid, to Al ions, and is then transferred to
deeper soil layers or to groundwater bodies. The chemical state
of the subsoil of the heath stand seems to be at the boundary of
the Al- and the exchange buffer range, since the equivalents of
released Ca and Mg are about equal to the released Al equivalents.
The observed release of Ca, Mg and Al from the soil agrees well
with results given by other others for soils treated with acid
rain (1,5,8,9).
Evaluating the sulfur balance of the soil, further information
about the time developement of the changes of sulfur storage
within the soil under beech and spruce can be obtained.

Tab. 5   Sulfur balance of the mineral soil under beech and
          spruce (kg . ha^{-1} . a^{-1})

	1969	1970	1971	1972	1973	1974	1975	1976	1977	1978	1979	1980
beech	+26.8	-4.6	+20.4	+13.8	+9.9	-9.5	+14.4	+11.7	+10.7	-3.8	-0.9	-8.3
spruce					+64.2	+37.1	+55.7	+15.9	-19.6	-40.8	-43.1	-69.0

For both stands, the accumulation of sulfur in the soil is not
continously throughout the period of investigation. Under beech
release of sulfur occures in 1970, 1974, 1978 and 1980.
Under spruce, the period of strong sulfur accumulation from
1973 to 1975 is followed by a period of low sulfur accumulation
(1976) and afterwards by an increasing sulfur liberation. At
present, the soil under spruce reacts as a source of sulfur,
indicating the instability of the chemical bounding.
Retention of sulfate in the soil may take place by precipitation
involving aluminium (presumable in the form of aluminumhydroxosulfate) or by adsorption on the exchange sites (17,4). Increasing soil acidification, which is to be expected in the upper
soil horizons, will effect the solubility of precipitated Al-
sulfates as well as the adsorption capacity cf sulfate and may
explain the observed release of sulfur under the spruce stand.
The developement of sulfur behaviour under beech seems to be
influenced by the rate of annual precipitation. In 1970 and 1974
the amount of annual rainfall was very high, so the increasing
water saturation of the soil may also effect the solubility of
Al-sulfates, an obervation, that corresponds to laboratory
studies (18).
The accumulation of sulfur and the connected accumulation of Al
ions seems to be the only buffer capacity of strongly acidified
soils (Al-buffer range), showing no more increase in exchangeable Al. The limitation of this buffer capacity becomes clearly

evident from the flux balance of the spruce stand. The
increasing sulfate concentrations in the soil solution since
1976 are followed by a drastic increase mainly of Al, but also
of Mg concentration (14), leading to heavy leaching losses of
these elements, which should have significant effects on ground
water quality as well as on microorganisms and the stand.
The sulfur balance of the soils covering oak and heath shows no
changes of sulfur storage, while high rates of sulfur release
occure under the pine stand. These may be explained by the
extreme high annual rate of precipitation in 1980 which may
have caused dissolution of aluminumhydroxosulfate as described
for the beech stand. Since only one year of measurement has been
evaluated so far, which was not representive with respect to the
water budget, final conclusions on the long term behaviour of
sulfur under these stands can not be drawn at present.
Summarizing the results of the flux balance for the soil of
different ecosystems, one can conclude, that the input of
protons leads to the release and leaching of Ca and Mg in soils
staying in the exchange buffer range, while simultaneously
acidity is accumulated in the form of hydroxo-Al-cations and
exchangeable Al. In soils having already reached the Al-buffer
range, acidity is mainly transferred in the form of Al ions into
deeper soil layers or to the groundwater and may be accumulated
in the form of aluminumhydroxosulfate.
Since the experimental approach included all major cations and
anions, the rates of ecosystem internal proton production can
be calculated from the changes of element storage within the
soil. This calculation presumes, that the excess release of
cation-equivalents from the soil must be balanced by the liberation of anions or by the net production of protons within the
soil. The results of this calculation are presented in Tab. 6
for the beech and spruce stand. The annual rates of proton
production given in this paper include the changes of soil
storage of organic anions which were not considered in previous
papers (22,23,13).

Tab. 6   Annual rates of ecosystem internal proton production
         (kg . ha^{-1} . a^{-1}

	1969	1970	1971	1972	1973	1974	1975	1976	1977	1978	1979	1980
beech	3.69	3.70	3.21	2.34	3.09	1.49	2.60	1.65	1.19	1.77	2.39	2.67
spruce				4.11	5.62	4.10	2.89	1.24	4.46	2.70	4.64	

Under both stands the annual rates of proton production show
great variation throughout the period of investigation, ranging
from 3.70 to 1.19 keq.ha^{-1}.a^{-1} under beech and from 5.62 to
1.24 under spruce. The average rates for both stands (2.48
beech; 3.72 spruce) are in the same order of magnitude than the
mean proton load from atmospheric deposition.
Calculating the changes of element storage within the soil from

the balance of the fluxes "total deposition", "output with
seepage water" and "increment storage", which could be done for
the oak, pine and heath stand, will lead to false conclusions
since the effects of humus accumulation and the formation of
organic acids within the humus layer are not considered.

As shown elsewhere (23,15) ecosystem internal proton production
is caused by the discoupling of the ion cycle. Discouplings of
the ion cycle may be caused by acid rain as shown in the case
of humus accumulation. Deterioration of decomposition will
therefore open ecosystem internal sources of protons and will
multiply the direct effects of acid rain leading to soil acidification and cation leaching.

4. LITERATURE

1: Abrahamsen, G., K.Bjor, R. Horntvedt and B. Tveite (1976):
    Effects of acid precipitation on coniferous forest. In:
    Impact of acid precipitation of forest and freshwater
    ecosystems in Norway. Ed.F.H. Braekke. pp. 38-63 SNCF
    project, Aas-NLR Norway.
2: Baath, E., B. Berg, U. Lom, B. Lundgren, H. Lundkvist, T.
    Rosswall, B. Söderström and A. Wiren (1980): Soil
    organisms and litter decomposition in a Scots Pine
    forest - effects of experimental acidification. In:
    Effect of acid precipitation on terrestrial ecosystems.
    E. T.C. Hutchinson and M.Havas, Plenum, New York,
    pp. 375-381.
3: Baum, U. (1976): Stickstoff-Mineralisation und Stickstoff-
    Fraktionen in Humusformen unterschiedlicher Waldökosysteme.
    Göttinger Bodenkundliche Berichte 38, 1-96.
4: van Breemen, N. (1973): Soil forming processes in acid
    sulphate soils. In: Acid sulphate soils. Ed. H.Dost,
    pp. 66-130, Wageningen.
5: Cronan, C.S. (1980): Controls on leachate from coniferous
    floor microcosms. Plant and Soil 56, 301-322.
6: Deuter Wetterdienst: Amtliche Auskunft.
7: Ellenberg, H. (ed) (1971): Integrated experimental ecology,
    methods and results of ecosystem research in the German
    Solling Project. Ecological Studies 2.
8: Lee, L.L. and D.E. Weber (1982): Effects of sulfuric acid
    rain on major cation and sulfate concentrations of water
    percolating through two model hardwood forests.
    J.Environ.Qual. 11, 57-64.
9: Likens, G.E., F.H. Borman, R.S. Pierce, J.S. Eaton and N.M.
    Johnson (1977): Biogeochemistry of a forested ecosystem.
    Springer Verlag.

10: Matzner, E. (1980): Untersuchungen zum Elementhaushalt eines Heidökosystems (Calluna vulg.) in Nordwestdeutschland. Göttinger Bodenkundliche Berichte 63, 1-120.

11: Matzner, E. and B. Ulrich (1980): The transfer of chemical elements within a heath ecosystem in Northwest Germany. Z.Pflanzenernähr.Bodenkd. 143, 666-678.

12: Matzner, E. and B. Ulrich (1981a): Effects of acid precipitation on soils. In: R.A. Fazzolare and C.B.Smith (ed): Beyong the energie crises. Third international conference on energie use management. Berlin Vol.II, 555-564, Pergamon Press.

13: Matzner, E. und B. Ulrich (1981b): Bilanzierung jährlicher Elementflüsse in Waldökosystemen im Solling. Z.Pflanzenernähr.Bodenkd. 144, 660-681.

14: Matzner, E., P.K. Khanna, K.J. Meiwes, M. Lindheim, J. Prenzel und B. Ulrich (1982): Elementflüsse in Waldökosystemen im Solling - Datendokumentation - . Göttinger Bodenkundliche Berichte 71, 1-267.

15: Matzner, E. and B. Ulrich (1982): The turnover of protons following mineralization and ion uptake in a beech (Fagus silvatica) and a Norway spruce ecosystem, this volume, 147.

16: Mayer, R. (1981): Natürliche und anthropogene Komponenten des Schwermetallhaushalts von Waldökosystemen. Göttinger Bodenkundliche Berichte 70, 1-292.

17: Meiwes, K.J., P.K. Khanna and B. Ulrich (1980): Retention of sulphate by an acid brown earth and its relationship with atmospheric input of sulphur to forest vegetation. Z.Pflanzenernähr.Bodenkd. 143, 402-411.

18: Prenzel, J. (1981): Ein bodenchemisches Gleichgewichtsmodell mit Kationenaustausch und Aluminiumhydroxosulfat. Göttinger Bodenkundliche Berichte, im Druck.

19: Seibt, G. (1981): Die Buchen- und Fichtenbestände der Probeflächen des Sollingprojekts der Deutschen Forschungsgemeinschaft. Schriften Forstl.Fak.Univ.Göttingen, Bd.72, 1-109.

20: Strayer, R.F. and M. Alexander (1981): Effects of simulated acid rain on glucose mineralization and some physicochemical properties of forest soils. J.Environ.Qual. 10,460-465.

21: Strayer, R.F., C.J. Lin and M. Alexander (1981): Effects of simulated acid rain on nitrification and nitrogen mineralization in forest soils. J.Environ.Qual. 10, 547-551.

22: Ulrich, B., R. Mayer und P.K. Khanna (1979): Die Deposition von Luftverunreinigungen und ihre Auswirkungen in Waldökosystemen im Solling. Schriften Forstl.Fak.Univ.Göttingen, Bd.58, Sauerländers Verlag Frankfurt/M., 1-291.

23: Ulrich, B., R. Mayer und P.K. Khanna (1980): Chemical changes due to acid precipitation in a loess derived soil in Central Europe. Soil Sci. 130, 193-199.

24: Ulrich, B.(1981): Theoretische Betrachtung des Ionenkreislaufs in Waldökosystemen. Z.Pflanzenernähr.Bodenkd. 144, 647-659

A MECHANISM FOR STORAGE AND RETRIEVAL OF ACID IN ACID SOILS

J. Prenzel

Institute for Soil Science and Forest Nutrition,
Goettingen

ABSTRACT

Acid soils transform part of incoming strong acids into Al-ions.
These cation acids are often bound to the exchange complex of
the soil. Another mode of storage for Al-ions in the soil is the
precipitation of basic aluminum sulfates. It is shown by calculated titration curves from simplified soil chemical equilibrium
models that substances like the mineral jurbanite ($AlOHSO_4$) can
be understood as acids, bound to the soil matrix, but still
rather soluble. Evidence from the Solling research area, for
probable accumulation of basic aluminum sulfate in that soils,
is presented.

1. INTRODUCTION

Strong acids can enter the soil by acid rain or by biological
processes (nitrification, ammonium uptake by plants in exess
of anion uptake). In either case the soil has four possibilities
in dealing with an input of H-ions: 1.) The strong acid is left
in solution and appears in the seepage. 2.) The H-ions are
stored by the soil matrix. 3.) The H-ions are transferred to the
conjugate base of a weaker acid and this appears in the seepage
or (4.) is stored by the soil matrix.
   In case 2 the mechanism is related to cation exchange. In a
first step the H-ions enter the diffuse electrical double layer
which counterbalances the negative charge of the cation exchange
material. An equivalent amount of other cations is released to
the soil solution. Part of the H-ions remain in this position,
where they can be displaced again by any other cations with

B. Ulrich and J. Pankrath (eds.), Effects of Accumulation of Air Pollutants in Forest Ecosystems, 157–170.
Copyright © 1983 by D. Reidel Publishing Company.

relative ease. But H-ions tend to penetrate to the surface or
into the interior of the exchange material where they are bound
at specific sites of the molecular structure thereby reducing
the exchange capacity (3: pp. 76-81, 16: 111-115). In these
positions the H-ions are not exchangeable against other cations
but they will be released to the soil solution if an attempt is
made to raise the pH of the latter. Organic cation exchangers
can be saturated with H-ions while clay minerals saturated with
H-ions are not stable. More and more H-ions penetrate into the
lattice structure and eventually part of the lattice dissolves,
releasing Al-ions which are then found to be exchangeably bound
instead. This process fulfills the definition of case 4 above –
storage of a weaker acid – and has to be dealt with below.

Case 3.): In calcareous and limed soils the base which
accepts the H-ions is provided by carbonates. Typically:

$$CaCO_3^S + 2 H^+ = Ca^{2+} + 2 HCO_3^-$$

In this example the H-ions have been transferred to a very weak
acid (pK = 10.3) and the strong acid has disappeared.

In soils where no carbonate minerals are present, the most
important source of bases is provided by aluminum bearing
minerals. H-ions can be transferred to the ligands of Al in the
mineral lattice while Al-ions are released to the solution.
Al-ions are not neutral but they are acids of moderate strength:
In aquaeous solution H-ions dissociate from the octahedral hydra-
tion shell of the Al-ions yielding Al-OH-complexes:

$$Al(H_2O)_6 = Al(OH)(H_2O)_5^{2+} + H^+$$
$$Al(OH)(H_2O)_5^{2+} = Al(OH)_2(H_2O)_4^+ + H^+$$
$$Al(OH)_2(H_2O)_4 = Al(OH)_3(H_2O)_3^o + H^+$$

The pK values of these 3 reactions are between 4 and 5 (7).
Hydrated Al-ions are called cation acids due to this feature.
Taking aluminum hydroxide as an example, the reaction with H-ions
can be split into a dissolution and an H-ion transfer step:

$$1^{st}: \quad Al(OH)_3^S + 3 H_2O = Al(OH)_3(H_2O)_3^o$$
$$2^{nd}: Al(OH)_3(H_2O)_3^o + H^+ = Al(OH)_2(H_2O)_4^+$$

In this way it becomes obvious that the H-ion is not transferred
to the weakest acid H2O but to an acid of moderate strength. In
transformations of this type acidity does not disappear but is
transferred to a weaker form. (As toxins these weaker acids are
often more hazardeous than the original strong ones: not only
aluminum but also many heavy metals form cation acids.) If not
stored by the soil matrix, acidity in the form of cation acids

will appear in the seepage.

Case 4.): One possibility for storage by the soil matrix is adsorption of Al-ions by the cation exchange complex. In case of H-ions which first enter the diffuse electrical double layer and later dissolve the lattice of clay minerals, cases 1 and 4 are linked.

A second possibility is the precipitation of an aluminum salt which is less soluble under the prevailing conditions than the dissolving mineral. One example of this is the formation of $AlOHSO_4$ which will be dealt with below.

The immediate effect of storing H- or Al-ions on the soil matrix is protection of soil life and seepage against hazard or pollution by these substances. In long term considerations three questions arise: 1.) If the load of acid entering the soil continues, what will happen? Will the capacity for storing acids be exhausted at some point? 2.) If the input of acid is stopped, what will happen? Will the stored amounts remain on the soil matrix or will they reenter the solution now? 3.) What are the possibilities and the costs of a soil melioration aiming at a non-acidified soil?

In order to answer questions of this kind, an understanding of the processes by which acids are stored and possibly released again (by the soil matrix) is required. In the course of acidification of a soil the storage of acid shows in an increasing part of the cation exchange capacity being saturated with Al and H (17).

After completion of this process - in an acid soil - how could further storage of acid occur? The purpose of the present paper is, to present one possible mechanism.

2. MODEL CALCULATIONS

2.1 Methods

Soil science is interested in numerous chemical species which are present in soils. These substances in general are not independent from each other but undergo reactions by which some of the species are produced at the expense of others. Therefore sets of chemical concentrations in soils form interconnected systems, where, as a rule, no member can be changed without influencing all others. Mathematical models which try to describe these dynamics, have to use simplifications in order to be manageable. One possible simplification is the assumption of overall chemical equilibrium (8,9).

A set of chemical reactions for which coupled equilibria are assumed lends itself to a mathematical model (13,14). The model consists of a set of non-linear equations together with specification of the total amounts of certain elements in the system (e.g. total Na, total SO_4, total Al). From these total amounts

Table 1:   Reactions considered in chemical equilibrium
           models A, B and C

present in model				pK
A	B	C		
x	x	x	$AlOH^{2+} = Al^{3+} + OH^-$	9.03
x	x	x	$Al(OH)_2^+ = Al^{3+} + 2\ OH^-$	18.7
x	x	x	$Al(OH)_3^o = Al^{3+} + 3\ OH^-$	27.0
x	x	x	$Al(OH)_4^- = Al^{3+} + 4\ (OH)^-$	33.0
x	x	x	$Al_2(OH)_2^{4+} = 2\ Al^{3+} + 2\ (OH)^-$	20.3
x	x	x	$Al_3(OH)_4^{5+} = 3\ Al^{3+} + 4\ OH^-$	42.06
x	x	x	$Al_{13}(OH)_{32}^{7+} = 13\ Al^{3+} + 32\ OH^-$	349.27
	x	x	$AlSO_4^+ = Al^{3+} + SO_4^{2-}$	3.2
x	x	x	$H_2O = H^+ + OH^-$	14.0
x		x	$Al(OH)_3^s = Al^{3+} + 3\ OH^-$	33.0
	x	x	$AlOHSO_4^s = Al^{3+} + OH^- + SO_4^{2-}$	17.7

the model calculates their respective distribution over the
different chemical species considered. If all major acids and
bases re included in the reaction system, the pH can also be
calculaed by the model.

In te following, three soil chemical equilibrium models are
presentd, which are by no means complete. The intention is,
rather, o show the effect of certain reactions more clearly by
isolatin them. The behavior of a more complete model would be
modifiedy additional features.

The deinition of the 3 models, named A, B and C, is based on
the reactons in table 1. It is assumed that Al-ions, in one
form or ther are the most important medium of storage of acids
in acid ols. So, Al-ions and -complexes have been included in
all thre mdels. The solids, which are assumed to be in equilibrium wih an aquaeous solution, are different for models A, B
and C: I A it is Al(OH)₃, in B it is AlOHSO₄ and in C it is a
combinaton of both.

The nthod of obtaining the mathematical model proper from
table 1 s the same as described elsewhere (13,14). In brief the
procedur is as follows: Under the assumption of equilibrium

each of the reactions in table 1 yields a thermodynamic equilibrium condition. In addition to these equations the condition of
electroneutrality and activity corrections according to an
extended Debye-Hueckel law are included. In model B an additional mass balance equation (soluble Al equals soluble SO_4) is
required. It is assumed that the only source of Al- and SO_4-ions
in the solution are the respective solid phases.
In the electroneutrality condition and in the ionic strength
equation, terms have been included to allow for an assumed input
of H- or OH-ions to the system (e.g. HCl or NaOH). This system
of equations has been solved by an iterative procedure for
different inputs of H- or OH-ions to give the titration curves
in fig. 1.

2.2 Results and Discussions

From thermodynamic stability it can be concluded that aluminum
hydroxide is controlling the solubility of Al in many soils
(3: pp. 113-118). A commonly accepted picture is (7,15: p.42)
that the Al-ions released from weathering parent material
precipitate as amorphous $Al(OH)_3$ or some Al-OH polymers which
enter clay mineral interlayers forming secondary chlorites. In
model A this group of solids is represented by an $Al(OH)_3$ with
pK=33, which is intermediate between crystallized and amorphous
forms (3: p. 122).
     For acid soils which are also rich in sulfate there is an
ongoing discussion whether aluminum hydroxy sulfates, especially
$AlOHSO_4$, could be controlling the soil solution composition
(4,5,10,11,12,13,14). A mineral jurbanite has been described,
which has the composition $AlOHSO_4$ and a solubility of pK=17.8 (12).
The value of 17.7 in model B has been taken from (13).

     The combination of both solids in model C results in new
features of the system. The equations

$$pAl + 3\ pOH \qquad\qquad = 33$$
$$pAl + \quad pOH + pSO_4 = 17.7$$

together with

$$pH + \quad pOH \qquad\qquad = 14$$

can be combined to give

$$pSO_4 + 2\ pH \qquad\qquad = 12.7 \qquad\qquad\qquad (2.1)$$

This means, system C maintains an activity product of H_2SO_4 in
a similar manner as would a system with solid Na_2SO_4 in equilibrium with its saturated solution the activity product of Na_2SO_4.
Hence system C is a perfect sink for H_2SO_4: Once the solution is
in equilibrium with both of the solid phases, an addition of
sulfuric acid would be digested by the system without any change

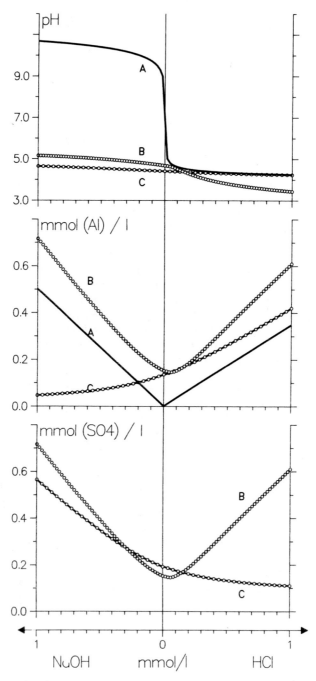

Fig. 1: Calculated response of models A (Al(OH)$_3$), B (AlOHSO$_4$) and C (A&B) to titration with base (l.h.s) or acid (r.h.s). The concentrations of Al and SO$_4$ include all soluble species.

in solution composition, including the pH. The only reaction
occuring would be

$$Al(OH)_3^s + H_2SO_4 = AlOHSO_4^s + 2 H_2O \qquad\qquad (2.2)$$

The reverse direction would occur if H_2SO_4 would be withdrawn
from the solution. Thus, system C is for sulfuric acid a system
of storage and retrieval of acid. The amount already stored is
equivalent to the amount of $AlOHSO_4$ accumulated. The capacity for
storing is equivalent to the amount of $Al(OH)_3$ available.
In soils eq. 2.1 can be interpreted as follows: At a given pH
there is a critical level of SO_4 above which reaction (2.2)
proceeds to the right and below to the left. For instance at
pH=4.2 the critical level is pSO_4=4.3.
      Fig. 1 shows results of a computer simulation where models
A, B and C have been exposed to titration with either NaOH or
HCl up to concentrations of 1 mmol/l, which corresponds to
approximate pH-values of 11 and 3. The line at zero (zero addition
of titrant) marks the concentrations calculated for water
brought in equilibrium with the respective solid phases.

      The solubility of Al in model A shows the well-known V-shape
(3: p. 115). The same solubility in model B is higher over the
full range of conditions applied. This means that in soils
$Al(OH)_3$ is more stable than $AlOHSO_4$, if no source of SO_4 is
present, as assumed in the simulation. But if SO_4-input occurs,
as in acid rain affected soils, $AlOHSO_4$ formation can take place
if the critical level is exeeded. As soon as the SO_4- (acid
rain-) input were stopped, $AlOHSO_4$ would dissolve again. The
accumulated acid would appear as H- and Al-ions in the solution.

      The largest relative difference in Al-concentrations between
models A and B is at zero: While $Al(OH)_3$ dissolves in water to
about 10^{-6} mol(Al)/l only, we find the corresponding concentrations about 10^{-4} mol(Al)/l in models B and C. Even after stoppage
of acid rain a soil which has accumulated $AlOHSO_4$ will maintain
an Al-concentration which is probably toxic (depending on the
Ca+Mg - level in the soil solution (18)).
      The nature of $AlOHSO_4$ as accumulated acid shows up also in the
pH-curves: On the addition of base, model A nearly adjusts to
the pH of the applied solution, while in models B and C the pH
remains low.
      The SO_4-concentration-curves show, that models based on preci-
pitation/dissolution of $AlOHSO_4$ can explain the extractibility
of sulfate from acid soils by alkaline solutions, which has
also been observed (11).

3. EVIDENCE FOR AL-SO$_4$-COMPOUNDS IN SOILS OF THE
   SOLLING RESEARCH AREA

The Solling research area is situated at 9° 30' east, 51° 40'
north, on the plateau of the Solling mountains, 500 m above sea
level. The soils are typic Dystrochrepts (acid "Braunerde")
developed from a 40-60 cm loess covering solifluction material
from weathered "Buntsandstein", a triassic sandstone sediment
(2,20). Some of the research plots within the area are covered
by beech (Fagus sylvatica: plots B1, B3, B4), and some by spruce
(Picea abies: plots F1, F2, F3, F4). The age of the stands
varies between 25 and 125 years (6).

It has been reported (13) that in soil samples taken from
near plot F1 the exchangeable cations could be extracted with
mere water nearly as complete as with a 0.1 N salt solution.
This was possible when a soil sample of 3 g was percolated with
50 ml of water, but not when a saturation extract of nearly 1 ml
water per 1 g soil was used. This indicates that a salt of medium
solubility was dissolving in the soil samples during the extraction. The amount of cations extracted by water-percolation can
be recalculated as 11 meq/kg(soil) or 0.66 meq/l(solution). The
above model calculations show that AlOHSO$_4$ dissolves 0.45 meq
(Al)/l in water, which is not too far from 0.66 taking into
account the uncertainty of the solubility constant and the
formation of Al-OH-complexes in the percolation experiment.
Hence AlOHSO$_4$ is more of a candidate for the unknown dissolving
salt than Al(OH)$_3$, which, according to model A above, dissolves
only 0.001 meq(Al)/l in water (fig. 1).

Soil material from the same site was used in a number of batch
experiment series (13). Each series consisted of several consecutive steps. In each step a soil sample was allowed to equilibrate
with a batch of offered solution. After 24 h the equilibrium
solution was removed and analized. In the next step the same soil
sample was equilibrated with the next batch of the same offered
solution.

Table 2 shows for several ions the sorption (removal from
solution) which occured in the single steps of a batch experiment
series with 1 mmol(MgCl$_2$)/l offered. Obviously cation exchange is
among the processes which can explain the observations: The
Mg-ions offered enter the soil matrix while other cations are
released to the solution. But there are much more equivalents of
cations released than sorbed. The main cation appearing in the
solution is Al while SO$_4$ is the main anion. Similar results
have been obtained with other offered solutions (13). These
findings look as if an Al-SO$_4$ compound is present and dissolves
during the experiments.

Fig. 2 shows results of similar batch experiment series with
the same soil material. An attempt was made (13) to explain the
observations by a chemical equilibrium model. The model was
similar to model B above, but more comprehensive. It included

Table 2:   Sorption ($\rangle$ O: retention by, $\langle$ O: release from the
            matrix) from offered 1 mmol(MgCl$_2$)/l solutions in the
            single steps of a batch experiment series

step	H	Na	K	Mg	Ca	Mn	Al	SO$_4$
			mikromol / kg					
1	-166	-177	-102	+352	-134	-139	-318	-494
2	- 41	- 23	- 47	+142	- 37	- 49	-177	-269
3	- 64	- 14	- 45	+ 68	- 26	- 30	-161	-207
4	- 59	- 13	- 37	+103	- 16	- 20	-127	-188
5	- 60	- 7	- 33	- 14	- 12	- 15	-116	-203
6	- 50	- 5	- 29	+ 25	- 9	- 12	-121	-146
7	- 45	- 17	- 30	+ 12	- 7	- 10	- 85	-146

cation exchange between several cations in addition to the
precipitation/dissolution of AlOHSO$_4$. The simulated lines in
fig. 2 show, that the model was able to explain the main
differences between Al- and SO$_4$-sorptions as well as between
treatments (offered solutions). This can be seen as an other
argument for the existence of AlOHSO$_4$ or a similar substance
in that soil.

This institute has analized soil samples from the Solling
research area since 1966. One method employed has been the
analysis of the chemical composition of the equilibrium soil
solution (ESS) which amounts to a saturation extract with less
than 0.8 ml (water)/g (fresh soil) and 24 h equilibration time.
Results of 474 ESS analyses have been reviewed. The soil
samples had been taken in 1966-1979 in different plots of the
Solling research area (B1, B3, B4, F1, F2, F3, F4) in different
depths (0-10, ..., 50-60 cm). Subplots which had recieved
fertilization have been included. The solutions had been
analized for pH, Na, K, Mg, Ca, Al, Mn, Fe, NH$_4$, NO$_3$, Cl, SO$_4$
and PO$_4$. A chemical equilibrium model, details of which are
given in table 3, has been used to calculate the activities of
Al, H, OH and SO$_4$. Fig. 3 shows the results.

For any solubility equilibrium of (basic) aluminum sulfate
we have

$$x \, pAl + y \, pOH + z \, pSO_4 = pK.$$

With

$$3x - y - 2z = 0$$

and

$$pH + pOH = 14$$

follows

$$x(pAl + 3 \, pOH) + z(pSO_4 + 2 \, pH) = pK + 28.$$

Therefore, in plots like fig. 3, equilibrium with any solid
Al$_x$(OH)$_y$(SO$_4$)$_z$ would force the points on a straight line, where
the slope would correspond to the formula and the intercept to

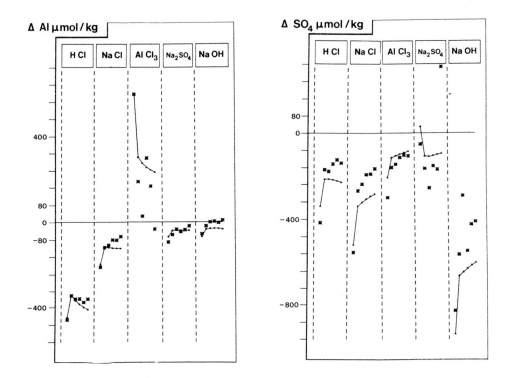

Fig. 2: Sorption ($>$ 0: retention by, $<$ 0: release from the
matrix) from offered solutions with 1 mmol/l of the
substance noted in the rectangle. Each strip gives
results from the single steps of one batch experiment
series (not cumulative). Stars mark experimental
results, lines are calculated from an equilibrium model.

the solubility product of the solid. The points in fig. 3 may
be interpreted as equilibrium of the solution with a solid
$AlOHSO_4$ with some uncertainty about its solubility. Other
workers have reported solubility products of 17.2 (4), 17.3 (5)
and 17.8 (12) for $AlOHSO_4$.
    There is a clear distinction between points from spruce and
beech sites in fig. 3. This is remarkable because several
different beech as well as spruce sites in different years and
soil depths have been included. In order to explain the scatter

Table 3:   Specification of an equilibrium model for calculation
            of activities from concentration measurements. For
            "elements" 1 to 12 analytical concentrations have to
            be supplied. Activities of "elements" 13 to 16 are
            calculated from a measured pH and an assumed equilibrium
            partial pressure of CO_2 of 0.3 mbar. The complexes are
            considered to have equilibrium constants for their
            disintegration into the "elements" as indicated by
            pK values. Sources for the pK are (1) for complex 1 to
            7 and (9) for all others. Activity coefficients are
            calculated from the Güntelberg-equation (9: p. 14).

i	Element	z	Komplex	z	pK
1	NA	1	(AL)(OH)	2	9.03000
2	K	1	(AL)(OH)2	1	18.7000
3	NH4	1	(AL)(OH)3	0	27.0000
4	CA	2	(AL)(OH)4	-1	33.0000
5	MG	2	(AL)2(OH)2	4	20.3000
6	FE	2	(AL)3(OH)4	5	42.0600
7	MN	2	(AL)13(OH)32	7	349.270
8	AL	3	(AL)(NO3)3	0	.124600
9	SO4	-2	(AL)(SO4)	1	3.19600
10	PO4	-3	(AL)(SO4)2	-1	1.89900
11	CL	-1	(CA)(HCO3)	1	1.12900
12	NO3	-1	(CA)(CO3)	0	3.15200
13	H	1	(CA)(CL)	1	-.997800
14	OH	-1	(CA)(CL)2	0	-.146600-002
15	HCO3	-1	(CA)(PO4)	-1	6.45900
16	CO3	-2	(CA)(H)(PO4)	0	15.1000
17			(CA)(H)2(PO4)	1	20.9500
18			(CA)(SO4)	0	2.30900
19			(K)(CL)	0	-.697200
20			(K)2(CO3)	0	-.146600-001
21			(K)(OH)	0	-.501500
22			(K)(SO4)	-1	.850400
23			(MG)(CL)2	0	-.307900-001
24			(MG)(CO3)	0	3.24000
25			(MG)(HCO3)	1	1.07000
26			(MG)(NO3)2	0	-.146600-001
27			(MG)(OH)	1	2.54800
28			(MG)(H)(PO4)	0	15.2600
29			(MG)(SO4)	0	2.22900
30			(MN)(CL)	1	.615100
31			(MN)(CL)2	0	.498500-001
32			(MN)(CO3)	0	-.703800
33			(MN)(HCO3)	1	1.81100
34			(MN)(OH)	1	3.41300
35			(MN)(OH)3	-1	8.00400
36			(MN)(OH)4	-2	7.70800
37			(MN)2(OH)	3	3.41300
38			(MN)2(OH)3	1	18.1100
39			(MN)(SO4)	0	2.26500
40			(NA)(CL)	0	-.733100-003
41			(NA)(CO3)	-1	1.26800
42			(NA)2(CO3)	0	.146600-001
43			(NA)(HCO3)	0	.249300
44			(NA)(OH)	0	-.200900
45			(NA)(SO4)	-1	.696500
46			(PO4)(H)	-2	12.3500
47			(PO4)(H)2	-1	19.5500
48			(PO4)(H)3	0	21.7000
49			(SO4)(H)	-1	1.97900

of the points in fig. 3 the following hypothesis may be helpful:
Solid solutions between $Al(OH)_3$ and $AlOHSO_4$ may be formed and
the concentration of the latter in the liquid phase would
increase with its concentration in that solid solution, that
means with an increasing amount of $AlOHSO_4$ relative to $Al(OH)_3$.
This would be the case at those sites where more $AlOHSO_4$ had
been formed due to higher input of sulfuric acid (see reaction
2.2). Such higher input has been observed under spruce (plot
B1) compared to beech (plot F1) in the Solling and has been
attributed to differences in the interceptive potential of the
respective canopies (19). Correspondingly, in the soil, higher
contents of sulfate (10) as well as of acidity (20) have been
found under spruce than under beech.

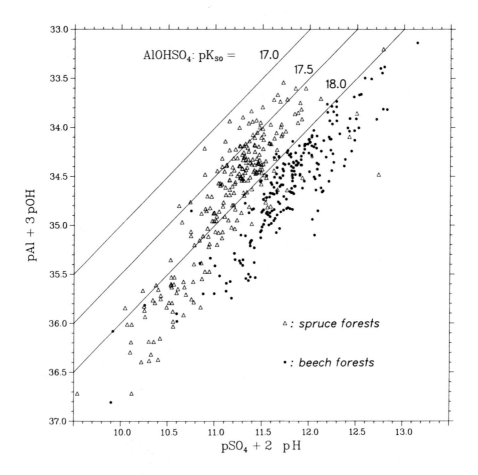

Fig. 3:   Activities in equilibrium soil solutions (ESS) from several
          years, depths and plots within the Solling research area.

4. CONCLUSIONS

Under the impact of acid rain, which consists largely of sulfuric acid, soils can accumulate $Al-OH-SO_4$-compounds similar to the mineral jurbanite ($AlOHSO_4$). This is a rather soluble form of acid, which, in contrast to H- and Al-ions adsorbed by the exchange complex, dissolves in water. Therefore it can be concluded, that after stoppage of acid rain immissions, these compounds would dissolve releasing H- and Al-ions into the soil solution. Meiwes (10) has found 952 $kg(SO_4)$/ha in the Solling research area under spruce. This can be understood to be $AlOHSO_4$ or a similar compound. Seepage is about 400 mm/a at that site, and the solubility of $AlOHSO_4$ in water is about 4 $mg(Al)$/l (see fig. 1). It follows, that this concentration can be maintained for 50 years in the seepage from dissolving $AlOHSO_4$ alone. In this way the impact of acid rain would be spread in time.

This solubility of Al has consequences for nutrient cations too. The soil is no longer able to truly hold its exchangeable stores of K, Mg and Ca. If not an atmospheric input would maintain sort of a steady state, the exchangeable stores would empty by exchange against Al-ions from dissolving $AlOHSO_4$.

5. REFERENCES

1. Baes, C.F. und R.F. Mesmer, 1976: The hydrolysis of cations. London: Wiley
2. Benecke, P. und R. Mayer, 1971: Aspects of soil water behavior as related to beech and spruce stands. In: H. Ellenberg (ed.): Integrated experimental ecology. Berlin-Heidelberg-New-York: Springer. pp. 153-163
3. Bolt, G.H. und M.G.M. Bruggenwert (ed.), 1976: Soil Chemistry, A, basic elements. Amsterdam: Elsevier. 281 pp. (Developments in Soil Science, Vol. 5A)
4. Breemen, N. van, 1973: Dissolved Aluminum in acid sulfate soils and in acid mine waters. Soil Sci.Soc.Am.Proc. 37 pp. 694-697
5. Breemen, N. van, 1976: Genesis and solution chemistry of acid sulfate soils in Thailand. Waageningen: pudoc. (Agricultural Research Report 848) pp. 1-263
6. Ellenberg, H., 1971: Introductory survey. In H. Ellenberg (ed.): Integrated Experimental Ecology. Berlin-Heidelberg-New-York: Springer (Ecological Studies.Vol.2). pp. 1-15
7. Hsu, P.H., 1977: Aluminum hydroxides and oxyhydroxides. In: J.B.Dixon and S.B.Weed: Minerals in Soil Environments. Madison (Wisconsin): Soil Science Society of America. pp. 99-143
8. Kittrick, J.A., 1977: Mineral Equilibria and the Soil System. In: J.B.Dixon and S.B.Weed: Minerals in Soil Environments.

Madison (Wisconsin): Soil Science Society of America.
pp. 1-25

9. Lindsay, W.L., 1979: Chemical Equilibria in Soils. New York:
   Wiley Interscience.

10. Meiwes, K.J., 1979: Der Schwefelhaushalt eines Buchenwald-
    und eines Fichtenwaldökosystems im Solling. Göttinger
    Bodenkundliche Berichte 60, pp. 1-108

11. Meiwes, K.J., P.K. Khanna und B. Ulrich, 1980: Retention of
    sulphate by an acid brown earth and its relationship with
    the atmospheric input of sulphur to forest vegetation.
    Z.Pflanzenernaehr.Bodenk. 143, pp. 402-411

12. Nordtstrom, D.K., 1982: The effect of sulfate on aluminum
    concentrations in natural waters: some stability relations
    ... Geochimica et Cosmochimica Acta 46, pp. 681-692

13. Prenzel, J., 1982: Ein bodenchemisches Gleichgewichtsmodell
    mit Kationenaustausch und Aluminiumhydroxosulfat.
    Gött.Bodenkundl.Ber. 72, pp. 1-113

14. Prenzel, J., 1982: Bodenchemische Gleichgewichtsmodelle:
    Berechnungsmethoden und Anwendungen. Mitteilgn.Dtsch.
    Bodenkundl.Ges., in press, 5 pp.

15. Scheffer, F., Schachtschabel, P., 1976: Lehrbuch der Boden-
    kunde. 9. Aufl.Stuttgart: Enke

16. Russel, E.W., 1973: Soil conditions and plant growth.
    10th ed. London, New York: Longman

17. Ulrich, B., 1981: Ökologische Gruppierung von Böden nach
    ihrem chemischen Bodenzustand. Z.Pflanzenernaehr.Bodenk.
    144, pp. 289-305.

18. Ulrich, B., Gefahren für das Waldökosystem durch saure Nieder-
    schläge. In: Immissionsbelastungen von Waldökosystemen.
    Recklinghausen: Landesanstalt für Ökologie, Landschafts-
    entwicklung und Forstplanung Nordrhein-Westfalen, Leib-
    nitzstraße 10, D-4350 Recklinghausen. (LÖLF-Mitteilungen,
    Sonderheft 1982) pp. 9-25

19. Ulrich, B., R. Mayer und P.K. Khanna, 1979: Deposition von
    Luftverunreinigungen und ihre Auswirkungen in Waldökosy-
    stemen im Solling. Frankfurt a.M.: Sauerländer. (Schriften
    aus der Forstlichen Fakultät der Universität Göttingen und
    der Niedersächsischen Forstlichen Versuchsanstalt, Vol.58)
    291 pp.

20. Ulrich, B., M. Ulrich und E. Ahrens: Soil chemical differences
    between beech and spruce sites. In: H.Ellenberg (ed.):
    Integrated experimental ecology. Berlin-Heidelberg-New
    York: Springer. pp. 153-163.

EFFECTS OF ATMOSPHERIC AMMONIUM SULFATE ON CALCAREOUS AND
NON-CALCAREOUS SOILS OF WOODLANDS IN THE NETHERLANDS

N. van Breemen and E.R. Jordens

Soil scientist and Visiting soil scientist,
Department of Soil Science and Geology
Agricultural University, Wageningen, the Netherlands

ABSTRACT

Throughfall and stemflow in a woodland in the Netherlands
had very high concentrations of ammonium sulfate, presumably
caused by dry deposition on the vegetation of NH_3 (volatilized
from animal manure) and SO_2 (from fossil fuels). Quick nitrification of the ammonium sulfate and of mineralized organic
nitrogen, even when soil pH dropped as low as 2.8-3.5, caused
strong soil acidification. In 1981 total acid inputs were
750 $kmol.km^{-2}$, which is more than double the acid inputs described for other areas with serious acid rain. The leaching
effect of nitric acid from nitrified ammonium was particularly
strong because plants took up little soil nitrate, probably
as a result of ample supply of atmospheric ammonia to their
leaves. In the calcareous soil HNO_3 mobilized and leached calcium, in the acid soils aluminium and calcium. Concentrations
of sulfate varied less with time and depth than those of nitrate, and sulfate was less important for leaching of cations.

INTRODUCTION

In 1980 a biogeochemical monitoring project was started
in a woodland near the 'Hackfort' estate, in Warnsveld, the
Netherlands. The aim of this project was to study ongoing soil
formation by tracing the fate of water and of a number of dissolved constituents moving through the atmosphere-vegetation-
soil system. Soon it became apparent that current soil formation
is determined to a large extent by inputs of atmospheric pollutants, mainly ammonium sulfate that is concentrated on vegetation

171

B. Ulrich and J. Pankrath (eds.), Effects of Accumulation of Air Pollutants in Forest Ecosystems, 171–182.

surfaces by dry deposition (9). In this paper we summarize the data on ammonium sulfate inputs for 1981 and discuss some of the soil chemical consequences of these inputs. Quantitative data are based mainly on one year of monitoring and need not be representative. Continuous monitoring will be done for at least two more years to study soil water movement and ionic activities in the aqueous solution.

MATERIALS AND METHODS

The study was done in a 3.2 ha woodland area, dominated by oak (Quercus robur) and birch (Betula pendula). The oak trees were coppiced until 1939, but except for occasional thinning of trees, the vegetation has not been disturbed since then. The area has been a strict natural reserve since 1968. Most of the surrounding land (about 75% of the area of the region) is used for intensive grass production for stall-fed dairy cattle.

Soil profile code	Classification	Remarks on soil genesis
A	coarse loamy, mixed, acid, mesic Aeric Haplaquept	decalcified to 80 cm depth, strongly enriched in iron supplied by groundwater
B	sandy, mixed, acid, mesic Umbric Dystro-chrept	decalcified to 130 cm depth, slightly enriched in iron, slightly podzolized
C	mixed, acid, mesic Aquic Udipsamment	reworked to about 120 cm depth, no calcium carbonate within 250 cm below the soil surface, slightly podzolized in the surface horizons
D	sandy, mixed, cal-careous, mesic Typic Haplaquoll	reworked to about 120 cm depth with the former calcareous subsoil now at the surface

Table 1   Classification of the soils studied -nearest equivalents at the family level of the Soil Taxonomy (8)-and remarks on their genesis.

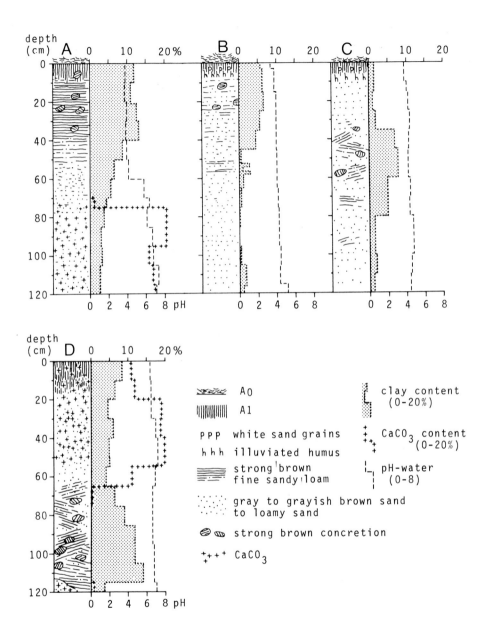

Fig. 1.   Some morphological and chemical characteristics of the
           soils used in this study.

The remaining land is mainly woodland similar to the area under
study.
     The parent material of the soils is sandy to loamy Pleisto-
cene Rhine sediment, presumably with admixture of coversand at
shallow depth, especially in slightly elevated areas. The river-
ine sediment was probably calcareous during deposition, but
the upper 80 to 250 cm of the soil has been decalcified.
Table 1 and Fig. 1 illustrate some of the soil properties.
Profile A represents the undisturbed soil in the lower part
of the area. Profile B is typical for 40-50 cm higher, dune-
like elevations in the terrain. The soils C and D have been
strongly disturbed, perhaps by mining activities for low-grade
iron ore. Profile C could have been formed from a profile similar to that of B, by turning the soil upside down. Profile D is
an upside-down version of profile A. So, strongly contrasting
soils occur close together under very similar environmental conditions: strongly weathered and acid, loamy (A) and sandy (B)
soils, a near-neutral calcareous soil (D) and a moderately acid
soil (C). No historical or archeological details are known about
the digging activities in the area. The well-developed A1 horizons and the apparently stable and well-adapted ground vegetation, however, indicate that the disturbance took place at
least several centuries ago.
     The selection of this area for biogeochemical research
was based mainly on the unique range of soil characteristics.
The same area has been used earlier (1957-1965) for studies
on decomposition of plant litter and the activity of soil fauna
and micro-organisms (4,10).
     At each of four 10 x 20 m^2 sites represented by the soils
A, B, C and D, throughfall (i.e. meteoric water collected below
the tree canopy), stemflow (water flowing down tree trunks) and
soil solution were collected and analysed regularly. Throughfall
was sampled fortnightly at 120 cm above the surface by 400 cm^2
polythene funnels into dark 5-litre polythene bottles. Six
throughfall collectors were placed at the apices of a randomly
located hexagon (of side 5 m) in each of the four sites. Samples from adjacent pairs of collectors were pooled to give
4 x 3 replicates. Stemflow was intercepted by silicon-sealed
polythene collars around six randomly chosen trees. Again, fortnightly samples of adjacent pairs of collectors were pooled to
give 12 replicates. Rainwater was sampled in duplicate from
collectors similar to those used for throughfall. The rainfall
collectors were placed at 4.5 m above the surface of pastures
adjacent to the woodland, about 400 m apart, and each approximately 100 m from the woodland.
     At each of the four sites soil solutions were sampled
monthly from Soil Moisture Equipment^c porous cups (with double
acces tubes) at 10, 20, 40, 60 and 90 cm depth, and from duplicate porous cups at 10, 40 or 60 and 90 cm depth. The sampled
points were located approximately 2 to 10 m apart, and 3-4 m

from the nearest trees. Vacuum tanks were connected to each set
of 3 or 5 porous cups and after 18 to 24 h the soil solution
collected inside the cups (normally 20 to 100 cc) was removed
by suction.
    Within one day of collection, all water samples were ana-
lysed for pH, electrical conductance, and total and inorganic
dissolved carbon and, in case of near-neutral soil solutions,
acidified. Within two weeks, all samples were analysed for Ca
and Mg (by atomic absorption spectrometry), Na and K (by atomic
emission spectrometry), NH_4 (colorimetrically with Nessler's
reagent), F, Cl, NO_3 and SO_4 (by ion chromatography) and H_2PO_4
(colorimetrically as phosphoric-molybdenic acid). Moreover, soil
solutions were analysed for Fe, Mn and Al (by atomic absorption
spectrometry), and for H_4SiO_4 (colorimetrically as the blue
silica-molybdic acid complex). Details of the the methods used
to analyse water and soil samples are given by Begheijn (2). We
anticipated that the ceramic cups used for sampling the soil
solution might release significant amounts of Al^{3+} in acid soil.
When flushed slowly with dilute sulfuric acid (pH 3-4), however,
the cups caused concentrations of $1/3$ Al^{3+} in the order of 0.01
to 0.15 mmol/l, which are small compared to those observed in
soil solutions in the same pH range (0.3-3 mmol/l).

RESULTS AND DISCUSSION

Atmospheric inputs

    The (inorganic) chemical composition of throughfall and
stemflow was dominated by ammonium and sulfate. Although the
concentrations of NH_4^+ and $1/2$ SO_4^{2-} varied strongly with time,
they were practically equal in most samples and hence were
strongly correlated, with r = 0.91^{xxx} for throughfall and
r = 0.95^{xxx} for stemflow. During 1981 concentrations of NH_4^+
(volume-weighted means ± st.dev.) were 0.47 ± 0.10 mmole/l
for throughfall, and 2.0 ± 0.5 mmole/l for stemflow. There
were no significant differences between the four sites in terms
of concentrations of ammonium and sulfate in meteoric water.
The volume-weighted mean ammonium concentrations in throughfall
and stemflow were respectively 4 and 17 times higher than in
rain water.
Van Breemen et al (9) explained the high ammonium sulfate levels in water collected beneath the tree canopy by dry deposition on the vegetation surfaces of NH_3 volatilized from domestic animal manure, followed by NH_3-enhanced oxidation and deposition of SO_2 and leaching of the resulting ammonium sulfate
by rain water. The total annual deposition of inorganic N in
throughfall and stemflow in 1981 was 460 ± 110 $kmol.km^{-2}$ (or 64
± 15 kg N per ha.yr). Of this, 75% was NH_4-N and 25% was NO_3-N.
Similar values were obtained in the southern part of the Nether-

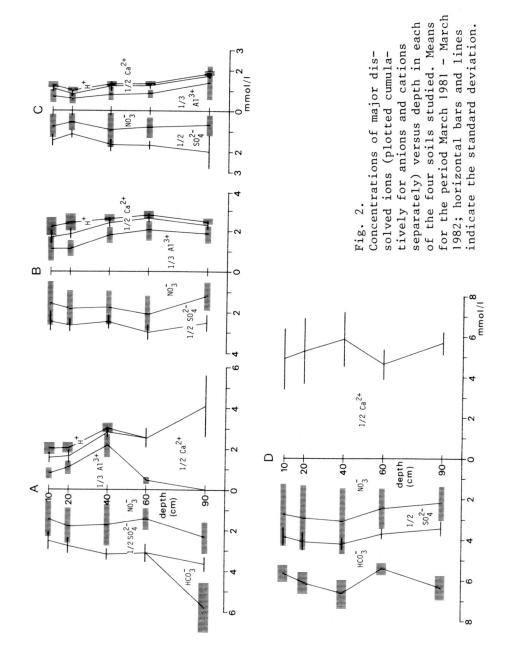

Fig. 2.
Concentrations of major dissolved ions (plotted cumula-
tively for anions and cations
separately) versus depth in each
of the four soils studied. Means
for the period March 1981 – March
1982; horizontal bars and lines
indicate the standard deviation.

lands in pine woods located at a greater distance (100 to 400 m)
from agricultural land (9). As will be discussed later, most of
the NH_4^+ was nitrified in the soil, leading to the production of
equivalent amounts of nitric and sulfuric acid:

$$(NH_4)_2SO_4 + 4O_2 \rightarrow 2HNO_3 + H_2SO_4 + 2H_2O \qquad\qquad (1)$$

In 1981 the input of potential acidity (from ammonium sulfate)
plus actual acidity (from H^+) amounted to 714 kmol.km^{-2}. These
are about double those for spruce forests near Göttingen (6),
and would be equivalent to the acidity delivered annually by
800 mm of precipitation of pH 3.0. In fact the volume-weighted
mean pH of throughfall and stemflow was about 4.6, so actual
acidity was about 3% of potential acidity. As will be discussed
in the next section the total potential acid inputs were probably still higher: part of the ammonia deposited on the vegetation was apparently taken up by the leaves and assimilated, and
would end up as nitric acid after mineralisation and nitrification of the organic N in leaf litter.

Soil solution chemistry

About 80% of the total ionic charge of the soil solutions
was due to the anions NO_3^-, SO_4^{2-} and (in calcarous soil) HCO_3^-,
and the cations Ca^{2+} and (in acid soil) Al^{3+} and H^+. Concentration profiles of these ionic species (Fig. 2) show a predominance of NO_3^- in most cases, of Ca^{2+} in calcareous soil, and
of Al^{3+} in acid soil. The good correspondence between summed
equivalent concentrations of inorganic anions and cations (assumed all dissolved aluminium to be trivalent) shows that partly hydrolysed and organically complexed dissolved aluminium
was quantitatively unimportant. Ammonium concentrations were
virtually nil in all soil solution samples from the calcareous
soil D, and were generally much smaller than those in throughfall for soil solutions from the acid soils A, B and C. So,
even at low pH ammonium is apparently nitrified rapidly according to equation 1. Moreover, except in soil C, the molar concentration of NO_3^- far exceeded that of 1/2 SO_4^{2-}, while equal
concentrations would be expected if nitrification of ammonium
sulfate were the only process involved.

Preliminary calculations indicate that 65% (in the acid
soils) to 30% (in the calcareous soil) of the dissolved nitrate
in the soil solutions came from ammonium in throughfall, so
that the remaining nitrate must have come from excess mineralization over NO_3^--uptake by plants. Because the year 1981 was
rather cool and wet, the excess nitrate formation was probably
not due to a temporarily increased decomposition of organic
matter which is often observed in warm, dry years (7), but
rather due to easily available nitrogen in the form of dry
deposited NH_3 on tree leaves, which would depress uptake of

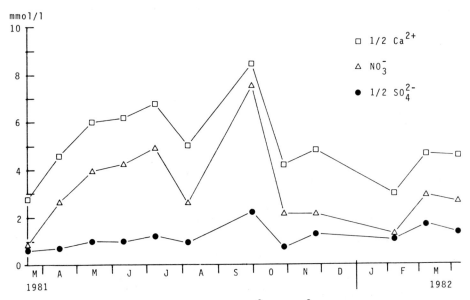

Fig. 3.   Concentrations of NO_3^{2-}, $1/2$ SO_4^{2-} and
$1/2$ Ca^{2+} vs. time in the soil solution at 10 cm
depth of the calcareous soil D.

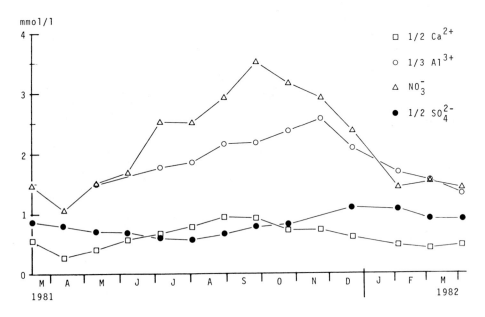

Fig. 4.   Concentrations of NO_3^-, $1/2$ SO_4^{2-}, $1/3$ Al^{3+} and $1/2$ Ca^{2+}
vs. time in the soil solution at 40 cm depth of the acid
soil B.

Soil	depth (cm)	$\frac{1}{2}Ca/NO_3$	$\frac{1}{2}Ca/\frac{1}{2}SO_4$	$\frac{1}{3}Al/NO_3$	$\frac{1}{3}Al/\frac{1}{2}SO_4$
A	0-40	0.37**	0.06	0.51***	0.39**
	60	0.75**	0.74**	0.89***	0.29
	90	0.53**	-0.07	-	-
B	0-90	0.71***	-0.17	0.56**	0.11
C	0-90	0.87***	0.02	0.70**	0.31*
D	0-90	0.84***	0.45***	-	-

* $P<0.05$ ** $P<0.01$ *** $P<0.001$

Table 2   Correlation coefficients for various cation-anion pairs in soil solutions collected from March 1981 to March 1982.

nitrogen from the soil solution.

During summer and early fall the pH values of the soil solution at 10 and 20 cm depth decreased with time and increased again during winter and early spring. The lowest pH values reached at 10 cm depth were 2.8-2.9 in profiles A and B, and 3.3-3.5 in profile C.

These changes in pH were paralleled by an increase in dissolved nitrate during summer and fall, and a decrease in winter (Figs. 3 and 4). There was a marked covariation between the concentrations of nitrate, calcium and aluminium (Table 2) indicating that nitric acid mobilized calcium and aluminium from solid phases. In the calcareous soils most of the Ca^{2+} must have come from calcium carbonate. In the acid soils, however, in addition to calcium-bearing (silicate) minerals, the exchange complex and mineralizing fresh organic matter could have been quantitatively important in supplying Ca^{2+} to the soil solution.

The source of dissolved Al^{3+} is not immediately obvious. Although the solubility of common Al-bearing soil minerals increases with decreasing pH below 6, aluminium concentrations in soils A and B were invariable lower in the soil solutions from very acid surface horizons (pH 2.8-3.4) than at higher soil solution pH (3.5-4) in subsurface horizons. Pending further research, a preliminary explanation is that dissolved aluminium came mainly from easily weatherable minerals in the sand and silt fraction (especially feldspars). Contents of sand- and silt sized aluminium increase with depth in the undisturbed soils A and B (Table 3), presumably because the most easily weatherable minerals had already been leached from the surface horizons in the course of soil development. By contrast, the contents of clay-size and free (dithionate-oxalate extractable) aluminium are higher in the surface horizons than in subsurface horizons.

The pattern of seasonally varying nitrate (and cation) concentrations was repeated at all depths, but at greater depth peak concentrations were found later. Peaks occurred at 10 and 20 cm depth in summer, at 40 cm depth in the fall, and at 90 cm depth in winter (Fig. 5). Apparently nitrate, together with calcium in the calcareous soil and with aluminium and calcium in the acid soils, was leached to the subsurface horizons and to the ground water. In profile A, dissolved Al^{3+} made way for Ca^{2+} at 60 and 90 cm depth, just above and in the calcareous subsoil. So, aluminium that was leached down was neutralized and precipitated by calcium carbonate, partly through intermediate steps involving dissolved (calcium) bicarbonate and exchangeable calcium.

In all soils sulfate concentrations were not only lower but also varied less with time and depth than nitrate concentrations. In most cases sulfate concentrations were poorly correlated with concentrations of calcium and aluminium (Table 2). These observations indicate that for leaching of cations sulfuric acid was less important than nitric acid. This difference in behaviour was probably related to little uptake of NO_3^- by plants, which

Soil	depth (cm)	$\frac{1}{3}Al^{3+}$ mmol/l	Al_2O_3, mass fraction (%) in fine earth		
			in sand & silt	in clay	free Al_2O_3
A	10	0.8	2.6	1.7	0.8
	20	1.1	2.5	1.9	0.9
	40	2.2	3.6	1.5	0.7
	60	0.5*	5.7	0.9	0.5
	90	0.0*	5.2	0.7	0.2
B	10	1.1	2.8	1.0	0.6
	20	1.1	2.4	1.1	0.6
	40	1.8	3.3	0.8	0.6
	60	2.1	5.1	0.3	0.4
	90	1.9	3.5	0.3	0.2
C	10	0.7	3.2	0.4	0.2
	20	0.6	2.9	1.0	0.6
	40	0.8	2.9	0.2	0.8
	60	0.8	3.3	0.9	0.5
	90	1.4	3.2	0.2	0.2

*   These low aluminum concentrations were due to high pH in the calcareous subsoil

Table 3   Dissolved Al^{3+} (mean concentration for March 1981- March 1982) and contents of soil aluminium in various fractions at different depths in soils A, B and C.

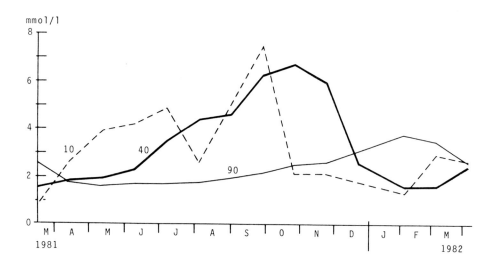

Fig. 5. Concentration of NO_3^- vs. time in the soil solution at
10, 40 and 90 cm depth of the calcareous soil D.

would increase the leaching efficiency of HNO_3, and to stronger
adsorption of SO_4^{2-} to the soil, at least at low pH.

The acid inputs reported here and their effect on the soil
solution in terms of pH and dissolved NO_3^- and Al^{3+} are extreme
in comparison with those reported from other areas influenced
by acid rain (3, 5, 6). Such extreme conditions may be common
in woodlands in the Netherlands (9) and in other areas with
intensive animal husbandry. Research will be continued to further clarify the effects of acid inputs, not only on soil but
also on vegetation. One aspect not encountered in the Hackfort
area (where below the groundwater table nitrate is removed by
denitrification) is the risk of a build-up of high nitrate
concentrations in ground water. Preliminary data by Appelo (1)
on recent samples of the upper 15 meter of the ground water body
in the ice-pushed sandy sediments of the Veluwe (used for domestic purposes on account of its excellent water quality) indicate
increased levels of nitrate and aluminium compared to samples
taken 5 to 10 years ago.

ACKNOWLEDGEMENTS

    We gratefully acknowledge the help of S. Slager and
A.G. Jongmans during describing and sampling of the soils.
E.J. Velthorst and N. Nakken-Brameyer did most of the
chemical analyses, Th. Pape helped with the data management,
G. Buurman drew the figures and Th. Neijenhuis-Reijmers typed
the paper.

REFERENCES

 1. Appelo, C.A.J.: 1982, "Acidification of ground water within
    quartz-rich sediments of the Netherlands Veluwe area",
    Working paper, 12th Annual Symp. Analyt. Chem. Pollut-
    ants, Amsterdam.
 2. Begheijn, L.Th.: 1980, "Methods of chemical analyses for soils
    and waters", Dept. Soil Science and Geology, Agric. Univ.,
    Wageningen, the Netherlands, 100 pp.
 3. Likens, G.E., Bormann, F.H., Pierce, R.S., Eaton, J.S., and
    Johnson, N.M.: 1977, "Biogeochemistry of a forested eco-
    system", Springer Verlag, N.Y., 146 pp.
 4. Minderman, G.: 1968, J. Ecol. 56, pp. 355-362.
 5. Overrein, L.N., Seip, H.M. and Tollan, A.: 1980, "Acid preci-
    pitation, effects on forest and fish". SNSF project,
    Ås, Norway, 175 pp.
 6. Ulrich, B., Mayer, R. und Khanna, P.K.: 1979, "Deposition von
    Luftverunreinigungen und Ihre Auswirkungen in Waldökosys-
    temen in Solling". J.D. Sauerlanders' Verlag, Frankfurt
    A.M., 189 pp.
 7. Ulrich, B., Mayer, R., Khanna, P.K.: 1980, Soil Science 130,
    pp. 193-199.
 8. U.S. Department of Agriculture, Soil Conservation Service: 1975,
    "Soil Taxonomy", US Government Printing Office, Washington
    D.C., 754 pp.
 9. Van Breemen, N., Burrough, P.A., Velthorst, E.J., Van Dobben,
    H.F., de Wit, T. Ridder, T.B. and Reijnders, H.F.R.: 1982,
    "Soil acidification from atmospheric ammonium sulphate in
    forest canopy throughfall". Nature 299, pp. 548-550.
10. Witkamp, M. and Van der Drift, J.: 1961, Plant and Soil, 15,
    pp. 295-311.

STUDIES OF PROTON FLUX IN FORESTS AND HEATHS IN SCOTLAND

Hugh G. Miller

The Macaulay Institute for Soil Research, Aberdeen,
AB9 2QJ, Scotland

Abstract

Models of nutrient uptake by tree roots and subsequent
release on litter decomposition show that, where ammonium is
the dominant nitrogen source, acidity as a result of excess
cation uptake is equivalent to that introduced in acid rain,
but because that in acid rain is introduced together with mobile
anions this is the more likely to reach streams, provided that
drainage over or through the soil is so rapid that neutralization
reactions are not completed. In polluted regions the neutralization process within a forest canopy appears to fail in late
winter, thus accentuating the episodicity of hydrogen ion input
to the soil.

INTRODUCTION

Although the evidence is largely circumstantial, there can
be little doubt that elevated acidity in the precipitation
reaching forests may lead to acidification of streams. Quite
how this small additional input, given the large production and
consumption of hydrogen ions within a forest ecosystem, reaches
streams without being neutralized within the soil is not at all
clear. Mechanisms have been postulated, but before definitive
explanations can be given more information is required on the
factors controlling the production, consumption and flux of
hydrogen ions within the ecosystem.

B. Ulrich and J. Pankrath (eds.), Effects of Accumulation of Air Pollutants in Forest Ecosystems, 183–193.

PRODUCTION AND CONSUMPTION ACIDITY

The role of the vegetation, and associated soil organic
layers, in the production and consumption of acidity is being
examined as part of a continuing study of element cycling in
forests and heathlands in Scotland and northern England.
Although in many areas only preliminary data is presently
available, it is clear both that tree growth may result in soil
acidification and that tree crowns may modify the pattern of
acid input to the soil, there being a complex interaction between
species, age and amount of received acidity.

Tree growth and hydrogen ion production

Studies of nitrogen mineralization in Scottish forest soils
suggest that at least three quarters, and often more, of the
nitrogen uptake by trees on these soils must be as ammonium,
rather than as nitrate ions (e.g. Williams *et al.* 17 & 18). This
conclusion is broadly supported by measurements of the amounts
of ammonium and nitrate in water extracted from the soil beneath
different tree species (Fig. 1), the proportion of nitrate being
particularly low beneath pine.

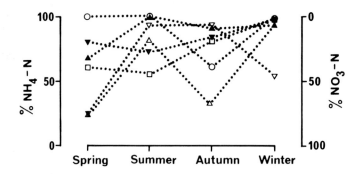

Fig. 1 Proportion of NH_4^+-N versus NO_3^--N in soil water beneath
the humus layer in stands of *Calluna vulgaris* (o) young (▼)
and old (▲) *Pinus sylvestris*, *Picea sitchensis* (▽), *Larix
kaempferi* (△) and *Betula* spp. (□).

If three-quarters of the nitrogen uptake by a pine crop is
as ammonium, then the trees accumulate an excess of cations over
anions, an excess that has to be balanced by a reverse flux of
H^+ from the roots into the soil. Plant growth in an ammonium

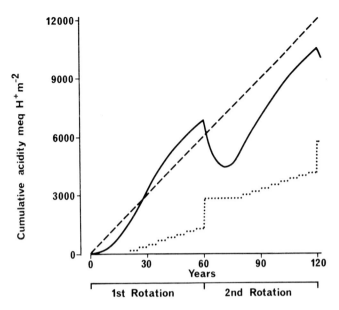

Fig. 2   Cumulative production of acidity as the net result of
root uptake and decomposition release of excess cations (———),
assuming that 75% of the nitrogen uptake is as NH_4^+, and the
proportion removed in harvested stems (·········).   The models are
for two 60-year rotations of regularly thinned and finally clearfelled pine, the first rotation being planted on bare soil with
no pre-existing soil organic layer (calculations as described in
Nilsson *et al*. (10) based on data of Miller *et al*. (8) and (9)).
Also shown (————) is the cumulative acidity that would result
from pH 4.0 rainfall at 1000mm per annum.

dominated environment, therefore, is acidifying.   The eventual
release of the excess cations on decomposition of the litter and
humus is the compensating neutralising step, but in a forest,
where large amounts of organic matter accumulate over many years,
there is a long delay between acidification and neutralisation.
Furthermore, an appreciable proportion of the accumulated cations
may be removed in harvested timber.   As shown in Fig. 2, therefore, there can be considerable acidification in the rooting zone
of fast growing trees, although some neutralisation occurs during
periods when organic matter decomposition exceeds net primary
production.

Hydrogen flux

        The hydrogen ion concentration in rainwater may be consider-
ably modified as it passes through the various layers of an ecosystem.   Eaton *et al*. (4) suggested that the reduction in rain

Table 1

Changes in amounts of hydrogen ion as rainwater passes through
the canopy of Sitka spruce at Fetteresso (F), Leanachan (L),
Elibank (E), Strathyre (S), Kershope (Ks) and Kilmichael (Kl)
during the year April 1976 - April 1977.

	Forest					
	F	L	E	S	Ks	Kl
Mean pH of incident rain	4.08	4.48	3.99	4.66	4.29	4.73
Equivalents H^+ ha^{-1} yr^{-1}:						
rain	910	350	1030	350	570	280
throughfall	240	10	240	250	530	100
stemflow	60	20	60	210	160	60
% change in H^+ beneath trees	-47	-91	-71	+31	+21	-43

water acidity they observed beneath the canopy of a northern
hardwood forest was due to cation exchange on leaf surfaces,
whereas Mayer and Ulrich (7) ascribed an increased acidity
beneath beech and spruce at Solling to the trapping of acidic
substances from the atmosphere on tree surfaces.

In studies in six stands of Sitka spruce (*Picea sitchensis*
(Bong.) Carr.), at locations ranging from the unpolluted Atlantic
coast to the relatively polluted eastern coast of Scotland,
throughfall water was generally, but not invariably, less acid
than the received rain (Table 1).   These changes in acidity were
not, however, mirrored by those found to result from the placing
of an inert trapping surface of polyehtylene coated wire mesh
above a rain gauge (Fig. 3), despite the fact that these "filter
gauges", as shown by the sodium results in Fig. 3, are extremely
efficient at scavenging the atmosphere.   Indeed, although the
increased sodium found in throughfall water is closely related
to the increase in the filter gauge water, the relationship for
hydrogen ions is very poor (Fig. 4).   In part this might be
explained by the uptake of acid-forming gases, particularly SO_2,
through leaf stomata, an uptake that does not occur on filter
gauges.   However, as the lack of a close relationship in Fig. 4
is generally due to a greater neutralisation of throughfall,
rather than acidification, this explanation would seem to be
inadequate.

More detailed examination of the neutralisation of throughfall water at Fetteresso forest, a polluted east coast site,
shows that the effect is distinctly seasonal (Fig. 5).   A
regular oscillation in pH is found in both throughfall and stemflow, although the pattern for stemflow appears to lag a month
or more behind that shown by throughfall.   As there is no such
regular variation in the pH of water collected beneath the filter
gauge, the observed pattern of neutralisation must be ascribed to
the presence of the living tree.   However, such oscillations
could not be detected in throughfall or stemflow water from

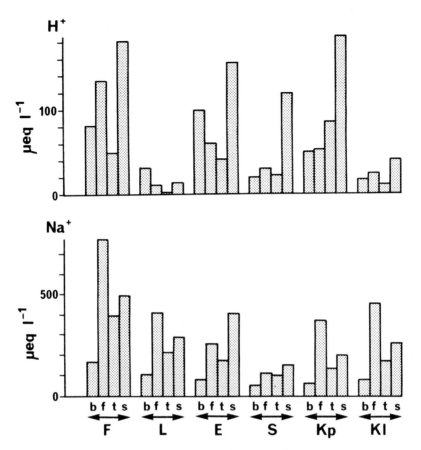

Fig. 3   Mean concentrations of H⁺ and Na⁺ in water collected
between April 1976 and April 1977 in bulk precipitation gauges
(b), filter gauges (f), throughfall gauges (t) and stemflow
gauges (s) in stands of *Picea sitcnensis* at Fetteresso (F),
Leanachan (L), Elibank (E), Strathyre (S), Kershope (Kp) and
Kilmichael (Kl).

Leanachan forest on the relatively unpolluted west coast (Fig. 6).
The difference between these forests appears to be that at
Leanachan, where the acid input is fairly low (Table 1), the
neutralisation process occurs consistently throughout the year,
whereas at Fetteresso there is a failure to neutralise the rainwater acidity arriving in winter and early spring.
      In a recently established study at Glen Tanar forest, not
far distant from Fetteresso and with a similar rainwater acidity,
bulk precipitation, filter, throughfall and stemflow gauges
together with water collecters beneath the ground vegetation,
below the humus layer and in the mineral soil, have been installed

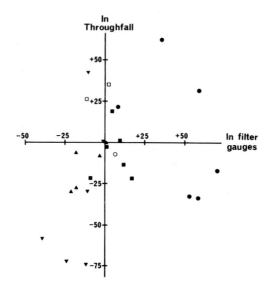

Fig. 4   Mean annual change in H$^+$, meq per litre, in comparison
to bulk precipitation for throughfall plotted against the
equivalent change in water collected in the filter gauge.   Data
from Fetteresso (•), Leanachan (▴), Elibank (▾), Strathyre (▪),
Kershope (▫) and Kilmichael (○).

in and around stands of *Betula* spp. (predominantly *pendula* Roth.),
Larix kaempferi (Lamb.) Carr., young (aged 45 years) and old
(aged 110 years) *Pinus sylvestris* L., *Picea sitchensis* and in a
Calluna vulgaris L. heathland.   Analyses are being made on
samples bulked into extended periods, but the data so far available again indicates that, like at the neighbouring Fetteresso,
the pH of throughfall under most tree species is depressed below
that of incident rain in late winter but is increased, except
beneath the old pine, in mid to late summer.
    Further changes in acidity occur as throughfall and stemflow
pass on through the ecosystem (Table 2).   When present, an
understory of shrubs or grasses generally increases the pH of
water passing over them, but by the time the water has drained
the surface organic layers the pH is once more lowered, only to
increase again in the mineral soil.   However, a striking feature
of the results from Glen Tanar is that at depth in the mineral
soil there remain considerable differences in soil water acidity
beneath the different species, with an appreciable difference
even being found between the young and old *Pinus sylvestris*
stands.

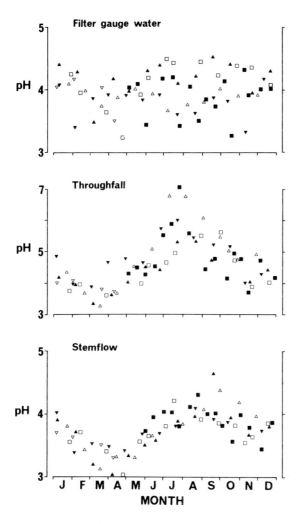

Fig. 5   Seasonal variations in pH of filter gauge water, throughfall and stemflow at Fetteresso, on the Scottish east coast, measured in 1973 (■), 1974 (□), 1975 (▲), 1976 (△), 1977 (▾) and 1978 (▽).

DISCUSSION

   The rate of acidification as a result of cation uptake by the modelled stand in Fig. 2 is broadly comparable to the acid brought in by an annual rainfall of 1000mm at pH 4.0.   Nilsson *et al*. (10), however, pointed out that the relative magnitude of rainwater acidity versus that from root uptake gives a very incomplete picture unless consideration is given to the likely

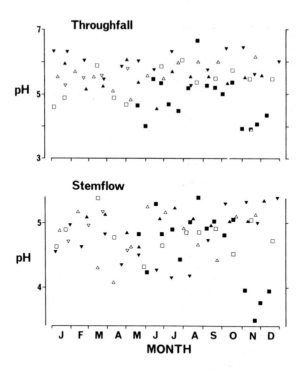

Fig. 6   Seasonal variations in pH of throughfall and stemflow
at Leanachan, on the Scottish west coast, measured in 1973 (■),
1974 (□), 1975 (▲), 1976 (△), 1977 (▼) and 1978 (▽).

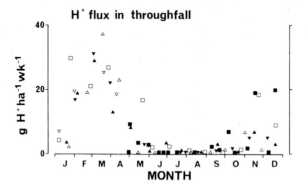

Fig. 7   Seasonal variation in the flux of hydrogen ions in
throughfall measured at Fetteresso in 1973 (■), 1974 (□),
1975 (▲), 1976 (△), 1977 (▼) and 1978 (▽).

Table 2

Mean pH of water at various points in different ecosystems for the year November 1980 to November 1981 at Glen Tanar in northeast Scotland.

pH of bulk precipitation 4.6
pH of filter gauge water 4.4

Water flux	Vegetation cover					
	Calluna vulgaris	Pinus sylvestris old	young	Picea sitchensis	Larix kaempferi	Betula spp.
Throughfall	–	3.9	4.7	5.4	4.8	4.8
Stemflow	'–	3.1	3.5	3.9	3.7	3.9
Beneath ground vegetation	4.5	5.5	5.5	*	4.7	6.4
Beneath humus	4.0	3.6	3.8	4.4	4.6	5.1
Below rooting zone	5.6	4.4	4.7	5.5	4.9	6.6

*No ground vegetation grew beneath the *Picea sitchensis*

fate of these hydrogen ions. The acidity from root uptake is a regular seasonal phenomenon, and because the hydrogen ions emerge in intimate proximity to the exchange surfaces or weathering minerals from which the nutrient cations were derived, they will almost certainly become attached to the exchange surface or be neutralised by weathering reactions. Furthermore, the transfer from root to exchange surface does not involve the movement of an anion. Rain introduced acidity, by contrast, is episodic, is often associated with large amounts of water with consequent rapid soil drainage, and is accompanied by mobile anions, such as sulphate and nitrate. It would seem, therefore, that rainwater acidity is more likely to pass through soils and acidify streams than is the biologically produced acidity resulting from root uptake.

Soil acidification early in the rotation has been noted by a number of workers (e.g. 16) and Popović (11) has recorded an increase in pH in the surface soil during the pronounced decomposition of logging remains that follows clear felling.

The acidity associated with root uptake is the result of transfer of basic cations within the ecosystem and as such may be supposed not to lead to a loss of H^+ from the system. However, as Rosenkvist (12) has pointed out, the input of neutral salts, including sea salt, may, through cation exchange and provision of a mobile anion, lead to leaching losses of naturally generated acidity. In addition, nitrate produced within the ecosystem may act as the mobile anion. Generally, nitrification is slow (Fig. 1) and most nitrate produced is retained by plant uptake. At times, however, such as during the accelerated decomposition

following clear felling, the quantity of NH_4^+ produced may be
such that considerable nitrification, with consequent release of
H^+, occurs (14).   Such acidity is counterbalanced by H^+ consumption during the ammonification step in organic matter decomposition
and by the efflux of OH^- from roots on uptake of NO_3^- (10).
However, as Ulrich *et al.* (13) have pointed out, if the rate of
nitrification outstrips plant requirements, then not only does the
acidity remain, but the accumulating NO_3^- provides a mobile anion
that enables the H^+ to be leached to greater soil depths, perhaps
eventually entering drainage water.

     It should be borne in mind, however, that weathering reactions
in all but the most impoverished soils should be more than adequate
to neutralise all acidity, whether introduced in rain or generated
internally (1).   Although observed acidification of streams has
generally been limited to regions with thin soils derived from
slow-weathering rocks (2, 3, 18) it has been necessary, even on
these, to implicate short contact, or reaction, time between
percolating water and the soil matrix in order to explain
incomplete neutralisation (5).   Short contact time may result
from channelled flow through the profile, or from overland flow,
during periods of episodically large inputs of water and acidity.
It would seem necessary to invoke some such mechanism to explain
the low pH values found at depth beneath some of the Glen Tanar
stands (Table 2).

     The variation in pH of throughfall and stemflow beneath the
spruce at Fetteresso (Fig. 5) may be relevant in this context.
Although the results from the filter gauge show that these changes
were not due to varying inputs, but rather to the transfer of H^+
or basic cations within the ecosystem, the net effect is that the
input of H^+ to the soil surface becomes markedly episodic (Fig. 7).
As this effect only occurs in polluted regions with high concentrations of sulphate in the rain, and at a time when precipitation
is moderately high but evapotranspiration low, there must be a real
possibility that some of the throughfall and stemflow acidity may
leave the site in drainage water before participating adequately
in soil neutralising reactions.   Clearly much more work is
required before this suggestion can be accepted, but if correct it
may go some way to explaining the observation of Harriman and
Morrison (6) that afforestation of open heaths and moors in acid
rainfall regions of Scotland is associated with an increasing
likelihood of streamwater acidification, with consequent loss of
fish.

ACKNOWLEDGEMENTS

     I am indebted to my colleagues J.D. Miller and C. Flower for
much of the data presented. The sites for the *Picea sitchensis*
experimental series were made available by the appropriate
Conservators of the Forestry Commission whilst that at Glen Tanar
is by kind permission of the owners.

REFERENCES

(1) Bache, B.W.: in press, Proc. Stockholm Conference on
     Acidification of the Environment, June 1982.

(2) Cronan, C.S. and Schofield, C.L.: 1979, Science 204,
     pp. 304-506.

(3) Dovland, H. and Semb, A.: 1978, SNSF IR 38/78, Oslo-Ås.

(4) Eaton, J.S., Likens, G.E. and Bormann, F.H.: 1973,
     J. Ecol. 61, pp. 485-508.

(5) Galloway, J.N., Scholfield, C.N., Hendrey, G.R., Peters,
     N.E. and Johannes, A.H.: 1980, pp. 264-265 in *Ecological
     Impact of Acid Precipitation* (eds. D. Drabløs and
     A. Tollan).   SNSF-project, Oslo-Ås.

(6) Harriman, R. and Morrison, B.R.S.: 1981, Scott. For. 35,
     pp. 89-95.

(7) Mayer, R. and Ulrich, B.: 1978, Water, Air and Soil
     Pollution 7, pp. 409-416.

(8) Miller, H.G., Miller, J.D. and Cooper, J.M.: 1980a,
     Forestry 53, pp. 23-39.

(9) Miller, H.G., Miller, J.D. and Cooper, J.M.: 1980b,
     *Tables of biomass and accumulated nutrients at different
     growth rates in thinned plantations of Corsican pine.*
     Macaulay Inst. for Soil Research, Aberdeen, Scotland.

(10) Nilsson, S.I., Miller, H.G. and Miller, J.D.: 1982, Oikos
     39 (1).

(11) Popović, B.: 1975, *Effect of clearfelling on the mobilisa-
     tion of soil nitrogen, especially nitrate formation.*
     Research Notes No. 24 Royal College of Forestry, Stockholm.

(12) Rosenkvist, I.Th.: 1978, The Science of the Total Environ-
     ment 10, pp. 39-49.

(13) Ulrich, B., Mayer, R. and Khanna, P.K.: 1980, Soil Science
     130, pp. 193-199.

(14) Wiklander, G.: 1980, pp. 226-233 in *Processer i kvävets
     kretslopp* (ed. T. Rosswall).   SNV PM 1213, Stockholm.

(15) Williams, B.L.: 1972, Forestry 45, pp. 177-188.

(16) Williams, B.L., Cooper, J.M. and Pyatt, D.G.: 1978,
     Forestry 51, pp. 29-35.

(17) Williams, B.L., Cooper, J.M. and Pyatt, D.G.: 1979,
     Forestry 52, pp. 151-160.

(18) Wright, R.F., Harriman, R., Henriksen, A., Morrison, B.
     and Caines, L.A.: 1980, pp. 248-249 in *Ecological Impact
     of Acid Precipitation* (eds. D. Drabløs and A. Tollan).
     SNSF-project, Oslo-Ås.

COMPOSITION OF PERCOLATE FROM RECONSTRUCTED PROFILES

OF TWO JACK PINE FOREST SOILS AS INFLUENCED BY ACID INPUT

Ian K. Morrison

Great Lakes Forest Research Centre
Canadian Forestry Service
Sault Ste. Marie, Ont. Canada P6A 5M7

ABSTRACT

Two soils, both acid glacio-fluvial sands from beneath mid-aged, natural jack pine *(Pinus banksiana* Lamb.) forest in northern Ontario, Canada, were reconstructed into 1-m-deep column lysimeters and subjected to dilute H_2SO_4 loadings over 4 1/2 years. Parameters measured were percolate volume, pH, conductivity, and concentrations of $SO_4^=$, K^+, Na^+, Ca^{++}, Mg^{++} and several trace metals. Results suggest a stage-by-stage process of element loss: first, soils exhibit considerable initial resistance, with $SO_4^=$ movement hampered by strong $SO_4^=$ adsorption; second, $SO_4^=$ adsorption capacity is saturated and bases move freely with $SO_4^=$ ions; third, when bases are depleted, H^+-ions increasingly dominate charge composition, and concurrently, there is substantial mobilization of trace metals.

INTRODUCTION

Jack pine *(Pinus banksiana* Lamb.) is one of the most widespread forest species in Canada, east of the Rocky Mountains. It is essentially a tree of the Boreal Forest and is, in fact, the principal Boreal Forest pine species of North America. In Ontario, Canada, it is second only to black spruce *(Picea mariana* [Mill.] B.S.P.), both as a component of the softwood primary growing stock and in terms of annual volume cut. It is used chiefly for the manufacture of Kraft pulp, but also sees extensive use as construction grade lumber, and as poles, piles, ties and pit-props. It is fast-growing in youth, slows down considerably in later years, and is not generally considered a high-yielding species. Well stocked stands on average sites yield

195

B. Ulrich and J. Pankrath (eds.), Effects of Accumulation of Air Pollutants in Forest Ecosystems, 195–206.

about 200 m^3/ha on a 60- to 70-yr rotation. Periodic annual increments in the fifth and sixth decades average about 5 m^3/ha/ yr gross, or 3-4 m^3/ha/yr net. The species is important to the forest economy of eastern Canada because extensive areas are stocked to it, chiefly as a result of natural seeding-in after fire.

Jack ·pine is the archtypical species of low nutrient requirement, and is characteristically associated with soils of low fertility. It is most commonly found in even-aged stands on low-base, acid outwash sands. Within the Canadian System of Soil Classification (1), profiles are typically classifiable as Podzols, or more specifically as Humo-Ferric Podzols (Orthic Podzol of FAO System) or as Brunisols (Cambisols of FAO), the latter an intergrade order of less-developed soils within the Canadian system.

Despite its low nutrient demand, jack pine commonly experiences nutrient limitation. Pawluk and Arneman (2), for example, studying growth on a range of soils in Minnesota and Wisconsin, USA, noted significant and highly significant correlations between Site Index at 50 years (SI_{50}) and total exchange capacity, exchangeable cations and percent base saturation of selected horizons and combinations of horizons. Chrosciewicz (3), on 'acid podzolized sandy soils' in northern Ontario, noted a general relationship between SI_{50} and content of 'basic intrusive and effusive rock particles' with the siliceous sand matrix. Further, on sites of generally low fertility, evidence from large numbers of field trials shows jack pine responding well to fertilizers, chiefly nitrogen (N) and, in many cases when N-demand is satisfied, to phosphorus (P) and potassium (K) as well (4, 5, 6, 7). Despite the general 'low base' reputation of many jack pine soils, however, there have been in eastern Canada relatively few 'forest liming' experiments, that is, experiments involving applications of either calcium (Ca) or magnesium (Mg), two elements which figure prominently in many acid rain/forest soils hypotheses. On one of the two soils of the present study, however, a Ca x Mg factorial trial was conducted between 1970 and 1980. Both interim and final results of this study showed some response to both elements, though only when NPK was also applied (8; also unpubl. data). Neither Ca nor Mg in this study could be considered 'most limiting'.

Finally, for some years, we have been concerned with mineral cycles in jack pine stands (e.g., 9, 10, 11, 12), although more recently we have been directing attention to other forest types. In our studies, we have attempted to quantify various reserves and various process rates. Against this background, then, the present study rests.

LYSIMETER EXPERIMENT

The experiment described herein is a continuing 'percolation' or 'leaching' study, using simulated 'acid precipitation' and soils contained in column- or monolith-lysimeters, within a greenhouse compartment. The experiment has been under way for approximately 4 1/2 years. An earlier report (13) described results to the end of 3 years; the present report covers the full 4 1/2 years. At the end of 5 years, it is planned to sample one replicate destructively for soils analysis. For the present, however, soils processes are inferred from percolate analyses. The overall approach has been to isolate and study a few individual processes, then to place these within a broader biogeochemical setting. Natural influences--plant uptake, root respiration, litter addition, wetting-drying cycles, freezing-thawing cycles, etc.--which cannot be accounted for properly, preclude simple extrapolation to natural systems.

Purpose

The purpose of the experiment, as far as the present report is concerned, is to compare products of leaching, chiefly bases: K^+, Ca^{++}, Mg^{++} and sodium (Na^+); plus hydrogen-ion (H^+) and sulphate-ion ($SO_4^=$) from two northern Ontario jack pine soils in relation to dilute sulphuric acid (H_2SO_4) loading. In design, the experiment includes two soils, four loadings and three replicates for a total of 24 lysimeters.

Soils

Both soils were from mid-aged pineries: one, an approximately 40-year-old, even-aged, close stand on a Site Class II sandy site in Wells Township, Ontario; the other, an approximately 50-year-old stand on a Site Class II site in Dupuis Township, 135 km to the north, near Chapleau, Ontario. In profile, the Wells Township soil is an Orthic Humo-Ferric Podzol of the Wendigo series, developed in deep, gravelly loamy-sand of glacio-fluvial origin. The Dupuis Township soil, on the other hand, is an Orthic Dystric Brunisol, unnamed as to series, developed in silt-loam over loamy sand. For convenience, the soils are hereinafter referred to as 'Wells' (the Podzol) and 'Dupuis' (the Brunisol). Physical and chemical properties are available elsewhere (11, 12). Generally, however, chemical properties expressed on a per-weight basis are similar: pH, 4.0-4.5 in the LFH, 5.0-5.5 in the mineral horizons; cation exchange capacity (CEC) and exchangeable cations low; base saturation about 10%. On a per-area basis, the Dupuis soil (the Brunisol) had more CEC and exchangeable cations, mainly because of greater depth of development. It is assigned to the Brunisolic Order, however, because it fails to meet the pyrophosphate

extractable aluminum (Al)-plus-iron (Fe) requirement for a <u>Pod-</u>
<u>zolic</u> B horizon.

Lysimeters

The soils were reconstructed into columns of clear acrylic
plastic, 100 cm high and 161.3 cm^2 in cross-sectional area, and
were set on perforated acrylic plastic lysimeter plates. Columns were foil-wrapped to exclude light. Lower mineral horizons
were air-dried, passed through a 7.7 mm screen, then placed in
the columns in appropriate order and thicknesses. Forest floor
and Ae horizon were excavated intact, trimmed to size, and in-
serted into the column top. All columns were then leached a
number of times with distilled water to promote settling.

Solutions

Solutions of pH2, pH3 and pH4 are prepared by appropriate
dilution of H_2SO_4. Distilled water serves as a fourth treatment
or 'control'. Volume is adjusted to the equivalent of 1000 mm
per annum, delivered in weekly 1-2 hour 'events', by a drip-feed
system. Columns drain freely by gravity, and effluents are recovered usually the day following application.

Measurements

Percolate volumes are recorded weekly. Other parameters
were initially measured weekly, but are currently measured only
every fourth week. These include: pH, by glass-electrode pH
meter; specific conductance by Radiometer Type CDM2e conductivity meter; $SO_4^=$, initially by a barium chloride titrimetric
procedure; latterly, according to a Technicon Auto-Analyzer II
method; major cations and some trace metals, by flame emission
or atomic absorption, as appropriate, with an atomic absorption
spectrophotometer. Where analytical procedures changed, conversion factors were developed to ensure comparability of results.

Results

On average, approximately 80% of solution volumes added are
recovered as percolates. Further, as far as percolate analyses
are concerned, only those of the pH2 series differed significantly in composition from those of the control series; therefore, comparisons are limited, for clarity, to those treatments
only. At the outset, mean specific conductance (Figure 1) of
percolates varied slightly between soils, but not between treatments within soils. At approximately 40 weeks, percolate conductivities associated with all lysimeters increased abruptly,
but by 80 weeks (or roughly 1 1/2 years) had returned to earlier
levels. Shortly thereafter, in the Dupuis/pH2 series, percolate

conductivity began to increase and reached its present level by
about 3 years. The Wells/pH2 series percolates followed, beginning about 2 1/2 years.

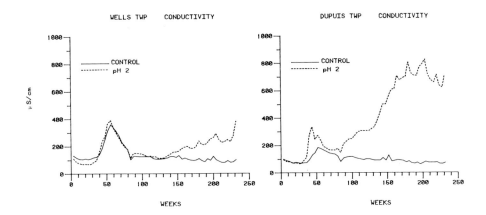

Figure 1. Change of electrical conductance (μS/cm) of percolates from Wells and Dupuis soils subjected to different acid treatments, over 4 1/2 years.

Major cation concentrations (Figure 2), expressed as the
sum of K^+, Ca^{++}, Mg^{++} and Na^+ concentrations, followed the pattern of conductance during the first 3 years, but declined
recently as H^+ replaced bases in the effluents. Regardless of
$SO_4^=$ concentration of the input solution, initial $SO_4^=$ output (Figure 3) remained low and differed little between treatments. After approximately 1 1/2 years for the Dupuis soil, and
2 1/2 years for the Wells soil, effluents from the pH2 series
showed increasing enrichment with $SO_4^=$. Regardless of input
pH (and despite the acid soils), the pH of all effluents (Figure
4) remained high, until approximately 3 years for the Dupuis/
pH2 series and 4 years for the Wells/pH2 series, whereafter it
declined abruptly. Finally, contemporaneously with the period
of pH decline in the effluent from each soil, significant quantities of trace metal cations appeared, e.g., manganese (Mn)
(Figure 5), with similar patterns being observed for other trace
metals as well.

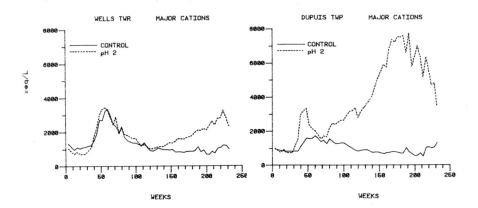

Figure 2. Change of total major cation concentration (μeq/L) in percolates from Wells and Dupuis soils subjected to different acid treatments, over 4 1/2 years.

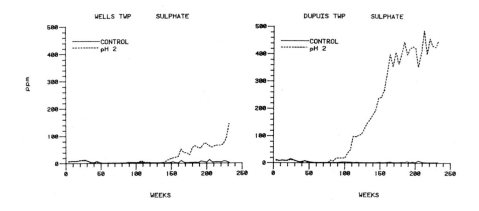

Figure 3. Change of sulphate-ion concentration (ppm) in percolates from Wells and Dupuis soils subjected to different acid treatments, over 4 1/2 years.

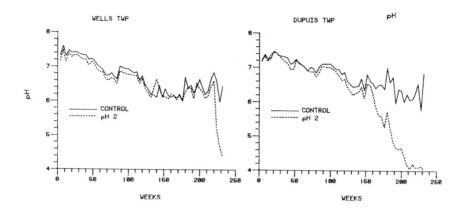

Figure 4.   Change   of   pH   of percolates from Wells   and   Dupuis
            soils subjected   to   different acid treatments,   over
            4 1/2 years.

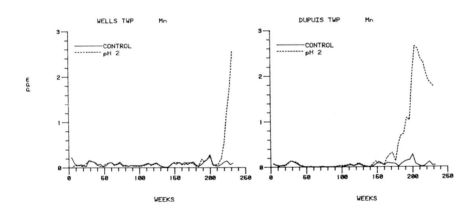

Figure 5.   Change of manganese concentration (ppm) in percolates
            from Wells   and Dupuis soils subjected   to   different
            acid treatments, over 4 1/2 years.

DISCUSSION

Both the Dupuis soil (the Brunisol) and the Wells soil (the Podzol) exhibited considerable initial resistance to dilute H_2SO_4 input: in the case of the former soil, 1 1/2 years of intense loading were required to produce a treatment-related effect on percolate 100 cm below the surface; in the case of the latter soil, 2 1/2 years were required. In addition to this 'initial resistance' phase, there appear to be several additional stages, viz., a phase in which $SO_4^=$ moves freely accompanied by substantial leaching of base cations, pH of the percolate remaining high during this period; then a phase in which H^+ comes increasingly to dominate the charge composition of the percolate, replacing the various soil bases; and contemporaneously with this, a phase of substantial mobilization of trace metals.

Explanation of the observed result must, at the minimum, account for the particular patterns of cation loss, for the fate of $SO_4^=$, and for the fate of H^+. Our work generally supports the hypothesis recently invoked by Johnson and Cole (14), Johnson and Henderson (15), Singh et al. (16), and others, involving anion regulation of base transport, and adsorption of $SO_4^=$ onto Fe- and Al-hydrous oxide coatings on soil particles within the illuviated zone of the soil. Indirect evidence in support of this also comes from the apparent greater retention of $SO_4^=$ by the Podzol as opposed to the Brunisol profile. As far as the present two soils are concerned, and with the pattern of results over the past 4 1/2 years, the exact nature of the process is not clear. Straightforward $SO_4^=$ displacement of hydroxyl-ions (OH^-) from positive sites on or in the hydrous oxide gels (with the $SO_4^=$ thus removed from solution no longer being available to transport bases, and the two OH^- released for every $SO_4^=$ adsorbed being free to neutralize the excess H^+) runs into the problem that, at pH 5.3 for the B_f horizon of the Podzol or pH 5.5 for the B_m horizon of the Brunisol, one would not expect total OH^- saturation of positive sites. Rather, a portion of the charge at least should be occupied by ions such as chloride (Cl) or nitrate (NO_3^-) that, when displaced, would be free to act as leaching counterions. While neither Cl^- nor NO_3^- were analyzed for NO_3 during the early period, there was little difference between control and treatment conductivities. This suggests little treatment-related transport at that time. It is to be hoped that soils analyses, particularly analyses of the intermediate treatments, will shed additional light on this problem.

It is again emphasized that results of this type of experiment should not be extrapolated directly to nature. Yet, it is the natural situation which is of ultimate interest. Even when reduced to its essentials, the interaction of acid precipitation with forest soil is complex. Distinct phases are passed through. Analyses of percolates--whether containing $SO_4^=$; whether the charge composition is base-dominated or H^+-dominated; to what degree mobilization of trace metals occurs--should give some indication of stage of advance.

After the $SO_4^=$-adsorbing capacity of the profile as a whole, or at least that portion occupied by tree roots, is saturated (which may take some time), and $SO_4^=$ moves freely, the cations-in-sum (not only bases, but H^+ and trace metals as well) will likely move in proportion to the anion throughput. The abruptness with which the major cation concentration, especially in the Dupuis/pH2 percolate, 'peaked out' (Figure 2), suggests mainly an ion-exchange process, with little enhanced weathering of K^+, Ca^{++} or Mg^{++} minerals (which would tend to prolong the period of base leaching). The opposite should be true with regard to trace metals.

Under natural conditions, inputs of bases from the atmosphere partially compensate for the two main avenues of loss: uptake and incorporation into the biomass, and loss into the groundwater. The balance is presumed to come from weathering of primary minerals which, in these soils, is slow (hence, their infertility). Forest productivity, then, is sustained by elements cycling amongst pools represented by the biomass (a portion of which turns over on an annual basis, another portion on a much longer basis), the forest floor (on these sites, about a five- to seven-year turnover time), and the mineral soil exchange complex.

Table 1. Potassium, calcium and magnesium contents (kg/ha) of trees, forest floor and mineral soil of Wells Township and Dupuis Township pineries.

Element	Wells Township			Dupuis Township		
	Trees[1]	Forest[2] Floor	Mineral[2] Soil	Trees[1]	Forest[2] Floor	Mineral[2] Soil
Potassium	105	71	539	150	38	421
Calcium	155	60	172	160	86	454
Magnesium	25	8	37	28	14	61

[1] Total
[2] Exchangeable

Relative sizes of K, Ca and Mg pools associated with the Wells and Dupuis Township stands are given in Table 1 calculated from data of Foster and Morrison (11), Morrison and Foster (12). Substantial depletion of any one pool, such as would occur by whole tree harvesting of the biomass could be damaging to productivity. Removals per rotation would be approximately as follows: K: 105 – 150 kg/ha, Ca: 155 – 160 kg/ha, and Mg: 25 – 28 kg/ha, with still further loss into the drainage. Additional leaching losses associated with acid precipitation, if superimposed upon these figures, would likely aggravate the situation, especially if not compensated for by enhanced weathering of primary minerals. For example, if 0.25 keq $SO_4^=$/year passed through the soil profile and carried with it K, Ca and Mg in rough proportion to their occurrence in the soil, excess losses over a 70-year rotation could be in the order of K: 20 kg/ha, Ca: 240 kg/ha, and Mg: 30 kg/ha. Thus, substantial amounts of both Ca and Mg could be lost to the trees. In relation to both harvest removal and acid leaching, Mg would appear to be the most critical element.

REFERENCES

1. CANADA SOIL SURVEY COMMITTEE. 1978. The Canadian System of Soil Classification. Can. Dep. Agric., Ottawa, Ont. Publ. 1646. 164 p.

2. PAWLUK, S. and ARNEMAN, H.F. 1961. Some forest soil characteristics and their relationship to jack pine growth. For. Sci. 7:160-172.

3. CHROSCIEWICZ, Z. 1963. The effects of site on jack pine growth in northern Ontario. Can. Dep. For., Ottawa, Ont. Publ. 1015. 28 p.

4. WEETMAN, G.F., KRAUSE, H.H., and KOLLER, E. 1976. Interprovincial forest fertilization program. Results of five-year growth remeasurements in thirty installations: fertilized in 1969, remeasured in 1974. Dep. Environ., Can. For. Serv., Ottawa, Ont. For. Tech. Rep. 16. 34 p.

5. WEETMAN, G.F., KRAUSE, H.H., and E. KOLLER. 1978. Interprovincial forest fertilization program. Results of five-year growth remeasurements in twenty-three installations: fertilized in 1971, remeasured in 1975. Dep. Environ., Can. For. Serv., Ottawa, Ont. For. Tech. Rep. 22. 27 p.

6. WEETMAN, G.F., KRAUSE, H.H., and KOLLER, E. 1979. Inter-
    provincial forest fertilization program. Results of
    five-year growth remeasurements in 17 installations fer-
    tilized in 1972 and remeasured in 1976. Dep. Environ.,
    Can. For. Serv., Ottawa, Ont. Inf. Rep. DPC-X-8. 27 p.

7. MORRISON, I.K. 1981. Assessment of the current state of
    forest fertilization research in Canada. p. 117-141 *in*
    K.M. Thompson, *(Ed.)*. An Industrial Assessment of For-
    estry Research in Canada. Vol. II. Present Status and
    Needs of Canadian Forestry Research. Pulp Pap. Res.
    Inst. Can., Pointe Claire, P.Q.

8. MORRISON, I.K., WINSTON, D.A., and FOSTER, N.W. 1977.
    Effect of calcium and magnesium, with and without NPK,
    on growth of semimature jack pine forest, Chapleau,
    Ontario: fifth-year results. Dep. Environ., Can. For.
    Serv., Sault Ste. Marie, Ont. Report O-X-259. 11 p.

9. FOSTER, N.W. and GESSEL, S.P. 1972. The natural addition
    of nitrogen, potassium and calcium to a *Pinus banksiana*
    Lamb. forest floor. Can. J. For. Res. 2:448-455.

10. FOSTER, N.W. 1974. Annual macroelement transfer from *Pin-
    us banksiana* Lamb. forest to soil. Can. J. For. Res. 4:
    470-476.

11. FOSTER, N.W. and MORRISON, I.K. 1976. Distribution and
    cycling of nutrients in a natural *Pinus banksiana* eco-
    system. Ecology 57:110-120.

12. MORRISON, I.K. and FOSTER, N.W. 1977. Fate of urea fer-
    tilizer added to a boreal forest *Pinus banksiana* Lamb.
    stand. Soil Sci. Soc. Am. J. 41:441-448.

13. MORRISON, I.K. 1981. Effect of simulated acid precipita-
    tion on composition of percolate from reconstructed pro-
    files of two northern Ontario forest soils. Dep. Envi-
    ron., Can. For. Serv., Ottawa, Ont. Can. For. Serv.
    Res. Notes 1(2):6-8.

14. JOHNSON, D.W. and COLE, D.W. 1976. Sulfate mobility in an
    outwash soil in western Washington. USDA For. Serv.
    Gen. Tech. Rep. NE-23. p. 827-835.

15. JOHNSON, D.W. and HENDERSON, G.S. 1979. Sulfate adsorp-
    tion and sulfur fractions in a highly weathered soil
    under a mixed deciduous forest. Soil Sci. 128:34-40.

16. SINGH, B.R., ABRAHAMSEN, G., and STUANES, A. 1980. Effect of simulated acid rain on sulfate movement in acid forest soils. Soil Sci. Soc. Am. J. 44:75-80.

SULPHUR POLLUTION: CA, MG AND AL IN SOIL AND SOIL WATER AND
POSSIBLE EFFECTS ON FOREST TREES.

Gunnar Abrahamsen

The Norwegian Forest Research Institute
1432 Ås-NLH, Norway

ABSTRACT

   Results from field plots (with Norway spruce, Scots pine or
silver birch) and lysimeters exposed to artificial rain are described. "Rain" was acidified to pH levels from 6 to 2.5 or 2
by means of H_2SO_4, and applied to field plots and 40 cm deep
lysimeters in quantities of 50 mm month^{-1} in the frost-free
period. Increased acidity of the "rain" increased the leaching
of Ca, Mg and Al. When "rain" acidity increased from pH 3 to
pH 2, Al concentration in the effluent increased from an average
of 80 µmol l^{-1} to an average of 1290 µmol l^{-1}. The Ca/Al mol
ratio decreased from 7 to 0.5. Significant and corresponding
changes in soil chemistry were observed. Neither the growth of
trees nor the chemical properties of the foliage indicate that
the trees suffered from Al-toxicity.

INTRODUCTION

   In some recent papers it has been suggested that acid precipitation mobilizes Al to the extent that forest trees are
damaged (1, 2, 3). Ulrich et al. (2) observed that Al concentration in the soil solution increased from 0.3-0.95 mg l^{-1} in April
of 1969 to values between 1.1 and 1.8 mg l^{-1} in November. Simultaneously the biomass of fine roots decreased from 2500 kg ha^{-1}
in May to 200 kg in August. On this basis they concluded that
"Aluminium concentrations of 1 to 2 mg Al/liter, as now existing
continuously, should therefore seriously damage root systems of
the trees especially in periods (seasons, years) where nitrification is favoured". They also concluded that "in central Europe

207

B. Ulrich and J. Pankrath (eds.), Effects of Accumulation of Air Pollutants in Forest Ecosystems, 207–218.

- central parts of West and East Germany, C.S.S.R., and Poland;
see OECD 1977 - the forests on most soils are highly endangered
by increasing soil acidification due to acid rain, if the site
is subject to substantial dry deposition of SO_2". In a later
paper (3) it was shown that the Al concentration in the seepage
water under beech during the years 1969 to 1979 has varied from
almost nil to about 6 mg l^{-1}. Under spruce the variation has
been from about 1 mg l^{-1} in 1972 to about 15 mg l^{-1} in 1976-79.

Ulrich also emphasizes the importance of the interaction
between Al and Ca in soil (4, 5). He assumes that Al injury is
likely when the mol ratio of Ca/Al in the equilibrium soil solution is one or lower.

These postulates must be evaluated on the basis of what is
known about Al-toxicity in trees, as well as consideration of how
much Al that may be mobilized due to the deposition of air
pollutants.

In a previous paper the Al-tolerance of trees was reviewed
(6). The conclusion of this review is that Al-tolerance appears
to be well correlated with acid soil tolerance, and that plants
classified as calcifuges (acid-soil plants) are more tolerant to
Al than those classified as calcicoles (calcareous-soil plants).
Because of this it is reasonable to expect that tree-species
like Norway spruce, Scots pine, beech and others which are widely distributed on acid soils have relatively high tolerance to
Al. Reference in this respect should be made to Rehfuess (7),
who also believes that Al-toxicity on acid-tolerant tree species
is not highly probable.

The next question, of how much Al that might be mobilized
due to deposition of air pollutants, is difficult to answer.
However, experiments with artificial acidification carried out
in Norway may give some indication. The intention of the present
paper is, therefore, to give results from lysimeter experiments
with particular reference to the leaching of Ca, Mg and Al.
These results are then compared with results on growth and foliar
properties of trees grown in field experiments with artificial
acidification.

MATERIAL AND METHODS

The lysimeters used in this experiment have been previously
described (8) and only a short description shall be given here.
The lysimeters were of the monolith type, 29.5 cm in diameter
and 50 cm high. They were made of fiberglass cylinders which
were forced/dug 40 cm into the soil. The lysimeters were then
turned upside down and a 5 cm layer of polyethylene pellets was
placed in the bottom before the bottom lid was glued on. In this
way neither the soil nor the ground vegetation was much disturbed.

The soil used in this experiment is an iron podsol
(U.S.D.A. Typic Udipsamment) at Nordmoen, approximately 45 km

north of Oslo. The lysimeters were dug out and placed in an area
where two field plot experiments were established, viz. A-2 and
A-3. The lysimeter experiment and field plot experiment A-3 were
from an area clearcut the winter of 1973. These two experiments
were started in July 1974. Experiment A-2 was started in 1972 in
a 16-year-old plantation of Norway spruce situated adjacent to
experiment A-3 and the lysimeters.

   These experiments - the lysimeters as well as the field
plots - are situated in the open and therefore not sheltered from
natural precipitation. Precipitation characteristics for Nordmoen appear in Table 1.

Table 1. Precipitation (mm yr^{-1}), volume weighted pH
of precipitation, and wet deposition (kg ha^{-1} yr^{-1}) of
S and N at Nordmoen.

	Nordmoen
Precipitation	830
pH of precipitation	4.3
Wet dep. of S	8.7
Wet dep. of N	7.7

Dry deposition which comes in addition to the wet deposition is
estimated to be about 30% of the total deposition (9).

   The treatments of the experiments included application of
lime ($CaCO_3$) as well as application of groundwater (50 mm month^{-1}
in the frost-free period) acidified to pH levels from 6 to 2 by
means of H_2SO_4. As natural rain comes in addition to the
"experimental rain" the bulk pH of the total amount of water
differ from that applied experimentally (10).

   Effects of lime will not be discussed in the present paper.
Table 2 gives a summary of the treatments discussed together with
the total amount of water and S applied experimentally to the
various treatments. More details about the field plot experiments are given previously (11).

Table 2. Total amount of water (mm) and SO_4^{2-} (mmol m^{-2}) applied
experimentally to the various treatments in the lysimeter- and
field plot experiments. Non-watered treatments were also
included in the experiments.

	Experimental period	No. of repli-cates	Amount of water applied	SO_4^{2-} mmol m^{-2}				
				pH 6	pH 4	pH 3	pH 2.5	pH 2
Lysimeters	1974-1979	4	1100 mm	45	240	710	–	5600
A-2	1972-1979	3	1100 mm	45	240	836	2165	–
A-3	1974-1979	4	1100 mm	45	240	715	1975	5600

The chemical analyses of soil and water are as described by
Ogner et al. (12, 13). Ca and Mg were extracted by 1 M NH_4OAc
at pH 7 and subsequently measured by an atomic adsorption spectrometer. Al was extracted by 1 M KCl and measured by an autoana-
lyzer. Soil pH was measured in a soil/water suspension of 1:2.5.

RESULTS

Figures 1-3 show the variation in Ca, Mg and Al concentrations in the effluent from the lysimeters. The figures cover the
period from April 1976 to June 1979. Data for the period before
1976 have been previously published (8). Increased acidity of
the artificial rain obviously increases the leaching of Ca, Mg
and Al. The pH 4 treatment is not included in the figures as the
concentrations in the effluent from these lysimeters are only
slightly higher than from the non-watered and pH 6-watered lysimeters. In contrast to Ca and Mg, relatively small differences
in Al concentrations were observed for treatments less acidic
than pH 3. The greatest difference in Al concentration is between
the pH 3 and pH 2 treatments. At pH 2 the Al concentrations
appear to have peaks in the spring and the autumn. From 1976 to
1978-79 there was a gradual increase in Al concentration from
about 1 mmol l^{-1} (27 mg l^{-1}) as the average for 1976 to about
1.4 mmol l^{-1} (38 mg l^{-1}) in 1978/79. The highest Al concentration
recorded was in April 1979 of 4.7 mmol l^{-1} (127 mg l^{-1}). At pH 3
the highest Al concentration recorded was 260 μmol l^{-1} (7 mg l^{-1}),
also in April 1979. At the less acidic treatments the average Al
concentration of the four lysimeters never exceeded 85 μmol l^{-1}
(2.3 mg l^{-1}). However, in the individual lysimeters, concentrations up to 110 and 160 μmol l^{-1} (3-4.4 mg l^{-1}) were observed.
The Ca/Al mol ratio in the effluent is shown in Figure 4.
Non-watered and "pH 6 lysimeters" generally have Ca/Al ratios
between 7 and 30. In the "pH 4 lysimeters" it varied between 3
and 47, and in "pH 3" between 1 and 30. At "pH 2" it varied
between 0.07 and 9, but during the last two years it never exceeded 0.2.
The average concentrations of Ca, Mg, Al and the average
Ca/Al ratio are given for the various treatments in Table 3.
The table also illustrates the sharp increase in Al concentration
and the sharp decrease in the Ca/Al ratio from the pH 3 treatment
to the pH 2 treatment.
Figures 5-7 illustrate the effects of the treatments on the
chemistry of the lysimeter soil. The content or concentration
of elements have been given as averages for the various soil
layers. In the B layer the average is given for each 5 cm layer.
Soil pH obviously has been reduced as result of the acid application (Fig. 5). The difference between treatments is greatest in
the top soil, declining somewhat in the mineral soil. Again it
was found that the greatest difference is between the pH 2 and

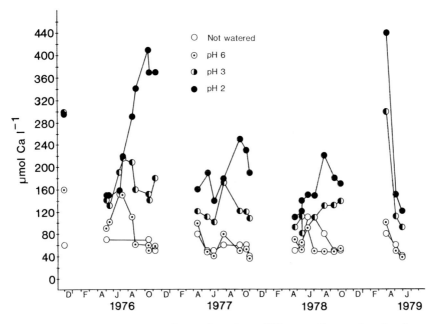

Figure 1.   Ca concentrations in the effluent from the lysimeters.
            Data are lacking during winter because of frost.

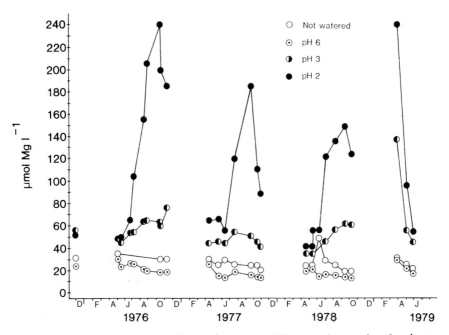

Figure 2.   Mg concentrations in the effluent from the lysimeters.
            Data are lacking during winter because of frost.

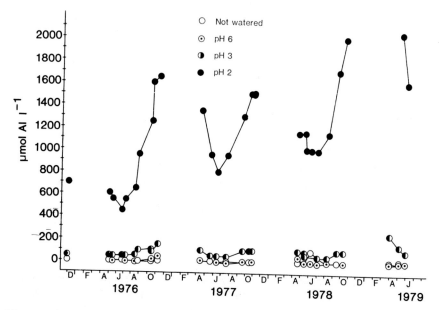

Figure 3.   Al concentrations in the effluent from the lysimeters
            Data are lacking during winter because of frost.

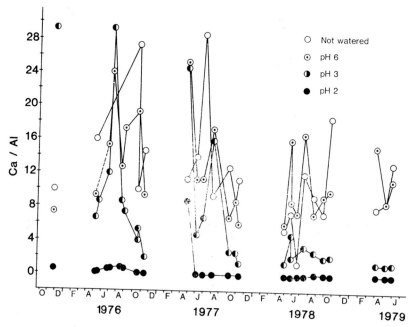

Figure 4.   Ca/Al mol ratio in the effluent from the lysimeters.
            Data are lacking during winter because of frost.

Table 3. Average Ca, Mg and Al concentrations ($\mu mol\ 1^{-1}$) and the
          Ca/Al mol ratio in the effluent of the different treat-
          ments. Averages are for the period 1976-1979.

Elements	n.w.	pH 6	pH 4	pH 3	pH 2
Ca	62	71	88	145	214
Mg	26	19	36	55	112
Al	13	12	23	83	1288
Ca/Al	11.9	12.8	15.4	6.9	0.52

the other treatments. At pH 2 the whole profile appears to be
acidified by about 0.6 pH units. Twenty-five cm into the B layer
pH is as low as 4.3, compared to almost 5.0 in the control treatments.

     The exchangeable Ca and Mg contents have been most influ-
enced in the raw humus, the bleached layer and the upper 5 cm of
the B horizon (Fig. 6). Deeper into the soil there were rather
small differences between treatments.

     The Al concentration is low in the top soil increasing to
a peak in the upper B horizon, declining again deeper into the B
layer (Fig. 7). The pH 2 treatment is the only treatment where
the exchangeable Al concentration was significantly influenced.
The concentration in the eluvial layer and the upper B layer
decreased, whereas it increased further down in the B layer.

     Figure 7 shows the mol ratio between exchangeable Ca and Al in
the soil. For the treatments less acidic than pH 3 the Ca/Al
ratio is greater than one in the raw humus layer. However,
further down into the soil the ratio is generally 0.1 or less.
The pH 2 treatment has significantly reduced the Ca/Al ratio in the
raw humus layer. Also further down in the soil the Ca/Al ratio
is lower than for the other treatments, but the great difference

Figure 5. pH_{H_2O} in the soil
           of the lysimeters
           at the end of the
           experiment
           (June 1979).

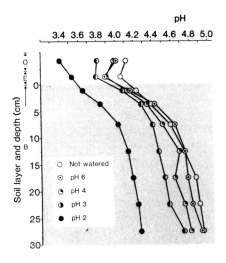

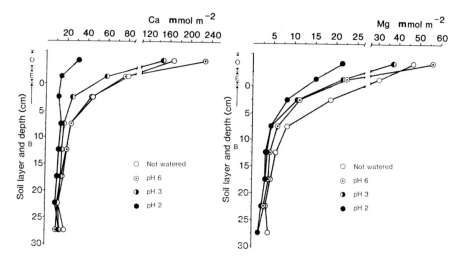

Figure 6.   Exchangeable Ca and Mg content of the lysimeter soil
            at the end of the experiment (June 1979).

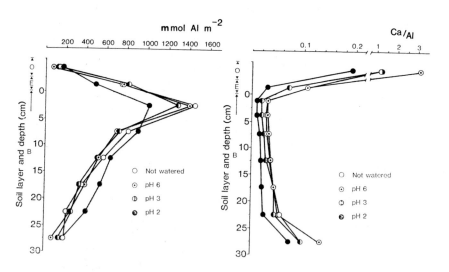

Figure 7.   Exchangeable Al and the Ca/Al mol ratio of lysimeter
            soil at the end of the experiment (June 1979).

observed in the raw humus layer is somewhat reduced.

    Finally we shall look at the tree growth observed in the
field plot experiments A-2 and A-3, which both are on the same
soil as within the lysimeters (Figures 8, 9). In experiment
A-2, the Norway spruce sapling stand, no statistically significant
effect of the treatments was found except for a reduction in the

A-2. Norway spruce

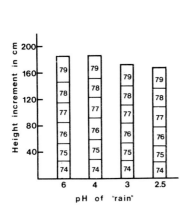
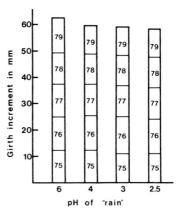

Figure 8.   Height and girth increment of Norway spruce in
           experiment A-2   (From 14).

Figure 9. Heights of
plants of birch, Scots
pine and Norway spruce
in experiment A-3 in Oct.
1979. Lime levels (CaO
kg ha^{-1})
1: 500, 2: 1500,
3: 4500.
Thin lines within
columns denote plant
heights in Oct. 1978
and Oct. 1977
(From 14).

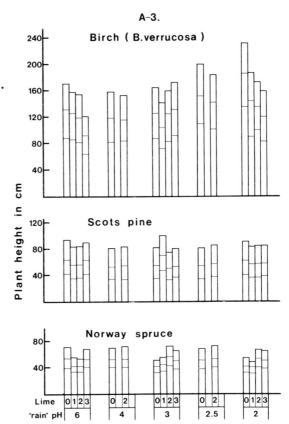

diameter growth in 1979 in plots supplied water of pH 4, 3 and
2.5. In experiment A-3 the only statistically significant
effect observed was increased growth in birch by increased
application of acid. There is also an indication of growth
reduction by increased lime application.

DISCUSSION

According to the literature on Al-toxicity on forest vegetation (see review paper (6)), Al-toxicity in the experiments
described can probably only be expected in plots supplied artificial rain of pH below 3. At the pH 2 treatment Al concentration generally was about 1.3 mmol l^{-1} (35 mg l^{-1}) and the Ca/Al
mol ratio generally about 0.5. During the last one and a half
years the Al concentration varied in the range of 1-4.7 mmol l^{-1}
(30-125 mg l^{-1}) and the Ca/Al ratio never exceeded 0.2. These
concentrations were measured in the effluent from 40 cm soil
depth. Regarding the soil chemistry data it appears likely that
Al concentrations were higher and the Ca/Al ratio lower further
up in the profile where the majority of the plant roots are
located. It is therefore possible that the risk of Al-toxicity
was in fact higher than indicated by the concentrations observed
in effluent from the lysimeters. The pH of the soil of the pH 2
lysimeters also indicates that Al predominantly should be present
as trivalent ions, unless it is organically complexed.

The tree growth experiments do not indicate that trees
suffer from Al-toxicity. We have not examined the tree roots,
as such studies would destroy parts of the field plots. Therefore we cannot disregard the possibility that roots have been
damaged. However, the foliar analyses carried out in experiment
A-3 and our other acidification experiments (15) do not indicate
abnormalities in the root function. In experiment A-2 the Mg
content of the current year's needles at the most acidic treatment was 0.07 per cent. In the other, less acidic treatments
the concentration was slightly higher (ca. 0.09 per cent). These
values are well above that is normally considered as deficiency
concentrations (16, 17). The Al concentrations of needles at the
most acidic treatment was 0.02 per cent and only slightly higher
than in the other treatments. The Al concentration in the
needles is probably not an adequate estimate of the content of
Al in the roots or of potential injury of the roots. Nevertheless, it may give some indication of a possible strain on the
root system. In this connection it is interesting to consider
the results from an acidification experiment carried out by Ogner
and Teigen (18). They applied water of pH 2.5 over a three-year
period to small plants of Norway spruce. There was no increase in
mortality during this treatment compared to the other treatments,
but some of the plants were slightly discoloured. The Al concentration of the needles was as high as 0.19 per cent, whereas the

control plants had an Al content of 0.04 per cent. Humphreys and Truman (19) mention that pine species growing on acid soils where Al is a major exchangeable cation have foliar Al concentrations up to 0.2 per cent. The results mentioned may indicate that Norway spruce, similar to pine species, has the ability to translocate Al from the roots to other parts of the plants. In this way extreme concentrations of Al in the roots may be avoided.

REFERENCES

1. Ulrich, B., Mayer, R. and Khanna, P.K.: 1979, Schriften aus der Forstlichen Fak. Univ. Göttingen und der Nieders. Forstl. Versuchsanstalt, J.D. Sauerländer's Verlag, Frankfurt am Main, 58, pp. 1-291.
2. Ulrich, B., Mayer, R. and Khanna, P.K.: 1980, Soil Sci. 130, pp. 193-199.
3. Matzner, E. and Ulrich, B.: 1981, in: Fazzolare, R.A. and Smith, C.B. (eds.). Beyond the energy crisis. Opportunity and challenge. Vol. II. - Pergamon Press, Oxford, New York, pp. 555-564.
4. Ulrich, B.: 1981, Z. Pflanzenernähr. Bodenk. 144, pp. 289-305.
5. Ulrich, B.: 1982, Paper presented at The 1982 Stockholm Conference on Acidification of the Environment, Stockholm, June 21-24, 1982.
6. Abrahamsen, G. and Tveite, B.: 1982, Paper presented at The 1982 Stockholm Conference on Acidification of the Environment, Stockholm, June 21-24, 1982.
7. Rehfuess, K.E.: 1981, Forstw. Cbl. 100, pp. 363-381.
8. Teigen, O., Abrahamsen, G. and Haugbotn, O.: 1976, SNSF-project, Norway, IR 26/76, pp. 1-45.
9. Joranger, E., Schaug, J. and Semb, A.: 1980, in: Drabløs, D. and Tollan, A. (eds.). Proc., Int. Conf. Ecol. Impact acid precip., SNSF-project, Norway 1980, pp. 120-121.
10. Abrahamsen, G.: 1980, in: Shriner, D.S., Richmond, C.R. and Lindberg, S.E. (eds.). Atmospheric sulfur deposition. Environmental Impact and Health Effects. Ann Arbor Science Publishers, Inc. Ann Arbor, Michigan, pp. 397-415.
11. Abrahamsen, G., Bjor, K. and Teigen, O.: 1976, SNSF-project, Norway, FR 4/76, pp. 1-15.
12. Ogner, G., Haugen, A., Opem, M., Sjøtveit, G. and Sørlie, B.: 1975, Medd. Nor. inst. skogforsk. 32, pp. 207-232.
13. Ogner, G., Haugen, A., Opem, M., Sjøtveit, G. and Sørlie, B.: 1977, Medd. Nor. inst. skogforsk. 33, pp. 85-101.
14. Tveite, B.: 1980, in: Drabløs, D. and Tollan, A. (eds.). Proc., Int. Conf. Ecol. Impact acid precip., SNSF-project, Norway 1980, pp. 206-207.
15. Tveite, B.: 1980, in: Drabløs, D. and Tollan, A. (eds.). Proc., Int. Conf. Ecol. Impact acid precip., SNSF-project,

Norway 1980, pp. 204-205.
16. Ingestad, T.: 1962, Meddn St. Skogforsk.Inst. Stock. 51(7), pp. 1-150.
17. Binns, W.O., Mayhead, G.J. and MacKenzie, J.M.: 1980, Forestry Commission, Leaflet 76, pp. 1-23.
18. Ogner, G. and Teigen, O.: 1980, Plant and Soil 57, pp. 305-321.
19. Humphreys, F.R. and Truman, R.: 1964, Plant and Soil 20, pp. 131-134.

SOIL PROPERTIES UNDER THREE SPECIES OF TREE IN SOUTHERN ENGLAND
IN RELATION TO ACID DEPOSITION IN THROUGHFALL

R.A. Skeffington

CEGB, Leatherhead, Surrey, KT22 7SE, UK

ABSTRACT

This paper reports a study of the pattern of soil pH, Ca
and Al round individuals of three tree species (*Pinus sylvestris,
Betula pendula, Quercus robur*) in an area subjected to acid
precipitation. All three species created a pH gradient extending
outwards to 5 m with the lowest pH close to the tree trunks.
Pine had more effect than birch which had in turn more effect
than oak. Acid deposition in throughfall under these tree
species showed a similar pattern, pine having the highest
deposition rate. However, Ca and Al are not behaving as
predicted, Al being lowest under pine and Ca in general highest.
Only under the pine canopy and in the surface layer close to the
trunk of the birches is the Ca:Al ratio greater than 1, yet there
is no above-ground visible injury which might indicate Al
toxicity to roots. Possible mechanisms for these effects
involve nutrient uptake and cycling, litter deposition and
throughfall deposition - these are discussed.

INTRODUCTION

This paper is concerned with the effects of three tree
species on the properties of the soil beneath them, in an area
where the input of acid from precipitation and dry deposited
gases is high by rural UK standards (1).

It has been postulated (2,28) that acid precipitation is
slowly acidifying soil, leading to the leaching of exchangeable
Ca and Mg and increasing the solubility of Al. Natural
acidification pulses, e.g. nitrification during warm dry years,

B. Ulrich and J. Pankrath (eds.), Effects of Accumulation of Air Pollutants in Forest Ecosystems, 219–231.
Copyright © 1983 by D. Reidel Publishing Company.

exacerbate this process, and the increasing concentrations of Al
and a declining Ca/Al ratio cause Al toxicity to the fine roots
of trees, leading to subsequent above-ground visible injury and
possibly death (13). It is also widely believed that the acid
input to forest ecosystems is greater than to low vegetation,
because of the "filtering effect" of a tree canopy which can
trap acidic gases, especially SO_2 (3). Therefore, the soil under
a tree canopy in a polluted area should have a lower pH and
[Ca] and a higher [Al] than the same soil outside, if the
hypothesis is correct. However, it is well known that certain
species of tree - particularly coniferous species - can acidify
soil even in areas remote from air pollution (see for example
the work of Nihlgård (4) for the effects of replacement of beech
by spruce forest).

    There are several mechanisms by which trees might affect
soil properties. Stemflow is concentrated close to the base of
the trunk and might be expected to have a radially diminishing
effect. Throughfall is clearly only important in areas under
the canopy. Root uptake and efflux affects an area which is
defined by the distribution of feeding roots, though extending
into the surrounding soil by transport phenomena. Litterfall
is also concentrated under the tree canopy, and the differential
effect of various litter types (e.g. bark versus leaf litter)
may be important. In forest stands these effects mingle and
may be impossible to distinguish. However, it may be possible
to differentiate them in a study of the effects of single
isolated trees on soil properties. This paper reports such a
study.

    Previous studies of the effect of individual trees on soil
properties have tended to show a depression of soil pH close to
the tree trunk (5,6,7,8,9). The effect of the trees on
exchangeable Ca has varied with species and investigator.
Gersper and Holowaychuk (5) demonstrated high exchangeable Ca
close to the trunks of *Fagus grandifolia* and *Acer saccharum*
except where the former were affected by large amounts of
stemflow. Zinke (9) found that exchangeable Ca and Mg were
highest under the canopy of a single specimen of *Pinus contorta*.
Lodhi (7) however found lower Ca levels than outside under the
canopy of various N. American species. Mina (8) was the only
investigator to measure Al - he found high levels of mobile Al
close to "pines". Only Haugbotn's study (6) took place in a
polluted area: he did not report Ca or Al levels around trees
though he showed base saturation to be less in areas affected
by stemflow. The present paper reports a statistically valid
study of the effect of isolated trees of three species on
soil pH, and on Ca and Al in the equilibrium soil solution.

METHODS

Site

This work was carried out as part of the Tillingbourne
Catchment Study, in which the influence of precipitation acidity
on stream acidity was investigated during the years 1977-1982.
The Tillingbourne Catchment is situated 7 km SW of Dorking,
Surrey in SE England, National Grid Reference TQ 140 435. The
area is a mixed coniferous-deciduous forest, largely self-sown,
the major tree species being oak (*Quercus robur* L.), birch
(*Betula pendula* Roth) and Scots pine (*Pinus sylvestris* L.). The
space between the trees is covered with a dense mat of bracken
(*Pteridium aquilinum (L)* Kuhn). For the work described in this
paper, an area where single isolated trees of the major species
were growing was chosen. Here the soil is a coarse loamy
humo-ferric podzol which has developed on a Cretaceous sandstone –
the Hythe Beds of the Lower Greensand formation (10). The soil
is naturally very acid and free-draining, variably stony, and
coarse loamy to sandy in texture. In the mineral layers, cation
exchange capacity and base saturation are both low, of the
order of 10 meq.100 g^{-1} and 5-15% respectively. Exchangeable
Al is high ($\sim$3 meq.100.g^{-1} oven-dry soil in M KCl) and exchangeable
Ca low ($\sim$0.3 meq.100.g^{-1} oven-dry soil in M ammonium acetate,
pH 7) Full details of the site, vegetation, soils and geology
are given by Skeffington (11).

Soil Sampling and Analysis

The soil round three isolated individuals of three tree
species (oak, birch, pine), growing at least 8 m from their
nearest neighbour in a gently-sloping area of the
Tillingbourne Catchment, was sampled with a soil auger. Tree
age, determined by counting annual rings in cores taken with an
increment borer, was 40-50 years, and heights varied between
11 and 18 m, but there were no significant differences between
tree species. Soil cores were taken at depths of 0-10, 10-20
and 20-30 cm at each site. Three cores were taken on each of
three transects, in the north, south west and south east
directions from each tree. The first core was taken 30 cm from
the bole, the second core 1 m from the bole (always under the
tree canopy) and the third between 4 and 6 m from the bole
(usually 5 m) ensuring this was outside the canopy. The number
of samples taken was thus 243. Further details are given
elsewhere (12). Samples were transported to the laboratory
in polythene bags, kept moist, and analysed within 4 days for pH.
Ten grams of each sample were shaken for 20 minutes in 20 cm^3
0.01 M calcium chloride solution, allowed to settle for 20
minutes, and the pH of the supernatant measured to ±0.01 units.
Sample water pHs were in general about 1 unit higher. Samples

were stored moist at 4°C and subsequently an equilibrium soil
solution (ESS) prepared using the procedure of Ulrich, Mayer
and Khanna (13). Ten g of soil were shaken with 8 cm^3
distilled water, allowed to stand for 24 hours and the
suspension filtered. The ESS was analysed for Ca and Al using
an inductively coupled plasma emission spectrometer.

Rain and Throughfall Collection and Analysis

Bulk precipitation was collected weekly at two sites inside
the catchment in glass funnels mounted 1.5 m above the ground
surface. Throughfall was collected weekly in 18 polythene
funnels distributed at random in a 2 ha area adjacent to that
described in this paper. Six collectors fell under oaks,
5 under birches, 4 under pines and 2 under bracken. Rain and
throughfall collectors were fitted with filters to reduce
particulate matter contamination. Each week samples were
returned to the laboratory and analysed immediately for pH
(electrometrically) and organic content by a uv absorbtiometric
method (WRC organic pollution monitor). Samples were stored
at 4°C and SO_4^{2-}, NO_3^-, Cl^- and NH_4^+ determined by autoanalyser
methods within 2 weeks. More details of rain and throughfall
collection are given elsewhere (14).

RESULTS

Effects of Trees on Soil pH

A visual impression of the effects of all three species on
soil pH can be gained from Fig. 1.

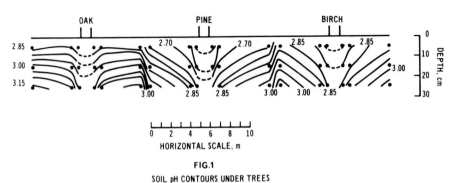

FIG.1

SOIL pH CONTOURS UNDER TREES

CONTOUR INTERVAL 0.05 pH UNIT

This shows averaged data from all replicates and orientations
as soil pH contours, and the lines for each species are joined
on the assumption that they are growing 11 m apart. The lowest
pHs are found close to the trunks and pHs increase radially
outwards. Soil pH also increases with depth. Pine has a
greater effect than birch, which has in turn a greater effect
than oak. In the case of pine and birch, the effect of the
tree can be demonstrated well outside the canopy.

    Analysis of variance showed the effects of tree species,
depth and distance from the trunk were significant (p = 0.001)
whereas the effect of orientation (compass direction) was not
significant. The effects of depth and distance operate in the
same way for all three species. Fig. 2 shows mean soil pHs
for the three species of tree plotted against soil depth
(a) and distance from the trunk (b).

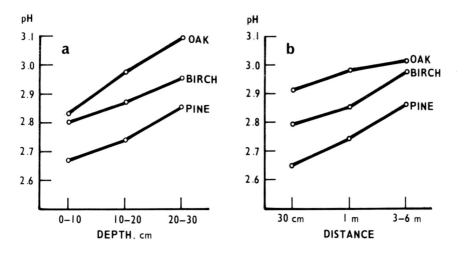

FIG. 2 EFFECTS OF DEPTH, DISTANCE, AND TREE SPECIES ON SOIL pH

Each point is thus the mean of 27 values. In both cases pine
acidifies more than birch which acidifies more than oak. In
all three species of tree, pH increases with distance from the
trunk, and with depth.

Effects of Trees on Ca and Al in the Equilibrium Soil Solution

Fig. 3 shows a similar plot to Fig. 2 for Al levels in the ESS.

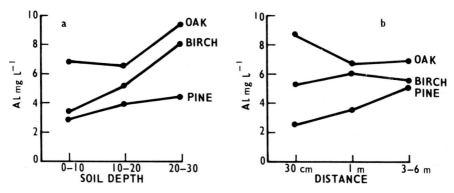

FIG. 3   EFFECT OF DEPTH AND DISTANCE FROM TRUNK ON Al IN ESS

Levels are high even compared to those published for the Solling (13). Perhaps surprisingly, Al is in general lowest under pine which had the lowest soil pHs. Al levels are significantly higher under oak than birch and pine ($p = 0.05$) in the upper soil layer and higher ($p = 0.05$) under oak than pine in the lowest soil layer (20-30 cm). Only under birch was there a significant effect of soil depths (Fig. 3(a)). Differences between species are greatest close to the tree (Fig. 3(b)) where all species differ significantly, oak and pine are still significantly different 1 m from the tree, but there are no significant differences 3-6 m out. Only in pine is there a significant change of [Al] with distance (Fig. 3(b)).

A similar plot of Ca levels in the ESS is shown in Fig. 4. Here the situation is the reverse of Al. Ca is in general highest under pine. There is significantly less Ca under birch than under pine and oak at soil depths of 20-30 cm, but not at other depths. Only under birch is the decline of Ca with soil depth significant ($p = 0.05$). Only at 30 cm from the tree is there a significant difference between the species in Fig. 4(b) (pine > birch; $p = 0.05$), and only under pine is there significantly more Ca close to the tree.

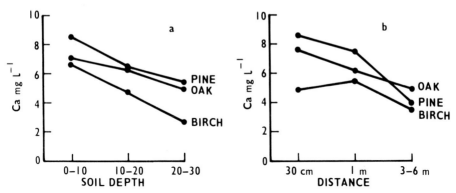

FIG. 4   EFFECT OF DEPTH AND DISTANCE FROM TRUNK IN Ca IN ESS

    Because of the importance of the Ca/Al ratio for toxicity
to seedlings in solution culture (K. Rost, pers. comm.), the
data are replotted to show the Ca/Al molar ratios in the ESS
(Fig. 5). Toxicity is expected where Ca/Al < 1.

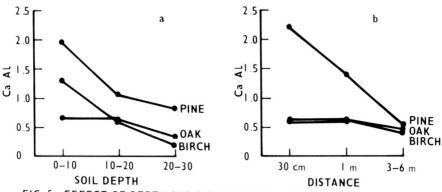

FIG. 5   EFFECT OF DEPTH AND DISTANCE FROM TRUNK ON Ca:Al MOLAR
         RATIO IN ESS

The Ca/Al ratio is less than the critical value of 1 under oaks
at all depths and distances and under birches exceeds 1 only
in the 0-10 cm layer. Only under pine is the Ca/Al ratio

greater than 1 under the tree canopy itself (Fig. 5(b)).

Acid Deposition in Precipitation and Throughfall

The deposition of acidic anions and NH_4^+ in bulk
precipitation and in throughfall under the various tree species
is shown in Table 1. Values are for 1981 only: the pattern in
the two previous years was very similar.

	Water mm	H^+	SO_4^{2-}	NO_3^-	CL^-	NH_4^+	Organic kg.ha^{-1}
		keq.ha^{-1}					
Rain	1082	0.64	0.90	0.38	0.92	0.45	31
Bracken	954	0.47	1.00	0.36	0.94	0.58	47
Oak	827	1.31	2.31	0.29	1.57	0.40	236
Birch	796	1.64	2.24	0.27	1.79	0.43	218
Pine	410	6.17	6.01	0.23	3.98	0.37	229

Table 1:   Deposition Rates at Tillingbourne, 1981

Deposition of H^+, Cl^- and SO_4^{2-} is least in bulk precipitation,
higher under oak and birch and very high under pine, whereas
NO_3^- deposition is less under trees than in bulk precipitation.
Interception loss reduces the water input to the soil especially
under pine. There is also a large flux of organic matter of
unknown nature under the trees.

DISCUSSION

Effects of Trees on Soil pH

Fig. 1 demonstrates that individual isolated trees markedly
depress soil pH even in this already acid soil. Each tree
creates a saucer of low pH around itself, the lowest pHs being
closest to the trunk. The effect extends down to at least 30 cm
in depth, and in the case of pine and birch out to at least
5 m from the trunk base (Fig. 2). Moreover there are significant
differences between species, consistent with the known
propensities of these species when growing en masse. There are
several possible explanations for this effect.

Firstly trees take up cations in excess of anions, to
satisfy their nutritional requirements. To preserve their

electrical neutrality they in effect extrude a hydrogen ion,
leading to soil acidification. To use Ulrich's terminology (2)
this is a discoupling of the ion cycle, which under natural
conditions is only temporary as the tree ultimately dies and
decays and the cations are recycled. However commercial forestry
involves harvesting the timber and in this case the
acidification is permanent. The fluxes of hydrogen ions
involved can be large. Rosenqvist (15) calculated that the H^+
introduced into Norwegian forests in this way was 0.1 to
0.6 $keq.ha^{-1}.yr^{-1}$. The effect varies with growth rate of the
trees and the stage in their life cycle. Nilsson, Miller and
Miller (16) calculated the effect for various species and growth
rates, the worst case considered being fast-growing pine between
the ages of 20-25 years, which introduced 4.45 $keq.ha^{-1}.yr^{-1}$.
Mayer (27) calculated that if all N uptake was as NH_4^+, about
7 $keq.ha^{-1}.yr^{-1}$ H^+ would be produced in the Solling beech forest,
whereas if all uptake was as NO_3^- 2 $keq.ha^{-1}yr^{-1}$ would be
consumed. The form in which N is taken up is thus critical to
H^+ production by the trees, though it is not important
considering the ecosystem as a whole, provided most of the N is
derived from mineralisation of organic matter. It seems likely
that the effect of the pines and birches on soil pH 5m from
their trunks is due to excess cation uptake, as the other
mechanisms would not be expected to extend so far.

The second mechanism is the production of acidic litter.
Pines in particular produce litter which decays to give quite
strong organic acids, such as malic and citric (17). Soil
acidification would be expected to be maximal in places with the
most litter deposition, though differences between tree species
and between litter types (e.g. bark versus leaf litter) might
be expected to be important. Zinke (9) reported that bark litter
was more acidic than leaf litter. Differences in both quantity
and quality of litter deposition might explain the species
differences observed in this study. The acidification produced
by organic acids is in a sense temporary, as ultimately they should
be respired to CO_2 and water, but where decomposition is slow this
can take a very long time. Litter deposition is maximal under
the tree canopy, explaining some of the variation in soil pH
with distance from the trunk.

The third mechanism is the hypothesis that trees act as
efficient filters of gaseous or aerosol pollutants, including
acidic species (3). Table 1 shows very different deposition
rates of H^+ in bulk precipitation and under the tree canopies,
the effect having the same direction as the observed
acidification of soil. Deposition of H^+, SO_4^{2-} and Cl^- under
pine is particularly high, and it is tempting to ascribe the
excess H^+ to washed-off dry-deposited SO_2. Dollard and Vitols
(18) showed in wind tunnel tests that SO_2 and sulphate aerosols

deposited on pine much more readily than on birch. However some of the SO_4^{2-} in throughfall may have been leached out of the needles and originated from the soil, and therefore should not be counted as an input. The relative amounts of crown leaching and dry deposition are still a matter of controversy. Methods which use an inert collecting surface to try and isolate the dry deposition component (4,19,20) tend to produce a lower proportion of dry deposition of S than the method of Mayer and Ulrich (3) which assumes that the gain in throughfall under beech during the leafless winter months is entirely dry deposition. Consideration of crude deposition velocities for SO_2 to forest with mean annual concentrations suggest the amount of dry deposition obtained by the Mayer and Ulrich method is reasonable. However, deposition velocity is known to vary depending on whether the canopy is dry or wet and the stomata are open or closed (see 21,22). Some of the absorbed SO_2 is known to be re-emitted in a reduced form and it is not known what proportion of SO_2 absorbed into the stomata is available for subsequent leaching. Application of the Lakhani and Miller method (19) to the Tillingbourne data yields the result that 65% of the gain of S in throughfall is due to crown leaching, though the proportion of this whose source was ultimately dry deposition is unknown. This leads to some anomalous results: crown leaching at Tillingbourne, under *Pinus sylvestris* is about 53 kgS.ha^{-1}.yr^{-1}, at Devilla in Scotland, 8 kgS.ha^{-1}.yr^{-1} (23) whereas Alcock and Morton (24) on a site only 40 km from Tillingbourne, found no SO_4^{2-} enrichment in Scots pine throughfall at all. Clearly much remains to be learnt about this subject, and all that can be said of the present data is that SO_4^{2-} deposition is increased under trees, and some of the increase is due to dry deposition of SO_2.

The values in Table 1, however, support the hypothesis that the pattern of soil pH observed under trees is at least partly due to the greater flux of acidity in throughfall, whatever the origin of that acidity. The species with the most acidified soil underneath (pine) has an acid deposition rate 6 times that of bulk precipitation (counting NH_4^+ as a Brønsted acid), oak and birch being intermediate. Nicholson et al., (23) found at Devilla that stemflow of Scots pine was even more acidic than throughfall, which may partially explain why pH of soil close to the trunk of the Tillingbourne trees is depressed relative to pH 1m away. Also notable in Table 1 are the high deposition rates for Cl$^-$. It is often assumed that Cl$^-$ input is accompanied by an equivalent amount of Na$^+$ or other sea-salt cations, but this is not necessarily the case. Chloride, like SO_4^{2-}, is the mobile anion of a strong acid, and its role in possible soil and water acidification should be more critically assessed.

Effects of Trees on Soil Ca and Al

It might be expected, from a knowledge of soil chemistry and the Ulrich hypotheses (2), that under trees with a low soil pH and a high acid deposition rate in throughfall, Al levels would be high and Ca levels low. Figs 3 and 4 show that this is not the case. Aluminium levels are lowest under pine which has the most acid soil and throughfall, and highest under oak, which has the least acid soil and throughfall. Some of this difference may be due to the slightly greater thickness of the Humus layer under pine, but even at 20-30 cm depth Al under pine is lower than Al in the 0-10 cm layer under oak (Fig. 3(a)). Calcium tends to be higher under the tree canopy and in the surface layers of the soil (Fig. 4). If the Ca/Al ratio is regarded as a measure of the potential for Al toxicity to roots, only under the pine canopy is the root environment still favourable (Ca/Al > 1, Fig. 5). These results suggest that the pines are able to modify the soil beneath by obtaining Ca either by more efficient aerosol capture or by recycling from deeper layers of the soil. Miller et al. (25) postulated that the enhanced input of atmospheric elements under the canopies of coniferous trees was a mechanism which helped them survive on nutrient-poor soils, and Kellman (26) showed nutrient enrichment had occurred under the canopies of certain shallow-rooted savannah trees (though not under young pine). Why Al levels should be lower under pine is not clear.

It is clear that at this site Al levels are high and the Ca/Al ratio and pH are low. These conditions are those postulated to cause Al toxicity damage to fine roots (13). Adult trees of all three species appear healthy, though it is noticable that only pine is regenerating freely on this soil type. The fraction of the soil Al which is in the less toxic organically-complexed form is not known. It will be interesting to note whether any damage occurs during the next warm dry year, when an "acidification push" might be expected.

This paper has demonstrated that oak, pine and birch are surviving without visible damage under conditions of high soil Al and low pH and Ca. The area has one of the largest inputs of acid precipitation in the UK (1), and would be expected to have a high SO_2 deposition rate, and a total deposition comparable to that in northern Germany. The forest damage occurring in Germany is not seen, however. The presence of trees is affecting soil pH, and some of this effect is probably due to capture of dry-deposited SO_2, but Ca and Al in soil are not reacting as expected. Calcium levels appear higher and Al lower, especially under pines, and it is clear that future studies should include not only the input and cycling of acidifying substances, but those of neutralising substances as well. Trees cannot be

regarded simply as dry deposition collectors: different species
have profoundly different interactions with the soil and the
atmospheric environments.

ACKNOWLEDGEMENTS

This work was carried out at the Central Electricity
Research Laboratories and is published by permission of the
Central Electricity Generating Board. I would like to thank
Patrick Whitehouse and Margaret Williams for help with soil
sampling; Brian Lewis and John Dedman for statistical analysis;
Derek Foster, Dave Hagger and John Ashford for chemical analysis,
and Geoff Cartwright for computing the rain and throughfall
fluxes.

REFERENCES

(1) Barrett, C.F., Fowler, D., Irwin, J.G., Kallend, A.S.,
    Martin, A., Scriven, R.A., and Tuck, A.F.: 1982, *Acidity
    of Rainfall in the United Kingdom - a Preliminary Report*,
    Warren Spring Laboratory, Stevenage, SG1 2BX
(2) Ulrich, B., this volume, p. 3; p. 127.
(3) Mayer, R. and Ulrich, B.: 1974, Oecologia Plantarum 9,
    pp. 157-168
(4) Nihlgård, B.: 1971, Oikos 22, pp. 302-314
(5) Gersper, P.L. and Holowaychuk, N.: 1971, Ecology 52,
    pp. 691-702
(6) Haughbotn, O.: 1976, Scientific Reports of the Agricultural
    University of Norway 55 (8), pp. 1-18
(7) Lodhi, M.A.K.: 1977, Am. J. Botany 64, pp. 260-264
(8) Mina, V.N.: 1967, Soviet Soil Science 10, pp. 1321-1329
(9) Zinke, P.J.: 1962, Ecology 43, 130-133
(10) Gallois, R.W.: 1965, *British Regional Geology: the Wealden
     District*, HMSO, London, 4th Edition
(11) Skeffington, R.A.: 1982, CECB publication TPRD/L/2278/N82
(12) Skeffington, R.A.: 1982, CEGB publication TPRD/L/2239/N82
(13) Ulrich, B., Mayer, R. and Khanna, P.K.: 1980, Soil Science 130,
     pp. 193-199
(14) Skeffington, R.A.: 1981, CEGB publication RD/L/2083N81
(15) Rosenqvist, I.Th.: 1976, *Contribution to an Analysis of the
     Buffer Effect of Geological Materials in Respect of Strong
     Acids in Rainwater*, Report to the Norwegian Council of
     Scientific Research. CEGB English Translation CE 5149
(16) Nilsson, S.I., Miller, H.G. and Miller, J.D.: 1982, Oikos
     in press
(17) Muir, J.W., Morrison, R.I., Bown, G.J. and Logan, J.: 1964,
     J. Soil Science 15, pp. 220-225

(18) Dollard, G.J. and Vitols, V.: 1980, In *Ecological Impact of Acid Precipitation*, Drabløs, D. and Tollan, A. eds, SNSF Project, Oslo, pp. 108-109

(19) Lakhani, K.H. and Miller, H.G.: 1980, In Effects of Acid Precipitation on Terrestrial Ecosystems, Hutchinson, T.C. and Havas, M. eds., Plenum Press, London, pp. 161-172

(20) Lindberg, S.E., Harris, R.C., Turner, R.R., Shriner, D.S., Huff, D.D.: 1979, Oak Ridge National Laboratory Environmental Sciences Division Publication No. 1299, Oak Ridge, Tennessee 37830, USA

(21) Fowler, D.: 1980, In *Ecological Impact of Acid Precipitation*, Drabløs, D. and Tollan, A., eds., SNSF Project, Oslo, pp. 22-32

(22) Granat, L.: 1983, this volume, p. 83.

(23) Nicholson, I.A., Cape, N., Fowler, D., Kinnaird, J.W. and Paterson, I.S.: 1980, In *Ecological Impact of Acid Precipitation*, Drabløs, D. and Tollan, A., eds., SNSF Project, Oslo, pp. 148-149

(24) Alcock, M.R. and Morton, A.J.: 1981, J. Applied Ecology 18, pp. 835-840

(25) Miller, H.G., Miller, J.M., Cooper, J.M. and Pauline, O.J.L.: 1979, Canadian Journal of Forest Research 9, pp. 19-27

(26) Kellman, M.: 1979, J. of Ecology 67, pp. 565-578

(27) Mayer, R.: 1979, in *Ecological Effects of Acid Precipitation*, M.J. Wood, ed., Report of a Workshop held at Gatehouse of Fleet, CERL, Leatherhead

(28) Ulrich, B.: 1981, Forstwissenschaftliches Centralblatt 100 (3-4), pp. 228-236

INFLUENCE OF SOIL REACTION AND ORGANIC MATTER ON THE SOLUBILITY OF HEAVY METALS IN SOILS

Gerhard Brümmer and Ulrich Herms

Department for Plant Nutrition and Soil Science, University of Kiel, Federal Republic of Germany.
Department of Soil Technology, Bremen, Federal Republic of Germany

ABSTRACT

The solubility behaviour of heavy metals determines mobility, leaching, availability, and toxicity of these elements in soils. In model experiments the solubility of Cu, Zn, Cd and Pb was investigated in equilibrium solutions of different soil samples under varying conditions.
The solution concentrations of heavy metals increase in the order Cd > Zn >> Cu > Pb with increasing total content of these elements and decreasing pH. The concentrations of Cd and Zn may already raise at pH values below 6, whereas those of Cu and Pb increase below pH 4 to 5.
At acid soil reaction the metal concentrations are reduced in subsoil samples with low content of organic matter by increasing clay content and CEC. In surface soils humic substances depress solution concentrations of heavy metals at strongly acid conditions more effectively than mineral soil components. The metals are immobilized by soil organic matter in the order Cu > Cd > Zn > Pb. At weakly acid to alkaline soil reaction these elements are mobilized in the same order by soluble organic substances. Especially additions of decomposable plant material to soils lead to a mobilization of heavy metals. Pb is least affected by these processes because of a strong fixation by mineral soil components.

1. INTRODUCTION

The humid climatic conditions of Germany lead to a degradation and acidification of soils. The influence of acid precipitation accelerates these processes (Ulrich et al. 1979). Many forest

B. Ulrich and J. Pankrath (eds.), Effects of Accumulation of Air Pollutants in Forest Ecosystems, 233–243.
Copyright © 1983 by D. Reidel Publishing Company.

soils show nowadays pH values in the range between 3 and 4 and
even values below 3 can be measured (Brümmer 1982).

Furthermore many results about heavy metal pollution of soils
are published. The environmental effects of heavy metals in
soils, like availability to plants and leaching in soils, are
strongly related to the solubility of these elements. The solubility is mainly influenced by the pH and composition of the
soils.

In order to get more informations about the influence of the
soil reaction and soil composition, especially of the content
and kind of organic matter, on the solubility of heavy metals in
soils, model experiments were carried out under controlled conditions with adjusted pH values, redox conditions and additions
of different kinds of organic matter. The details of these model
experiments were described by Herms and Brümmer (1980) and Herms
(1982).

2. RELATION BETWEEN TOTAL CONTENT AND SOLUTION CONCENTRATION OF HEAVY METALS

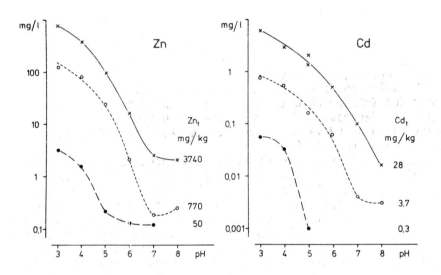

Figure 1. Concentration of Zn and Cd in equilibrium solutions of three soil samples (A horizons) with different total content of these elements in relation to pH.

In Fig. 1 the concentrations of Zn and Cd in the equilibrium solution of three soil samples with different content of heavy metals are given in relation to adjusted pH values between 3 and

8. Two of the soils taken for these experiments were highly contaminated by industrial pollution. The concentration of Zn and
Cd in the equilibrium solutions strongly increases with decreasing pH and increasing content of total metals in the soils.
In the same way the solution concentrations of Cu and Pb were influenced by both factors. The total content of heavy metals is
of importance, because mainly adsorption-desorption processes of
heavy metals with different soil components determine the solution concentration of these elements.

In order to get more detailed information about the influence
of the soil reaction and soil composition on the solubility of
heavy metals, it is therefore necessary, to carry out experiments
with different soil samples but constant total content of heavy
metals.

3. INFLUENCE OF PH ON SOLUTION CONCENTRATION OF HEAVY METALS

Five samples from A-horizons of different not-contaminated
soils were adjusted to comparable total contents of metals by
additions of 100 mg/kg Zn, Cu, and Pb and 15 mg/kg Cd. The
added metals were equilibrated with the soil samples by alter-
nating drying and wetting during eight weeks. The pH dependent
solubility of these elements is presented in Fig. 2 .

The solution concentrations of Zn and Cd are generally low at
pH 7 to 8 but strongly increase below pH 6. The Cu and Pb concentrations show distinct minima between pH 5 and 6 and increase
below pH 5 and 4, respectively. From the pH dependent solubility
of these elements, it can be concluded, that the mobility and
availability of heavy metals in soils is high at strongly to extremely acid soil reaction. Therefore a high input of heavy metals in acid forest soils may lead to toxic environmental effects
of the accumulated elements.

The influence of pH on the solution concentration of heavy
metals is modified to some extent by the content and kind of organic matter in soils. In some soil samples with higher content
of organic matter (Fig. 2: Podsol, A horizon, 4.4 % o.m.; Rost-
braunerde, A horizon, 8.0 % o.m.) the Cu and Pb concentrations
also increase at pH values above 6 or 7, because soluble metal
organic complexes are formed in this pH range. In soil samples
with low content of organic matter (Fig. 2: Kalkmarsch, A horizon, 2.6 % o.m.; Parabraunerde, A horizon, 2.8 % o.m.) the
solution concentrations of Cu and Pb show only a slight increase
at neutral to alkaline soil reaction.

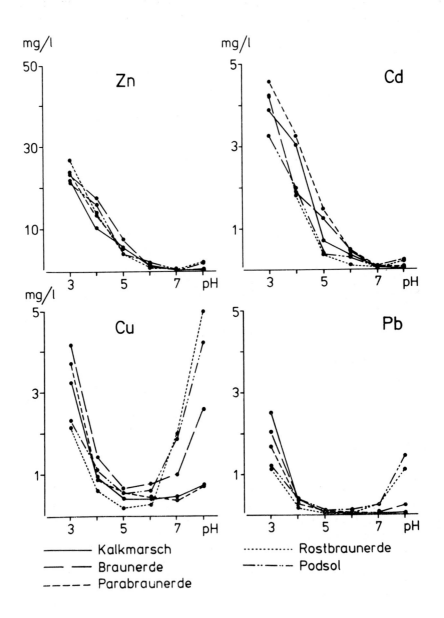

Figure 2. Concentration of Zn, Cd, Cu and Pb in equilibrium solutions of five soil samples of different composition but comperable total content of heavy metals (soil samples: A horizons from a calcic gleysol, sandy eutric cambisol, loamy luvisol, sandy dystric cambisol, sandy gleyic podsol).

4. EFFECT OF DIFFERENT CLAY CONTENTS

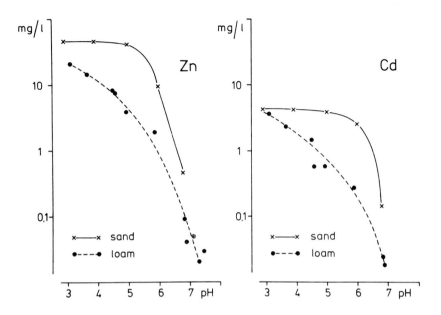

Figure 3. Zn and Cd concentrations in equilibrium solutions of sandy and loamy subsoil samples of comparable total Zn and Cd content

Besides the influence of organic matter also differences in clay content modify the pH dependent solution concentration of heavy metals in soils. In Fig. 3 the Zn and Cd concentration of equilibrium solutions are given in relation to pH for two subsurface samples with very low content of organic matter, comparable total content of Zn and Cd, but different clay content (sand: 2 % < 2 μm; loam: 24 % < 2 μm). Below pH 6 high concentrations of Zn and Cd were measured for the sandy samples and lower concentrations for the loamy ones. Because of the higher cation exchange capacity (CEC) of the loam (15.3 meq/100 g) in relation to the sand (2.5 meq/100 g) more unspecific adsorption sites for Zn and Cd as well as for other cations are available at acid soil reaction, so that the solution concentration of both metals decreases with increasing CEC. At neutral to alkaline soil reaction Zn and Cd are bound by specific adsorption and precipitation processes, which are mainly related to hydroxylated surface sites of Fe, Al, and Mn oxides and are independent of the CEC of soils (Tiller et al. 1979, Brümmer et al. 1982). Therefore at high pH values the solution concentration of heavy metals is low and differences between sandy and loamy samples diminish. Therefore toxic effects of heavy metals in polluted soils can be minimized, if the pH is adjusted close to neutral conditions.

5. INFLUENCE OF DIFFERENT CONTENTS AND KINDS OF ORGANIC MATTER

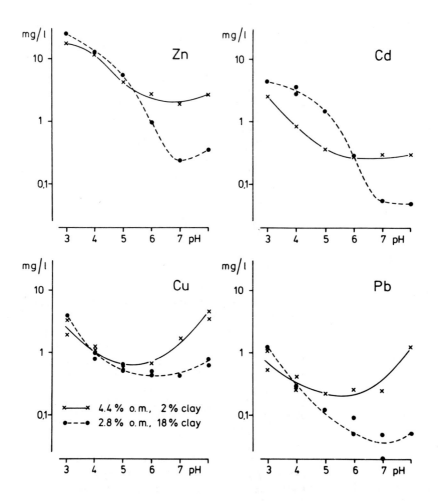

Figure 4. Heavy metal concentration in the equilibrium solution of two soil samples of different organic matter and
clay content but comparable total content of heavy metals.

In order to get more informations about the influence of soil
organic matter on the solubility of heavy metals, the data of
two soils with higher and lower content of organic matter (Podsol: 4.4 % o.m.; Parabraunerde: 2.8 % o.m.) were taken from
Fig. 2 and represented separately in Fig. 4. At acid soil reaction, especially the concentration of Cd is much lower in the
equilibrium solution of soils with higher content of organic

matter. Therefore soil organic matter can reduce toxic effects of Cd in contaminated soils at pH values below 6. Humic substances obviously contain a much higher bonding capacity for Cd at acid soil reaction than mineral substances. Therefore Cd is often accumulated in the humic layer of forest soils (Heinrichs and Meyer 1980, Schwertmann et al. 1982).

Also the Zn concentration is depressed at pH values below 5.5 by higher contents of organic matter. Because of the logarithmic scale in Fig. 4 the Zn curves of both soils are close together at low pH values, but the differences in the solution concentrations are quite clear. Also for Cu and Pb soil organic matter decreases the solution concentration at pH values below 4.

On the other side at pH 6 to 8 higher concentrations of Zn and Cd were measured for soil samples with higher content of organic matter because of a complexation of heavy metals by soluble organic substances (Fig. 4). Cu and Pb form much stronger complexes with organic substances than Cd and Zn. Therefore the Cu and Pb concentrations already raise at a lower pH of about 5. So the organic matter of the soils immobilizes heavy metals at strongly acid conditions and mobilizes metals at weakly acid to alkaline reaction by forming insoluble or soluble organic metal complexes, respectively.

The organic matter of soils consists of numerous compounds with different composition, structure, molecular weight and functional groups. Therefore it is difficult to determine, which organic substances increase or decrease the solution concentration of heavy metals. However, to get more general informations about the influence of different kinds of organic matter, further experiments were carried out with loamy and sandy subsoil samples with additions of 5 % peat or hay as model substances for raw humus and fresh organic matter (Fig. 5 and 6).

Additions of 5 % peat to loamy subsoil samples with very low content of soil organic matter (0.4 %) depress the Cu concentration of the equilibrium solutions remarkably in the whole pH range in relation to mineral soil samples without peat additions (Fig. 5).

The immobilizing effect of peat for heavy metals decreases in the order Cu > Cd > Zn > Pb. The solution concentration of lead is hardly affected by the addition of peat, because this metal is bound very strongly by mineral components of soils.

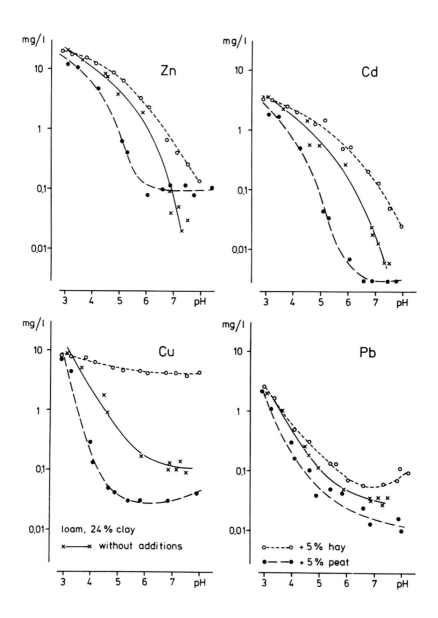

Figure 5. Heavy metal concentration in the equilibrium solution of loamy subsoil samples with additions of 5 % peat
or hay in relation to pH.

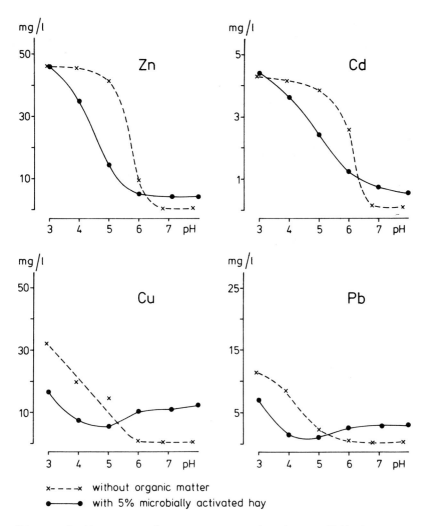

x – – – –x   without organic matter

●———●   with 5% microbially activated hay

Figure 6. Heavy metal concentration in the equilibrium solution of sandy mineral soil samples with and without
additions of 5 % microbially fermented hay (sterile conditions).

Additions of fresh organic matter (hay) to the loamy samples lead
to a strong   mobilization of Cu at all pH values above 3. The
Cd and Zn concentrations are also raised considerably at pH 4
to 8. The solution concentration of Pb is again hardly affected
(Fig. 5). The mobilization processes also change the metal concentration in the order Cu > Cd > Zn > Pb. The mobilizing effect
of fresh organic matter is due to the complexing ability of soluble organic acids and other organic substances delivered from

all kinds of plant residues like leaf litter and plant roots.
Such compounds may also originate from metabolic processes of
microorganisms. The results of other experiments show, that soluble metal organic complexes can be transformed by microbial
transformation processes to insoluble compounds (Herms 1982), so
that phases of heavy metal mobilization are followed by immobili-
zation phases in relation to the vegetation cycle and growing
conditions of microorganisms.

    The effect of organic substances on the solubility of heavy
metals in soils is modified to some extent by the kind of mineral components present in soils, because competitive effects
exist between mineral and organic substances. Therefore the
heavy metal concentrations of sandy samples without soil organic
matter but with additions of 5 % microbially fermented hay are
quite different from those of the experiments with loamy samples,
although the total content of heavy metals in both sets of
samples is comparable. In the sandy samples with very low adsorption capacity even additions of fresh organic matter depress
the concentrations of Zn and Cd at pH values below 6 and of Cu
and Pb below pH 5 (Fig. 6). Because the metal concentrations in
the equilibrium solutions of the sandy samples are much higher
than those in the loamy samples (cp. Fig. 5), the effect of the
additions of fresh organic matter is quite different in both
experiments.

6. CONCLUSIONS

    In order to assess the mobility, availability and potential
toxicity of heavy metals in soils the total content of the
different elements, the soil reaction and the amount and kind of
organic and mineral substances in soils have to be taken into
account. These soil properties determine the solution concentration and ecological efficiency of the different heavy metals.

ACKNOWLEDGEMENTS

    The authors are grateful for the able assistance of
Mrs. S. Kneesch, who carried out the analysis and prepared the
diagrams.

REFERENCES

Brümmer, G.: Einfluß des Menschen auf den Stoffhaushalt der Bö-
    den. Christiana Albertina 17 (im Druck) 1982.
Brümmer, G., Tiller, K.G., Herms, U., and Clayton, P.: Ad-
    sorption-desorption and/or precipitation-dissolution processes
    of zinc in soils.
    In preparation for Geoderma, 1982.
Heinrichs, H. and Meyer, R.: The role of forest vegetation in
    the biogeochemical cycle of heavy metals.

J. Environ. Qual. 9, 11-118, 1980.
Herms. U.: Untersuchungen zur Schwermetallöslichkeit in Abhängigkeit von pH, Redoxbedingungen und Stoffbestand von Böden und Sedimenten. Dissertation Agrarw. Fakultät, Kiel 1982.
Herms, U. and Brümmer, G.: Einfluß der Bodenreaktion auf Löslichkeit und tolerierbare Gesamtgehalte an Nickel, Kupfer, Zink, Cadmium und Blei in Böden und kompostierten Siedlungsabfällen. Ldw. Forsch. 33, 408-423, 1980.
Schwertmann, U., Fischer, W.R., and Fechter, H.: Spurenelemente in Pedosequencen.
Z. Pflanzenernähr., Bodenkde. 145, 161-196, 1982.
Tiller, K.G., Nayyar, V.K., and Clayton, P.M.: Specific and non-specific sorption of cadmium by soil clays as influenced by zinc and calcium.
Aust. J. Soil Res. 17, 17-28, 1979.
Ulrich, B., Mayer, R. and Khanna, P.K.: Deposition von Luftverunreinigungen und ihre Auswirkungen in Waldökosystemen im Solling. J.D. Sauerländer, Frankfurt, 1979.

Topic 4:

Effects on Biological Soil State and on Animals

MICRO-MORPHOLOGICAL CHARACTERISTICS OF HUMUS FORMS AS INDICATORS
OF INCREASED ENVIRONMENTAL STRESS IN HAMBURG'S FORESTS

Max-W. v. Buch

Institute of World Forestry, Federal Research Centre
for Forestry and Forest Products, Hamburg-Reinbek, FRG,
presently at Chair of Soil Science, University of Hamburg

ABSTRACT

Micro-morphological investigations of some humus forms, decomposition sequences and traces of soil animals in soils of periurban
forests of the city of Hamburg, in combination with field investigations and a number of chemical analyses, allow the interpretation that the conditions for the decomposition of organic matter
have become one-sided, i.e. have deteriorated with a decrease of
biological activities. Even under consideration of the particular
visitor stress to which periurban forests are subjected, it can
be concluded that the deterioration can be largely attributed to
immissions, i.e. to acid precipitation during the last decades.
In addition, the comparison with analytical results obtained from
a raw-humus site 30 years ago, indicate that soils have become
more acid and that in some stands penetration of the H- and Ah-
horizons by roots has strongly decreased.

Forest humus form is a comprehensive term for all humus-like topsoil horizons of forests. Within a forest ecosystem they represent typical associations of litter decomposers and the related
processes of decomposition. These associations are generally very
sensitive systems, furnishing a number of macro- and micro-
morphological indicators, characteristics and properties of development processes and detrimental influences.
In the area of investigation, the "peri-urban" forests of the
city of Hamburg, rawhumus and "moder" predominate. Humus forms
are characterized by slow to impeded decomposition of material

247

B. Ulrich and J. Pankrath (eds.), Effects of Accumulation of Air Pollutants in Forest Ecosystems, 247–255.
Copyright © 1983 by D. Reidel Publishing Company.

with destabilizing influences on the decomposer chain which have
been active for a long time. Their formation took place and is
still taking place under the following characteristic variables:
- poor in weatherable minerals and in nutrients;
- humid climate;
- silvicultural preference for coniferous species;
- severe past interference through timber removal, forest fire,
  forest pasturing as well as litter and turf removal, particularly on the poorer sites.
Additional stress is caused by its function as a "peri-urban"
recreational forest: intensive pedestrian use cause top-soil
compaction, people and dogs leave their excrements;
- the entire area is subject to immissions from central European
  industries, particularly acid-rain, of increasing significance
  during the last 20 years;
- immission stress from the adjacent industrial and city regions;
- as a consequence of several adverse effects blowout of litter
  may occur.
The present investigations have been carried out on small sample
areas by means of field observations and microscopic-micromorpho-
logical methods with soil-thin-slades and polished embedded soil
samples for assessment of the organic substance and related
decomposition processes. Particular weight has been laid on the
identification of traces of animal activity, i.e. frass patterns,
fecal aggregates and forms of subterranean passages of those
groups of soil animals dominating the decomposition of matter and
which provide valuable indications of ecological conditions and
development of humus forms. A number of chemical analyses
provided supplementary information and controls of micro-morphological diagnoses.

One example each of rawhumus and acid mull will be described.
In the Hausbruch forest district podsols and rawhumus prevail
on old diluvial, glacial sands of low silica content. Here the
opportunity existed to link the results obtained in 1981 with
those of fine root intensity measurements carried out by W.
GROSSKOPF in 1949. During the subsequent years ([20],12,3,4) conducted further investigations of humus forms at this site.
Since the forest stands and their structure remained intact over
the past 30 years, a comparison of results is possible (Tab. 1).
The following changes took place since 1950:
- the thickness of the humus layers has increased;
- the properties of humus have changed;
- the acidity values increased considerably;
- penetration of the humus layers and the top mineral soil layers
  by fine roots has been considerably reduced.
These changes are particularly clear in the spruce rawhumus no.1,
but a similar picture is obtained with spruce raw-humus no.5
(Tab. 2), which has been additionally sampled because of apparent
crown damage.

The following are considered possible causes for the increase of soil acidity:
- Warm-dry years which according to B. ULRICH may cause long-lasting acidity phases because of sudden mineralization of N-containing organic matter.
  The year 1947 had been extremely dry, the first samples were taken in 1948.
  Measurements in 1981 had been preceded by several cool-humid years which, according to B. ULRICH, provide a regeneration phase, for soils with a related decrease of acidity.
- Humus accumulation phases during the time of stand development which in spruce monoculture are particularly marked in dense pole timber and small trees cause phases of increasing acidity and decoupling processes in the decomposition chain.
  This situation could be applicable to the dense spruce stand age 74 years of 1950, but not to the spruce stand of 105 years age, opened-up through fellings, sampled during 1981.
- Special interference to which peri-urban forests are particularly subject - i.e. excrements of people and dogs, result predominantly in a reduction of soil acidities.
Hence, the stated increase of acidity from 1950 - 1981, similar to the results obtained by BUTZKE in Westfalia, can only be attributed to acid immissions.

Micro-morphological characteristics of the spruce raw-humus no 1:
In each normal raw-humus, as also stated at this site in the year 1950, dominate, in addition to fungi and collembols, particularly the enchytraeids, living from bacterial mucus, detritus, fecal rests, strongly decomposed plant material and hyphae. They are not primarily decomposers of falls. Generally they produce dark, unregularly formed fecal aggregates of up to 100 m diameter, which act as a bonding agent in the L- and upper F-horizons, glueing needles, leaves and plant rests together. In the enchytraeids feces frequently the fecal aggregates of collembols and rests of plants can be recognized. Enchytraeic aggregates fall apart easily, their substrate being generally resistant against decomposition, it is not further processed and is enriched within the H 2 and Ah-horizons. The present humus state is essentially characterized by the fact that enchytraeids predominate soil faunal development because of the disappearance of other soil animals and an increase of fungal activity. Within the L- and F-horizons plant rests on particular sites are embedded net-like in fungal hyphae while in the H 2 and Ah-horizons fungal sklerotiae are frequent. This predominance of fungi is less attributable to conditions more favourable for their development rather than by the circumstance that animal groups feeding on mycelia are missing or are only weakly represented.
In addition, this increase of fungal growth indicates markedly wet periods.

Table 1: Changes of humus form, acidities and fine-root formation in the Hausbruch forest district (30 year comparison)

compartm. No.	1950 after GROSSKOPF					1981				
	stand description	humus form	thickness (cm)	pH KCl	root development in F – H	stand description	humus form	thickness (cm)	pH KCl	root development in F – H
1 47b	spruce site cl. II,5 74 years no soil cover vegetation	typic. raw humus	11	L 4.0 F 3.6 H 2.9 A 2.9 B 3.0	intensive	spruce site cl. III,5 105 years no soil cover vegetation	typic. raw humus	16	L 3.6 F 2.9 H 2.7 A 3.1 B 3.3	little
2 45b	oak site cl. II,5 108 years Blueberry	"moder"	4.4	L 4.4 F 4.3 H 3.9 A 3.9 B 3.5 BC 4.5	moderately dense	oak site cl. III,5 139 years Blueberry	moder. raw humus	6.5	L 4.1 F 3.7 H 3.2 A 3.2 B 3.3 BC 4.1	moderately dense
3 47c	pine site cl. II,5 104 years no soil cover vegetation	raw humus	8.4	L 4.0 F 3.8 H 3.2 A 3.4 B 3.8	moderately dense	pine site cl. II,5 135 years understory Blueberry	moder. raw humus	6.5	L 3.9 F 3.1 H 3.0 A 3.1 B 3.3	moderately dense
4 49a	pine	–	–	–	–	pine site cl. III 145 years no soil cover vegetation	typic. raw humus	11	L 3.3 F 3.0 H 2.7 A 3.0 B 3.2	little

Site class determination according to 1966 management plan

Table 2: 30 year comparison (1950 - 1981) of C, N, P. contents in
soils of 3 different sites in the Hausbruch forest district

No. 1 Compartment 47 b, spruce (Picea abies) 105 years, slope, no
soil covering vegetation, fine-humus rich raw humus;

No. 2 Compartment 45 b, oak (sessiliflora) 139 years, Vaccinium myrt.,
some Deschampsia fl., "moder"-raw humus;

No. 5 Compartment 47 d, spruce (Picea abies), 130 years, upper slope,
N-NW exposition, no soil covering vegetation, fine-humus rich
raw humus.

No of sample	horizon	depth (cm)	pH H₂0	pH KCl	C %	C/N	C/P
1	L₁	0.5	4.8	3.8	51.4	28	540
	L₂	0.5	4.2	3.5	50.5	25	462
	F	3	3.7	2.9	50.3	29	680
	H₁	7	3.6	2.7	50.5	30	992
	H₂	7	3.6	2.7	46.7	30	1454
	Ahe	0-17	3.8	3.1	0.8	42	1430
	Bh	17-21	3.9	3.3	1.4	45	175
2	L	1	4.9	4.1	45.6	27	408
	F	3	4.3	3.7	41.5	20	370
	H	2	3.7	3.2	26.6	25	473
	Ah	0-2	3.8	3.2	16.1	25	487
	Ahe	2-4	3.7	3.2	3.0	37	589
	B	4-9	3.9	3.3	2.2	37	328
5	L	1	4.5	3.9	47.0	24	482
	F	5	3.7	3.1	47.4	25	554
	H₁	7	3.5	2.9	29.0	37	1085
	H₂	7	3.5	2.9	29.3	37	1511
	Ah	0-10	3.6	3.1	3.2	45	1410
	Bh	48-58	3.7	3.2	1.3	42	33

In comparison to enchytraeids aggregates the feces of other animal groups are rare:
Feces of collembols can only be found in the L- and upper F-horizons, only sporadical occurrence of fecal aggregates of dipters
insect larvae of up-to 400 my-size and the feces forms (up-to
500 my-size) of the litter inhabiting earthworm species
Dendrobaena.
Very rare are also the oval feces of 30 to 60 my diameter of
oribatids, found in frass caverns of weakly decomposed needles
and rests of wood. Since many oribatid species are known to be
fungiphag their weak occurrence can be related to the strong
increase of fungi micelium. More frequent are, however, bark and
wood rests transformed abiologically by chemical and mechanical
decomposition to the so-called "red moder material", which is
extremely resistent against further decomposition.
Living roots in significant quantities are only found in the H 1-
horizon, they are almost exclusively bigger roots of bad vitality,
while fine-roots are almost non-existant.
Root surfaces are frequently cracky, with the bark coming-off,
lateral roots are rare, those present give a cracky and decomposing picture. Similar descriptions have been made by A. HÜTTER-
MANN 1982 for damaged roots.
Traces of dead fine and corse roots, however, are numerous in the
H-horizons, easily identified on thin slades by the black Mycorrhiza rings, which are well conserved when the roots have already
decomposed leaving no morphological traces.
In 1950 W. GROSSKOPF reported for this site high fine-root intensities particularly in the F-H-horizons, no mention whatsoever
had been made of root damage.
Apart from effects on stand development the missing roots have
far reaching consequences for humus formation. Fine roots, of
which even under steady ecological conditions a large part is
dying annually, form an important link as root litter within the
decomposer chain, i.e. in the food supply of soil animals. In
addition, within the dense compact material of the H-horizon
roots form a drainage network. Where they are absent, local formation of stagnant water may temporarily occur. It is not certain
whether or not suppressed root formation leads to the development
of hitherto unknown humus forms.
In the moder-raw humus forms no. 2 under oak and no. 3 under pine
with a broadleaved understorey (Tab. 1) enchytraeids predominate
with the quantity of their fecal deposits, while fungal hyphae
and sklerotiae are found only infrequently and fecal pellets of
collembols, dendrobaena, dipters and oribatids are much more
frequent than in the spruce raw humus no. 1 samples and no. 5. In
particular in the moder-raw humus, the border between H and Ah
horizons is not well defined due to the activity of tipulids which
is well expressed by the gradation of C-contents as shown in no.
2, Tab. 2.
The micro-morphological characteristics of raw humus and moder-raw

humus are well confirmed by the C/N and C/P relationships (21)
in Tab. 2.
The C/N-values for these humus forms are relatively low. Whether
the comparatively high N-contents can be attributed to N-immis-
sions (B. ULRICH found in the Solling-mountains 29 kg N/ha/year)
and/or to the high number of visitors of city forests has not yet
been analysed.
In the forest district Wohldorf at the northern city limits of
Hamburg soils are mainly composed by predominantly young-diluvial
material with higher clay fractions and higher silica proportions.
Moder humus of all types predominates. Rather infrequent an
"acid F-mull under mixed broadleaved forest", with a dense soil
cover of Glechoma hederaceae, Anemona nemorosa, Aspidium spinolo-
sum, Polygonatum multiflorum and Sambucus nigra, can be found and
is described as example in the following.
The highest acidity values in the top soil amount here to pH
4,0 (H_2O) and pH 3,5 (KCl).
Mixture and decomposition of litter function relatively fast. The
humous mineral soil Ah (0 - 18 cm) is loose and friable with a
favorable fine root situation. Soil aggregates, however, decompose easily and the transition to the Bv-horizon is markedly
offset.
Earthworms do occur but litter decomposition is essentially
performed by isopods, dipters and diplopods while voles perform
the intermixture of substrates.
A census of soil animals performed by U. GRAEFE by means of
Formalin-eviction produced several earthworm species known to
live near the soil surface and several juvenile specimen of
Lumbricus terrestris, however, only a few adults which are the
only worms of those found here which could perform intensive bioturbation and deeper soil penetration.
Important for the intended evaluation of traces is the question
whether the activity of Lumbricus terrestris has been higher in
former times than today. For the evaluation procedure the
following indicators can be analysed:
1) Fecal aggregates, particularly in the top soil, up-to several
   mm large, frequently with several phases sticking together.
   Aggregates of all earth worm species contain larger longitudinal
   plant rests.
   Worm aggregates with a high mineral particle content and leaf
   fractions or entire leaves surrounded by feces are characteristic
   of adult Lumbricus terrestris.
2) Larger worm exit funnels of L. terrestris at the soil surface
   frequently with leaves which have been partly drawn into the
   soil.
3) Worm passages and passage coatings.
   Passages of adult L. terrestris in the humus top-soil are
   inclined, from the lower Ah they ramnificate radially toward
   the soil surface. From the Al- or B-horizon respectively down-
   wards the passages are more or less vertical in swinging curves.

Oblique and vertical passages have coatings composed spread of
earthworm excrements. These coatings contain fragments of
leaves and plants and within the upper portions of the passages
also leaves and larger leave-particles pulled-in by the
earthworms.
In older passages which have not been used for 1 or 2 years
the coatings become rough and cracky and finally fall off.
With even older passages the coatings are completely missing,
the originally smooth walls of the passages become rough and
uneven.
Fecal aggregates and exit funnels, typical for the adult L. terrestris occur at the site only in a limited number. Bioturbation
of larger closed soil volumes as observed in the humus form
L-mull does not occur at this site.
Of 12 worm-passages followed up to a depth of 100 cm only two
could be identified as recent, while 9 were definitely very old:
without coatings, the rough and cracky walls partly collapsed,
several having been completely penetrated by roots.
The relationship of old to recent earthworm passages, traces of
other animals and the quality of humus form indicate that
earthworm activity must have been greater and extended to greater
depths formerly. Increase of acidity and decrease of earthworm
activity, however, must have taken place during the last 20 years,
obviously as a consequence of acid precipitations.

REFERENCES

1 Babel, U. (1965). Die Ansprache von Pflanzenresten im mikro-
       skopischen Präparat von Humusbildungen. Z.Pflanzenernähr.
       Bodenkunde 109, H. 1: 17-26.
2 Babel, U. (1972). Moderprofile in Wäldern. Morphologie und
       Umsetzungsprozesse. Hohenheimer Arbeiten Bd. 60, Ulmer
       Stuttgart.
3 Babel, U. (1981). Humusmorphologische Untersuchungen in Nadel-
       holzbeständen mit Wuchsstörung. Mitt.Verein Forstl.Stand-
       ortskunde u. Forstpflanzenzücht. H. 29: 7-20.
4 Buch, M.-W.v. (1962). Vergleichende chemische und mikromorpho-
       logische Untersuchungen bei der Extrahierung von Huminstof-
       fen aus Waldböden. Z.Pflanzenernähr.Bodenkunde 97 (142),
       H.3: 255-265.
5 Buch, M.-W.v. (1981). Einfluß von Schwefelimmissionen auf Wald-
       humusformen am Beispiel eines Buchenmischwaldes südlich
       Hannover. Forstarchiv 52, 1: 15-18.
6 Buch, M.-W.v. (1982). Humusformen umweltbelasteter Bestände
       Hamburger Waldungen. Forstarchiv 53, 2: 46-51.
7 Butzke, H. (1981). Versauern unserer Wälder ? Erste Ergebnisse
       der Überprüfung 20 Jahre alter pH-Messungen in Waldböden
       Nordrhein-Westfalens. Forst- u.Holzwirt 36, 1: 542-548.
8 Graefe, U. Bodenzoologische Indikatoren der Versauerung von

Waldböden von Hamburg. (1982) (In Bearbeitung). Gutachten an die BBNU - Landesforstverwaltung, Hamburg .

9 Großkopf, W. (1950). Bestimmung der charakteristischen Fein-wurzelintensität in ungünstigen Waldbodenprofilen und ihre ökologische Auswertung. Mitt.Bundesforschungsanst.f.Forst-u. Holzwirtschaft 11: 1-19.

10 Hüttermann, A. (1982). Frühdiagnose von Immissionsschäden im Wurzelbereich von Waldbäumen. Mitt. LÖLF Sonderheft, S.26-31.

11 Knabe, W. (1981). Immissionsökologische Waldzustanderfassung. Allg.Forstzeitschr. 26: 641-643.

12 Meyer, F.H. (1959). Untersuchungen über die Aktivität der Mikroorganismen in Mull, Moder und Rohhumus. Archiv f. Mikrobiologie 33: 149-169.

13 OECD (1977). The OECD-programme on longrange transport of air pollutants OECD, Paris.

14 Rehfuess, K.E. (1981). Über die Wirkungen saurer Niederschläge in Waldökosystemen. Forstwiss.Centralbl.100, 6: 363-381.

15 Rusek, J. (1975). Die bodenbildende Funktion von Collembolen und Acarina. Pedobiologia, Bd. 15: 299-308.

16 Ulrich, B., Mayer, R. und Khanna, P.K. (1979). Deposition von Luftverunreinigungen und ihre Auswirkungen in Waldökosy-stemen im Solling. Schriften Forstl.Fakult.Univ.Göttingen, Bd. 58, Sauerländer Verl. Frankfurt a.M.

17 Ulrich, B. (1981). Zur Stabilität von Waldökosystemen. Forst-archiv 52, 5: 165-170.

18 Ulrich, B. (1981). Destabilisierung von Waldökosystemen durch Akkumulation von Luftverunreinigungen. Forst- u.Holzwirt 36, 21: 525-532.

19 Ulrich, B. (1982). Gefahren für Waldökosysteme durch saure Niederschläge. Mitt.LÖLF, Sonderh. 9-25.

20 Zachariae, G. (1964). Welche Bedeutung haben die Enchytraeen im Waldboden. in: A.Jongerius, Soil Micromorphology (ed.), Elsevier, Amsterdam.

21 Zachariae, G. (1965). Spuren tierischer Tätigkeit im Boden eines Buchenwaldes. Forstwiss.Forschungen, H. 20.

22 Zezschwitz, E.v. (1980). Analytische Kennzeichen typischer Humusformen westfälischer Bergwälder. Z.Pflanzenernähr. Bodenkunde 143: 692-700.

BIOCHEMICAL REACTIVITY IN FOREST SOILS AS INDICATORS
FOR ENVIRONMENTAL POLLUTION

A.Hüttermann,B.Fedderau-Himme,K.Rosenplänter

Forstbotanisches Institut der Universität
Göttingen, Büsgenweg 2, D-3400 Göttingen,
W.-Germany

Abstract: In order to get a picture about the state of
the physiology of the soil, we measure the activity
of four soil enzymes and the rate of nitrification
with respect to different substrates. The four enzymes
are from four different areas of metabolism: ß-glucosidase, phosphatase, phosphodiesterase and aminopeptidase. The activities of these enzymes are a) measured
in the presence of either buffer or water (= soil solution) and b) determined in the three upper horizons
of the soils: in the organic top layer (L and H-Horizon, if present) and the two next horizons in the
mineral soil (usually A_h and the following) down to a
depth of about 30 - 50 cm. Both patterns of enzyme
acitivities can be used for an evaluation of the physiological state of the soil.

1. INTRODUCTION

Already in the 1950s, a correlation was postulated between the biochemical agricultural properties of soils,
i.e. enzyme activities, and plant yield (for a review
see 1). The hope in these days even was to be able to
get to a "fertility index" via the determination of
special "Key enzymes" (2). Today, it is obvious that
these far ranged goals of soil enzymology cannot be
achieved. On the other hand, it is altogether clear,
that the measurement of soil enzymes provides useful
information with regard to the physiological state of
the soil.

B. Ulrich and J. Pankrath (eds.), Effects of Accumulation of Air Pollutants in Forest Ecosystems, 257–270.

Very few measurements have been performed so far on
enzymes in forest soils and even less were done under
the aspect of evaluation of polluting effects. The
most intensive study so far published is the one of
Greszta et al. (3) who measured the effect of dusts
emitted by metal smelters on soil enzymes, soil microflora and selected tree species. These authors found
a decrease of soil enzymes with an increase of heavy
metal-dusts in the soils, especially for urease.
These observations were confirmed by Zwolinski (4),
for smelter emissions, and by Maier et al. (5) for
in vitro studies, too. No attempt was made in these
studies, to determine the pattern of enzymic activity
along a soil profile and no variation was employed in
the condition of the assay.
In this study, we will report on the first results of
our study on biochemical aspects of soil reactivity,
which so far includes forest soils of the area of the
German Federal States of Hamburg, Nordrhein-Westfalen
and Niedersachsen. Our results indicate that both the
enzymic reactivity under different assay conditions
and along the soil profile provides useful data for
the evaluation of the physiological state of the soils
under study.

1. MATERIALS AND METHODS

1.1 Soil sampling and preparation for enzyme analysis
The soil samples were taken from the different locations either in the area around Göttingen or in the
Haard, a small hilly area north from the Ruhr-Gebiet.
After sampling, the soil probes were collected in
plastic bags and transferred within less than two days
to the Institute and stored in a cold-room at $4^{\circ}C$
prior to analysis. After a medium storage time of
about one week, the soil samples were air-dried at the
open air, and samples of 5 g each were weighed and
put into 20 ml plastic bottles and kept in the cold
again until subsequent enzymic analysis.

1.2 Determination of enzymic activities
To each of the 5 g soil samples, 10 ml of substrate
solution was added, either dissolved in buffer (comp.
Table 1) or in water.

Table 1 Conditions for the enzyme assays

Enzyme	Incubation mixture	Terminating solution
acid phosphatase	0.1 M Tris-maleate (pH 5.5) 0.01 M 4-nitrophenylphosphat	1 M Na_2CO_3 (6)
ß-glucosidase	0.05 M Tris-maleate (pH 5.5) 0.005 M 4-nitrophenyl-ß-D- glucopyranoside	1 M Na_2CO_3 (7)
phosphodiesterase	0.05 M Tris-maleate (pH 7.0) 0.001 M bis-4-nitrophenyl- phosphate, Na-salt	1 M Na_2CO_3 (6)
aminopeptidase	0.1 M Tris-acetate (pH 8.0) 0.003 M L-alanyl-4-nitroanilide	dimethyl- formamide (8)

The samples were then incubated at 30°C for 180 min.
The reaction was then terminated by the addition of
the stopping solution (Table 1) and aliquots were
transferred to Eppendorf-reaction vessels. They were
spun for 3 min at 14 000 x g and the appropriate OD
in the supernatant was determined in a semiautomatic
Eppendorf-Photometer (PCP 6121) with Sampling Table.
The units of activity were directly printed according
to the program which was fed into the photometer
according to the standard calculations. The values
are given in nKatalg^{-1} of the soil.

1.3 Determination of the nitrification activities
The method was adapted from (9): To each of the soil
samples, 10 ml of 10 mM substrate solution in water
was added, the different substrates being: either
$(NH_4)_2SO_4$, aspartic acid, casamino acids, and no addition for measuring the internal rate of nitrification.
The samples were treated as described above, and nitrite and nitrate was determined in the supernatant of
the centrifugation according to the standard procedures, the OD being measured at 628 nm, the nearest
filter available in that region of the spectrum.
The values are given in nMolh^{-1} g^{-1} of the soil.

2. RESULTS

2.1 Changes in absolute biochemical reactivity with
regard to soil treatment.
For a test of our system, we had access to the following series of experiments carried out by J. GEHRMANN
(Institut für Bodenkunde und Waldernährung, Göttingen)
(10). For a study on the reasons for failure of natural regeneration in beech-forests in the Haard, Nordrhein-Westfalen, he applied to following three treatments to the soils at three different sites (Arnsberg,
Haard, Wuppertal) in beech-forests.

1. he did not change anything (= control site)
2. the litter and humus layer was mixed with the
   mineral soil down to a depth of about 30-50 cm, a
   procedure which is known in German Forestry as the
   "Gahrenberg-Verfahren"
3. he did the same treatment, bud added lime
4. he removed the soil completely down to a depth of
   50 cm and replaced it with an artificial soil mixture, rich in organic material, especially peat,
   which is used in plant nurseries. This substrate is

called "Einheitserde", although this name may be mis-
leading, since the composition is not identically in
all places where it is supplied.
These four variations were applied at the three sites.
Therefore we were able to study four controlled treatments of the soil in three parallel experiments. The
first round of analysis was made after 100 days after
the application of the treatments mentioned above.

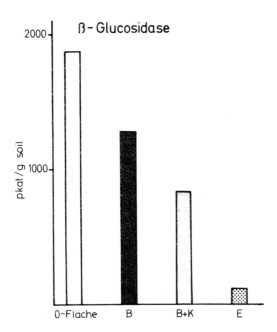

Fig. 1 Acitivty of ß-glucosidase in relation to the
applied treatments to the soil. O: no treatment,
B: mineral soil and top soil mixed, BK: mixing done
with added lime, Einheitserde: artificial substrate
(used for nurseries)

       In our enzyme and nitrification analysis, all
three sites behaved perfectly the same and identical
pattern were observed with regard to enzyme and nitrification activity in each of the four different
types of soils. In addition, rather drastic changes
of both types of reactivity were observed, indicating,
that the enzymes which we had been measuring, are
indeed sensitive probes for measurement of the physiological state of the soil. Fig. 1 shows the activity pattern of one enzyme, ß-glucosidase, in buffer,
for the four types of soils at one site. As can be

seen, the disturbance of soils due to the different
treatments resulted in rather dramatic decreases of
the enzymic activity, however with different relative
rates of decline.

The same connection between soil treatment and
biological reactivity was found in the nitrification
assay. In addition (Fig. 2), it can be seen, that the
artificial soil "Einheitserde" did not yet reach a

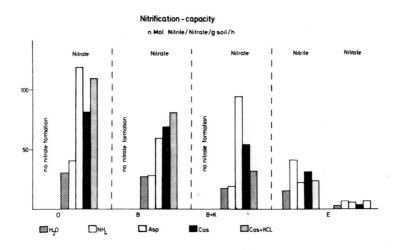

Fig. 2. Nitrification capacity in relation to the
applied treatments of the soil and with different sub-
strates. (for the soil types see Fig. 1)
Water: Incubation with water only
NH$_4$: Incubation with (NH$_4$)$_2$SO$_4$ dissolved in water
Asp   : Incubation with aspartic acid in water
Cas   : Incubation with casamino acids (Difco) in water
Cas + KCl: Incubation with casamino acids in 10 mM
              KCl-solution

balanced microflora with regard to nitrification.
Whereas in all the several hundred soil samples, which
were studied according to our scheme so far, almost
no nitrite was found under our assay conditions, only
in all three "Einheitserde"-samples development of
nitrite was observed, at rates considerably higher
than nitrate formation. This result does not necessary
mean, however, that high levels of nitrite are to be
expected when "Einheitserde" is used as a substrate
for plant growth. It may be that nitrite appears only
temporarily under the conditions of our assay and will
be eventually removed by the activity of nitrite-

oxydizers in the soil. These values indicate, however, that a different microflora is present in the different types of soils, a fact which is sufficient for a further use of this type of assay in evaluation of the physiological state of soils.

2.2 Differences in the enzymic activity with regard to different assay conditions and types of soils.

In our routine assay, we measure the enzymic activities in two different assay systems: 1. in a buffer which is considered to be the optimal buffer system for most of the respective soil enzymes (cf. Table 1), 2. in water, with no added buffer. For a test whether the two different assay systems reveal additional and useful information, again the experiment of Herrn Gehrmann (10) was exploited. The results for one site are shown in Fig. 3, identical results were obtained

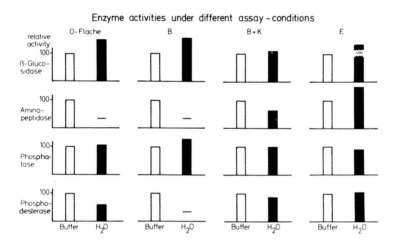

Fig. 3. Activities of enzymes measured under different assay conditions in the same type of soil, either in buffer or in water.

again in the other two repetitions of this experiment. Whereas in assays with the buffer systems, activity was observed in all soil types, this was not the case in the assay using only water as solvent for the substrate. Here one enzyme was not active in the untreated soil and another one disappeared in the sample where the top layer was mixed with the mineral soil. Both enzymes were active in the soil which was treated with lime and in the Einheitserde.

The results presented in Fig. 3 are compatible with the data reported by Zantua and Bremner (11) who found

remarkably high differences in soil urease determinations dependent on the presence or not of an appropriate buffer in the assay. In very acidic soils,
these authors too found no or only little urease in
the non buffered assays as compared to the ones with
the buffer present. The authors conclude that the
non-buffer method provides "a very good index of the
ability of soils to hydrolyse urea under natural conditions" (11). On the basis of the data we have collected so far, we feel confident to extend this view
on the biochemical activity of soils in general.
In addition, the differences between the data obtained
in buffered assay and the ones in the non-buffered
assay are, at least for the two enzymes phosphodiesterase and aminopeptidase, indicators for the physiological state of the soil. From our data, we are not
able to draw conclusions on the mechanisms of the in-
hibition of enzyme activity in the non-buffered system.
It is yet quite clear, that this effect is not due to
possible differences in the very pH in the two types
of assay.
Another interesting result of the experiment reported
in Fig. 3 is the very high activity of ß-glucosidase
in Einheitserde in the non-buffered assay. This indicates again the totally different physiological state
of this artificial soil and the additional wealth of
information which can be obtained by the comparison
of the two assay systems.

2.3.   Distribution of enzymic activity within a
       soil profile

Most enzymic activities in soils decrease with soil
depth (for a review see Speir and Ross, 12). It is
in general agreed upon that this decrease in enzymic
activity with increase in soil depth can be mainly
attributed to the diminuition of biological activity
down the profile.
In our studies on the activity of enzymes and nitrification activity along soil profiles, rather different
patterns were observed in profiles from different sites,
ranging from rather little or almost no decrease in
activity down to about 30 cm depth in calcareous soils
to nearly complete disappearance of biological activity within this depth-range in soils from polluted areas.
An example for a soil exhibiting relatively little
decline in reactivity is given in Fig. 4a and b.

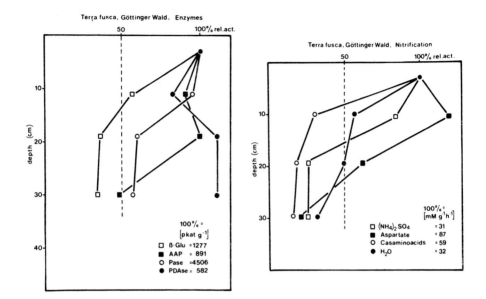

.Fig. 4. Relative distribution of enzyme (=a, left figure) and nitrification (=b, right figure) activities within a soil profile from the Göttinger Wald, terra fusca.

It is a profile from a "terra fusca" from the Göttinger Wald. Starting with rather high values for the enzymic and nitrifying activities in the top layer, about 50 % are still measurable at about 30 cm soil depth. Similar activity patterns were observed in an additional "terra fusca"-profile and two "rendzina"-profiles taken from the Göttinger Wald, too. In the latter profiles, no decrease in reactivity at all was observed down to a depth of 25 cm.

A different picture was observed in more acidic soils from Spanbeck in Niedersachsen (Fig. 5a and b). Here the absolute activities in the top layer were much less than the ones in the soil samples from the Göttinger Wald and a rather steep decline in reactivity takes place.

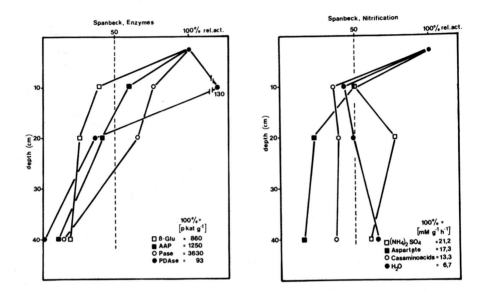

Fig. 5. Relative distribution of enzyme (=a, left figure) and nitrification (=b, right figure) activities within a soil profile from Spanbeck, Niedersachsen, brown earth.

The most extreme profiles which we have measured so far came from the Haard, Nordrhein Westfalen (Fig. 6 and 7), a small hilly wood area directly north from the Ruhr-Gebiet, which was under heavy immission for several decades. Here reactivity is measurable practically only in the top layer of the soil, with almost no activity in the mineral soil below 15 cm of depth. Life has retreated here almost completely from the mineral soil and natural regeneration, e.g. beech, dies off here within few years (comp. 10).

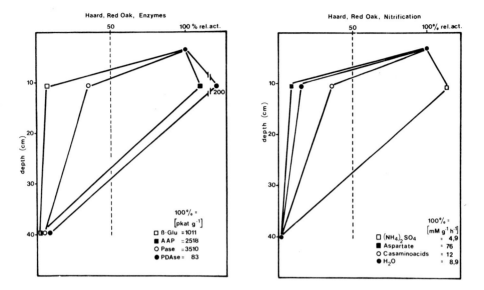

Fig. 6. Relative distribution of enzyme(=a,left figure) and nitrification (=b,right figure) activities within a soil profile from a Red Oak forest in the Haard, Nordrhein-Westfalen, brown earth.

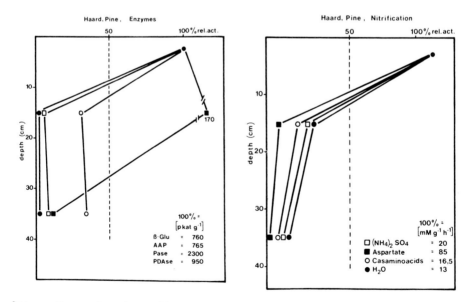

Fig. 7. Relative distribution of enzyme (=a,left figure) and nitrification (=b,right figure) activities within a soil profile from a pine forest in the Haard, Nordrhein-Westfalen, brown earth.

3. DISCUSSION

Chromogenic substrates for the assay of soil enzymes
have been used only rather little, only aryl-sulfa-
tase (c.f. 13) is determined routinely this way.
This is rather surprising, because there is no syste-
matical reason known to us and other authors for this
lack of use of these very convenient substrates (13).
Apparently historical developments are responsible
for this situation: the assays for the "classical"
soil enzymes were worked out before the chromogenic
substrates have been developed to a great extent.
In our experience, the use of chromogenic compounds,
especially when nitrophenole or nitroaniline are the
products of the enzymic reaction, are ideal substrates
for the study of soil enzymes. The danger of artifacts
is rather low, since the product of enzymic reaction
does not react further with soil components and can
be easily extracted from the reaction mixture. In
addition, it allows to make use of the sophisticated
procedures and machinery which has been developed in
the last years for clinical enzymology, thus permit-
ting the analysis of many samples with many replica-
tions in a reasonably short time. Our assay procedure
which is described in the "Methods-section" is a compromise between two apparently uncompatible demands:
the processing of large numbers of samples with high
precision. In introductory experiments it was found,
however, that with these conditions, both requirements could be met, and although each soil enzyme is
determined in quadruplicates, using four individual
soil samples of 5 g each, the maximal deviation which
was found by us is still about $\pm$ 10 % of the absolute
value, which is sufficient enough for such studies.
The same is true for the study of the nitrification
activity.
    A common problem for all studies of soil enzymes
is the use of the appropriate buffer systems. Theo-
retically it would be necessary to determine the optimal buffer system for each enzyme in each soil
sample individually. In our screening program, this is
far out of any consideration. We therefore decided to
use those buffer systems which have been applied suc-
cessfully by us for the determination of the same enzymes in pure cultures of different soil organisms
(fungi, slime molds, myxobacteria) in prior studies
(c.f. 6,7,8,14). This approach, however, appears to
be reasonable in view of the rather broad pH-optimum
curves which were observed for soil reactivity in
other studies (c.f. 15).

The use of the additional assay parameter (substrate dissolved in water only) gives additional information on the state of the actual enzymic activity in the soil (comp. Fig. 2). If therefore, the buffer for some reasons is not in the optimum, this can be cross-chequed in the water system. In most of the soils which have been studied so far, the buffer values have been either higher or more or less the same as the ones observed for the water-assay. Only the values for ß-glucosidase in the Einheitserde-samples (Fig.2) differed more than 100 % of the buffer-value. This, however, indicates the very big differences in the microflora of the individual soils in that field experiment.

A very critical parameter in our view is the pattern of enzymic activity along a soil profile. Unfortunately, very little work has been done with regard to this aspect of soil enzymology, although the information which can be achieved by these measurements is a rather valuable indicator for the physiological state of the soil under study. The pattern of enzymic and microbial activity within a soil profile is a direct indication for the activity of the higher soil organisms. In soils with high activity of insects and lumbricides, the decline of enzymic activity is rather low (Fig. 4). Here an active mixing of the top soil components into the soil takes place, resulting into very active transformation of the litter. The other extremes are the soils from the Haard (Fig. 6 and 7) where almost no biological activity is measurable in soil depths below 15 cm.

Between these two extremes, all kinds of transitions were found in our measurements as well as in literature-data which we have calculated according to our scheme. For instance, the soils studied by Dutzler-Franz (16,17) fall between the two extremes of our study. The interpretation of the data presented here and others which are not yet published by us are compatible with the conclusions which have been drawn by Dutzler-Franz(17) on the correlation between soil enzymes and other soil factors. They show in addition, that the activity pattern of enzymes and microbial reactivity in a soil profile can be used as an indicator of the physiological state of the soil.

Acknowledgement:
The work was supported by grants from the Federal-States Nordrhein-Westfalen (Pilotprojekt "Saure Niederschläge") and Niedersachsen (Niedersächsisches Zahlenlotto).

References

1. Hofmann, E.: 1955, Z.Acker-u.Pflanzenbau 100,
   pp. 31-35.
2. Skujins, J.: 1978, in R.G.Burns(ed.) Soil Enzymes,
   pp. 1-49. Academic Press.
3. Greszta, J., Braniewski, S., Marczynska-Galkowska,
   K., and Nosek, A.: 1979, Ekol.Pol. 27, pp. 397-426.
4. Zwolinski, J.: 1980, Papers presented to the Symposium on the effects of air-borne pollution on
   vegetation, U.N., Economic Commission for Europe,
   p. 159.
5. Maier, R., Maier, F., and Maier, G.: 1981, Ber.
   Deutsch.Bot.Ges. 94, pp. 709-718.
6. Hüttermann, A., and Volger, C.: 1973, Arch.Mikro-
   biol. 93, pp. 195-204.
7. Hüttermann, A., Porter, M.T., and Rusch, H.P.:1970,
   Arch.Mikrobiol. 74, pp. 90-100.
8. Hoffmann, W., and Hüttermann, A.: 1975, J.Biol.Chem.
   250, pp. 7420-7427.
9. Hanson, R.S., and Philips, J.A.: 1981, in Philip,
   G.(ed.) Manual of Methods in General Bacteriology,
   pp. 328-364, ASM, Washington.
10. Gehrmann, J., and Ulrich, B.: 1982, Sonderheft
    LÖLF-Mitteilungen, pp. 32-36.
11. Zantua, M.I., and Bremner, J.M.: 1975, Soil Biol.
    Biochem. 7, pp. 291-295.
12. Speir, T.W., and Ross, D.J.: 1978, in R.G.Burns
    (ed.) Soil Enzymes, pp. 197-250. Academic Press.
13. Roberge, M.R.: 1978, in R.G.Burns(ed.) Soil enzymes,
    pp. 341-370. Academic Press.
14. Guntermann, U., Tan, I., and Hüttermann, A.: 1975,
    J.Bacteriol. 124, pp. 86-91.
15. Duddridge, J.E., and Wainwright, M.: 1982, Water
    Res. 16, pp. 329-334.
16. Dutzler-Franz, G.: 1977, Z.Pflanzenernaehr.Bodenkd.
    140, pp. 329-350.
17. Dutzler-Franz, G.: 1977, Z.Pflanzenernaehr.Bodenkd.
    140, pp. 351-374.

MERCURY – ACCUMULATION IN GAME

S. Bombosch

Institut für Forstzoologie, Universität Göttingen

Hg occurs naturally everywhere in the countryside, in addition man has spread considerable amounts on agricultural land as fungicide. A further source is the airborne Hg. As opposed to agricultural areas, Hg in forest land has its origin only in natural sources and in air pollution. However, forest soils (9) exhibit in their upper strata approximately ten times higher concentrations than agricultural soils (Fig. 1). The plants also, for example spruce (1), are clearly contaminated with Hg (Tab. 1).

age of the needles	Hg in ppb
1 - year old	$56,4 \pm 19,2$
2 - year old	$88,3 \pm 19,6$
3 - year old	$106,8 \pm 22,6$
4 - year old	$131,9 \pm 23,9$

Tab. 1: Mercury content of spruce needles

Therefore one has to expect that forest animals take up Hg either directly with their food plants (f. example red deer, roe deer, hare) or via food chains (f. e. wild boar, fox). The question is, whether there is an enrichment of Hg in deer, and if so, to what extent.

In the years 1975 to 1979 the Hg-content in kidneys, liver, and muscle of approximately 2000 game animals was examined (Tab. 2).

B. Ulrich and J. Pankrath (eds.), Effects of Accumulation of Air Pollutants in Forest Ecosystems, 271–281.

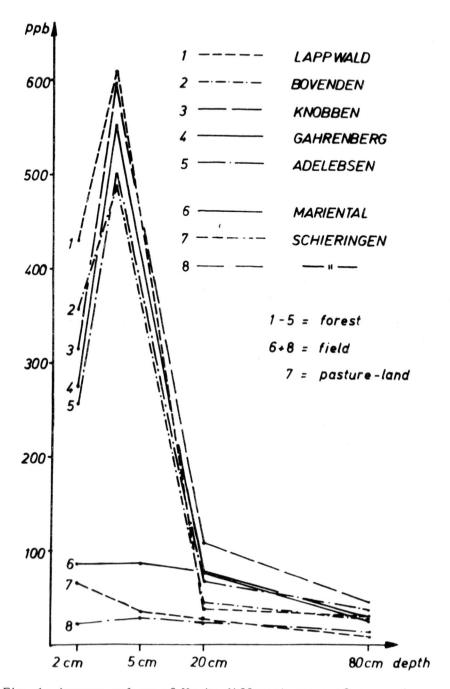

Fig. 1: Average values of Hg in different areas of research

species	muscle	liver	kidney
red deer	230	225	225
fallow deer	243	237	235
roe deer	374	366	368
wild boar	680	696	663
hare	314	327	326
rabbit	108	111	110
fox	71	70	71
S.	2.020	2.032	1.998

Tab. 2: Number of samples examined

Hg was determined by means of an atomic absorption spectrometer
(6) in duplicates for every organ after it had been disintegra-
ted by acid. The results cannot be expressed as mean values,
since the data which were gained in this way differed widely
between "not detectable" and 160 000 ppb. For this reason 4
classes were formed (1. up to 100, 2. up to 1.000, 3. up to
10.000, 4. greater than 10.000 ppb Hg per sample) and the mean
value was formed from simple values within each range. The
results are shown in table 3.

The lowest contamination was found with red deer. The values
which were registered in the muscle and in liver correspond
roughly to the natural Hg-content of deeper soil strata. Even
the concentrations in the kidneys are only increased insigni-
ficantly. Apart from a few exceptions fallow deer as well is con-
tamined only to a small degree. Larger amounts of Hg are found
frequently in the kidneys of roe deer, however as in red and
fallow deer, the values are very low in liver and muscle. In
contrast to the species already mentioned, wild boar is heavily
infested wich Hg. Even in the muscle of this game very high values can be noted. Still greater concentrations are registred
in hares. Low contaminations are detected in rabbit and fox,
both species exhibiting similar values.

If one tries to find out the reasons for these considerable
differences in the Hg-accumulation one can exclude a different
Hg-burden of the biotops of the individual game species since
the differences occur also when various species inhabit the same
area. It seems much more likely that specific qualities of every
game species are responsible for infestation taking place and
for the extent of same by the Hg-content of its biotop. This may
be explained wich the data of my collaborator Teuwsen (8) concerning wild boar.

By screening the data it became apparent that areas exist with

species: red deer

organ	total n	1	class %	of contamination 2	%	3	%	4	%
muscle	230	8	99	269	1				
liver	225	13	99	236	1				
kidney	225	45	86	211	13	3.099	1		

species: fallow deer

organ	total n	1	%	2	%	3	%	4	%
muscle	243	13	96	212	4	4.180	1		
liver	237	19	89	380	10	3.747	1		
kidney	235	45	60	316	32	1.946	8	57.677	1

species: roe deer

organ	total n	1	%	2	%	3	%	4	%
muscle	374	25	94	196	5	1.886	1		
liver	366	32	89	186	10	4.016	1		
kidney	368	60	20	315	71	2.393	8	21.647	1

species: wild boar

organ	total n	1	%	2	%	3	%	4	%
muscle	680	28	79	345	14	2.626	6	14.996	1
liver	696	46	63	322	24	3.078	12	16.797	1
kidney	663	64	26	287	49	3.663	14	42.181	11

species: hare

organ	total n	1	%	2	%	3	%	4	%
muscle	314	50	45	272	48	1.746	6	18.732	1
liver	327	55	4	462	47	2.270	48	34.010	1
kidney	326	49	2	473	31	3.509	54	26.171	13

species: rabbit

organ	total n	1	%	2	%	3	%	4	%
muscle	108	23	87	200	13				
liver	111	55	60	282	35	1.672	5		
kidney	110	61	20	259	65	2.690	13	17.156	2

species: fox

organ	total n	1	%	2	%	3	%	4	%
muscle	71	34	85	168	15				
liver	70	53	37	362	54	1.739	9		
kidney	71	72	24	411	56	2.293	20		

class   of contamination   1 = up to   100 ppb
                             2 = up to   1.000 ppb
                             3 = up to   10.000 ppb
                             4 = greater than 10.000 ppb

Tab. 3: Mercury content in the organs of different game species

constantly high values and other ones with constantly low values
of Hg in wild boar. There is a third group where the Hg-content
of this game shows very strong variations from one year to the
other (Tab. 4). It was not possible to explain these differences

locality	investiga-tions car-ried out in	kidney class of contamination							
		1	n	2	n	3	n	4	n
Forest district	1975			443	7	3.163	10		
Bleckede and	1976	61	13	444	10	2.992	3		
neighbouring	1977			305	7	3.030	5	55.447	16
districts of	1979			370	4	4.755	13	12.330	4
private owners									
Forest district	1975	89	6	135	2				
Lappwald	1976	84	4	263	25				
	1977			155	2	4.626	5	35.646	17
	1979	85	2	248	18	3.194	3	55.088	1
Solling =	1976	72	10	111	3				
forest distr.	1977	67	11	259	34	1.983	3		
Seelzerthurm	1978	81	2	218	2				
Fürstenberg	1979	36	9	112	1				
Winnefeld									
Hardegsen									
Knobben									
Uslar									

class of contamination   1 = up to   100 ppb
                          2 = up to   1.000 ppb
                          3 = up to   10.000 ppb
                          4 = greater than   10.000 ppb

Tab. 4: Mercury content in the kidneys of wild boar

by immission or waves of immission. All efforts, to find a
correlation to the food sources of the wild boar (mushrooms, insects etc.) were without result. Being unsuccessful with this
approach Teuwsen tried to investigate the problem by analyzing
the faeces at regular intervals and to find out in this way the
time of Hg-uptake. As can be seen very clearly from figure 2,
small amounts of Hg are found in the faeces from wild boar of
the Solling mountains thoughout the year while those from the
North german heath exhibit a strong increase in spring and autumn. In the Solling the wild boars live in a spacius enclosure

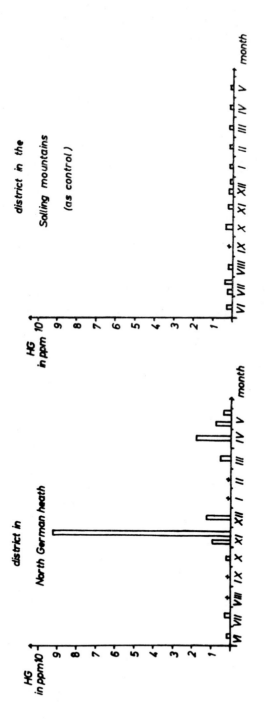

Fig. 2: Average mercurycontent in the dry mass of faeces of wild boar (1981/82)

without constant access to fields. In contrast the ones in the North german heath can visit the fields all over the year. There they prefer in spring and autumn the areas with germinating, Hg-treated seed. Therefore one has to look for the source of the contamination in the treated seed which leads to high Hg-values in wild boar and thus can explain also the difference of the Hg-concentration of wild boar in the Solling and the heath. However the question cannot be answered why the wild boar in the area of Bleckede is constantly, in the area of Lappwald only temporarily infested with Hg although both places are comparable regarding the proportions of field and wood. Hg is found in large amounts in fields of both areas. It is probable that a solution to the problem can be attained by comparing the wood species and

Forest district with low percentage of hard wood:

hunting year	intensity of mast oak	beech	Ø-Hg-content ppb	number n
1975	++	O	2040	18
1976	+++	+++	550	29
1977	O	O	32300	28
1978	O	O	no data	
1979	+	O	5360	24
1980	+	O	9890	45

Forest district with high percentage of hard wood:

1975	++	O	100	10
1976	+++	+++	240	30
1977	O	O	26230	26
1978	O	O	no data	
1979	+	O	2890	25
1980	O	O	11420	17

Enclosure with high percentage of hard wood:

1975	++	O	no data	
1976	+++	+++	230	41
1977	O	O	380	67
1978	O	+	150	4
1979	+	+	1070	25
1980	O	+	260	70

(Explanation of symbols: O = mast failure, + = scattered mast, ++ = medium crop, +++ = full mast)

Tab. 5: Mercury contents of Kidneys of wild boar (average values) in forest districts with different percentage of hard wood

age composition in both places. Ohne finds that the share of
hard wood which is able to produce seed is much smaller in
Bleckede than in Lappwald (Tab. 5). If one takes into conside-
tation further that much seed was produced by oak and beech in
1975 and 1976 it becomes apparent that the high supply with food
in Lappwald kept in wild boar in the forests. In Bleckede however the supply of acorns and beech-seeds was low and not sufficient to keep the animals off the fields. This led to the different Hg-values of the wild boar population in both areas. In
1977 and the following years no seed was produced by the forest
trees which caused the wild boar in both areas to feed on the
fields. As compared to 1976 this led to a strong increases of
the Hg-values in Lappwald bringing them to a similar level as
those in Bleckede.

In other words: The Hg-content of wild boar does not reflect the
Hg-content of its biotop since the degree of uptake of trated
agricultural seed is geared by the food supply in the forests.
One the contrary it is - or was - a good indicator for the
amount of food available to the wild boars in the forests.

Even if the use of Hg-containing fungicides will no longer be
permitted in future these results point out the possibilities for
counteracting other potential dangers resulting from agricultural
management by temporarily frequent feeding of wild boars in the
forest.

A connection between treating seed with fungicides and Hg-contamination of hares is suspected by austrian colleagues (7). Our
results do not support these suppositions (2), since in all areas
of research a high Hg-concentration has been found which is independent of the individual percentages of field and forest. The
generally very low Hg-concentrations in red deer and fallow deer,
expecially in the kidneys, indicate that mercury is taken up only
in very small amounts by the guts of these game species. The few
exceptions to this rule may be explained by exceptional conditions which might be caused by trated cereal seed offered at the
feeding place. The Hg-values of the other game species show
clearly that they take up mercury and store it in different organs. No definite statement can be made at the moment if this
enrichment is a direct reflection of the Hg-burden of their biotop. Research on trouts and carps show how careful one has to be
in constructing such a conclusion. Starving individuals of both
fish species excrete only minute amounts of the absorbed Hg. Well
fed carps exrete up to 80 %, trouts ca. 50 % (3, 5). Thus one has
to take into consideration that under natural conditions unexpec-
tedly high Hg-contents can be reached when little food is available (4, 6).

Not yet finished research work of the author concerning cadmium
and lead in 161 wild boars and 137 red deer revealed in some
classes very high concentrations (Tab. 6). The results seem to

lead

species organ	total n	1	%	2	%	3	%	4	%
wild boar									
muscle	99	46	49	224	48	3.832	3		
liver	141	38	8	341	86	2.051	6		
kidney	150	70	8	321	85	1.970	7		
red deer									
muscle	64	49	53	305	47				
liver	103	41	13	395	66	2.185	21		
kidney	111	53	4	320	68	2.864	28		

cadmium

species organ	total n	1	%	2	%	3	%	4	%
wild boar									
muscle	104	48	50	255	42	2.018	8		
liver	129	67	12	327	67	2.528	21		
kidney	131			617	13	3.566	80	20.550	7
red deer									
muscle	68	24	85	349	13	1.703	2		
liver	119	49	46	342	46	1.766	8		
kidney	122	85	4	362	43	3.452	50	18.372	3

Tab. 6: Content of lead and cadmium in organs of wild boar and
red deer

confirm the well known relationships between contamination and
stress on the environment. In addition as far as the uptake of
Cadmuium is concerned the age of the game plays an important
role. Numerous exceptions however indicate that further factors,
as yet unknown, influence the incorporation of these heavy
metals. Therefore a generalization upon one old research results
is very problematic.

Our results until now demonstrate clearly that heavy metals can
be accumulated in our game species. This may even occur in large
areas. Further they showed that, for example as far as mercury
is concerned, a direct correlation does not always exsist between Hg concentrations in the organs of the game species and
the ones of their biotops. By the elucidation of the conditions

which lead to the incorporation we were able in addition to indicate ways of avoiding strong contaminations (f.e. wild boars)
and to elaborate criteria for judging the impact of a heavy
metal on its environment (f. ex. fishes). It follows that the
significance of a heavy metal in a given environment can be
estimated only when we know the pathways of its incorporation.
it seems therefore very urgent to intensify such kind of research.

SUMMARY

Investigation of the mercury content of spruce needles and forest
soil revealed such high values that one has to take into account
an enrichment of this heavy metal in several game species. As a
result of samples of the kidneys, liver, and muscles of approximately 2000 individual animals, the lowest Hg-contamination was
found with red deer, the highest values were recorded with wild
boar and hare. In contrast to the homogeneously contaminated
hares, very different values were found in wild boars which were
related both to different districts and to different years in the
same district. It could be shown that these differences originated in Hg-treated cereal seeds which were taken up when the
food sources in the forests were small. If enough food is supplied in the forests by seed production of oak and beech trees the
wild boars do not visit the fields and therefore contain less Hg.
High values are registered when this food is missing and the
wild boars take up larger amounts of Hg-treated cereal seed in
the fields. Thus the Hg-content of wild boar does not reflect the
Hg-content of their biotop but of their food supply in the forests. If the food basis is limited, trouts and carps excrete
only very small amounts of mercury, with sufficient amounts of
food the rate of excretion is up to 80 %. These examples demonstrate clearly that the importance of Hg in an ecosystem can be
estimated only when the pathways are known by which Hg is incorporated in the body of higher organisms. The same seems to be
true for cadmium and lead which were registered in remarkable
amounts in game of certain areas.

References

1. Altrogge,D.: 1977, Untersuchungen über den Quecksilbergehalt in Fichtennadeln (Picea abies (L.)) (Karst.). Diplomarbeit.

2. Bombosch,S.: 1981, Schadstoffe im Wildpret. Wild und Hund, 84, pp. 461-465, 492-494.

3. Branz,H.: 1978, Über den Einfluß der Futtermenge auf die Hg-Speicherung in verschiedenen Geweben des Karpfen (Cyprinos carpol.). Diplomarbeit.

4. Gämisch,R. and Sander,D.: 1976, Vergleichende Untersuchungen von gefütterten und ungefütterten Regenbogenforellen (Salmo gairdneri) auf Quecksilberanreicherungen. Semesterbegleitende Arbeit im Fach Biologie.

5. Kaufmann,R.: 1978, Über den Einfluß der Futtermenge auf die Hg-Speicherung in verschiedenen Geweben der Regenbogenforelle (Salmo gairdneri, RICHARDSON 1936). Diplomarbeit.

6. Peters,L.: 1977, Über das Quecksilbervorkommen in gering bealsteten Ökosystemen. Dissertation.

7. Tataruch,F. and Onderscheka,K.: 1981, Belastung freilebender Tiere in Österreich mit Umweltschadstoffen (III). Gehalt an Quecksilber in Organismen von Feldhasen. Zeitschr. Jagdwiss. 27, pp. 266-270.

8. Teuwsen,N.: 1982, Zu den Ursachen von Quecksilberkontamination von Schwarzwild. AFZ im Druck.

9. Uffelmann,W.: 1977, Untersuchungen über den Quecksilbergehalt verschiedener Böden. Diplomarbeit.

Topic 5:

Effects of Soil Acidification and Accumulation of
Air Pollutants on Plants

AIR POLLUTANT DEPOSITION AND EFFECTS ON PLANTS

Theodor Keller

Swiss Federal Institute of Forestry Research,
CH-8903 Birmensdorf

Abstract

Plants are often more sensitive to air pollutants than man.
They may accumulate air pollutants in parts of their shoots without showing any injury at all. This has the advantage of making
pollutants accessible to chemical analysis even if they occur
only temporarily or in such low concentrations as to make their
detection very difficult. The presence of a substance does not necessarily imply its phytotoxicity. Therefore the terms "bioindicator" and bioindication" should be distinguished. Bioindicators
are convenient if they exhibit typical and distinct symptoms
(effect). A toxic effect in the invisible range, however, is only
shown by bioindications, i.e. by the physiological, biochemical
or ecological behavior of the plant. Examples are given for accumulative bioindicators as well as for bioindications.

It has long been known that vegetation filters the air and
accumulates pollutants. Plants, however, are often more sensitive
to air pollutants than man. Thus trees may die in the presence of
SO_2 at concentrations which do not impede man's activity. Ever-
green forests with big crowns which reach into air layers with
increased wind speed are particularly efficient. This filtering
and accumulating action for gases which involves the stomata is
considered active in comparison to the merely passive deposition
by sedimentation.

Man has made use of vegetation as an indicator of certain
site aspects for decades. Because air pollutants recently became
part of our environment, it is quite natural that mankind should
try to extend the use of vegetation as a bioindicator of this new

B. Ulrich and J. Pankrath (eds.), Effects of Accumulation of Air Pollutants in Forest Ecosystems, 285–294.
Copyright © 1983 by D. Reidel Publishing Company.

ecological factor.
     Such a use, however, makes a distinction between "bioindica-
tor" and "bioindication" necessary. Figure 1 seeks to distinguish
between these two terms by accentuating the fact that a plant may
be under stress in the range where no visible symptoms of injury
occur. As bioindicators man mainly uses particularly sensitive
plant strains which show symptoms at an early stage (range "a"
of figure 1). Such a use of sensitive strains is particularly
convenient because the visible symptoms show that the plant has
not only accumulated the compound(s) but has also reacted. This
reaction of the plant as a biological entity becomes obvious and
demonstrates much more than the mere presence of a compound,
which is not necessarily proof of toxic action. This reaction of
the type "a" occurs, however, only in the visible range.

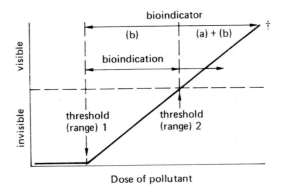

Figure 1: Diagram of the relationship between bioindicator and bioindication as affected by the pollutant dose in the visible and invisible range (the
thresholds are ranges rather than exact values).

     Bioindicators are also often used as "accumulator plants"
(range "b" of figure 1). This use extends into the range of invisible injury. It utilizes the plant's ability to store and accumulate compounds which pollute the air only temporarily or in
such low concentrations that their detection in the air becomes
very difficult. The determination of the compound by chemical
foliar analysis, however, presupposes the following five points:
1. The compound is chemically relatively easy determinable;
2. The compound is not destroyed or changed by the plant;
3. The compound is not easily transferred to other plant organs
   or exuded by the roots or easily leached by rain;
4. The compound is normally present in the plant in minute amounts

only and is not e.g. a major essential element;
5. The compound is easily taken up or adsorbed by the shoot and foliar uptake greatly exceeds root uptake.

In spite of these points we have to bear in mind that the mere presence of a compound does not prove its toxicity if the latter has not yet been established.

For detection of the toxicity to the plant we often have to rely on bioindications in the invisible range. I call this "latent injury" (1), which means a biochemical, physiological, pathological or ecological reaction of the plant even when visible symptoms of injury are lacking. Of course such reactions may also be used as evidence of toxicity in the range of visible injury although there they are usually replaced by the symptom occurrence. The term "latent" injury was chosen because the term "latent" deficiency of nutritive elements is used in plant nutrition.

In forestry it is particularly important to recognize the existence of a plant stress or a restriction in plant reaction to any stress before symptoms become visible. We need an early warning of a biologically dangerous situation, because it is often too late to react once a tree has started to show symptoms or has even died! After all, (6) pointed out more than a hundred years ago that a tree as a whole may be much more sensitive to air pollutants than its assimilatory tissue.

All to often people fail to realize that vegetation may be under stress before symptoms appear and may resort to a mere dilution of pollutants, e.g. by building high chimney stacks. This is believed to provide sufficient environmental protection because there are no visible symptoms of injury in the vicinity. Threshold 2, which separates the invisible from the visible injury, has therefore attracted much attention.

When utilizing bioindicators or bioindications we have to take into consideration that any plant reaction is governed by many internal and external factors. One important internal factor in plants is their genetical make-up, which modifies their responses. This fact is often neglected in forestry, where each tree is an individual. Figure 2 shows one reaction (CO_2 uptake is considered a very sensitive bioindication) of three different spruce clones, under identical conditions, to a constant but low SO_2 concentration. One clone showed neither statistically significant reduction of CO_2 uptake nor visible symptoms of injury within eight weeks. Another clone, however, reacted after only two weeks with significantly reduced CO_2 uptake and after eight weeks even with visible symptoms of injury. The third clone showed an intermediate reaction. Similarly, not all the trees in a forest will react alike to air pollution, not even when belonging to the same tree species.

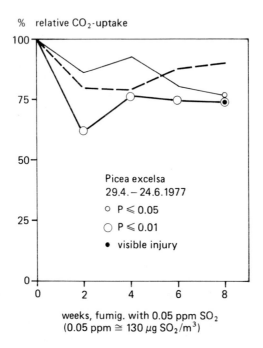

% relative CO_2-uptake

Picea excelsa
29.4. – 24.6.1977
○ $P \leqslant 0.05$
◯ $P \leqslant 0.01$
● visible injury

weeks, fumig. with 0.05 ppm SO_2
(0.05 ppm $\cong$ 130 μg SO_2/m^3)

Figure 2: The effect of the genetic make-up (3 spruce clones) on the relative CO_2 uptake of the plants as affected by a prolonged SO_2 fumigation (each value is the average of 5 replicates).

In order to make the distinction between bioindicator and bioindication clearer, the following examples are given. Figure 3 shows the accumulation of chloride by beech foliage near a garbage incinerator in relation to the height above (below) the chimney mouth (Δ H_K). In all years sunleaves of the uppermost crown of the same, marked trees were sampled in order to allow comparisons (2). The trees are part of a protective forest on a steep slope. From 1971 to 1973, before the incinerator began its operation, the chloride contents were very low and practically uniform at all heights. In the years 1975/77 the highest chloride contents were detected, whereas in 1978/80 lower (but still much increased) chloride contents were detected. Although fortunately no visible symptoms of injury occurred in this forest, the data demonstrate that the flue gases reach the forest in spite of a chimney height of 60 m. The foliage of the trees acts as an accumulative bioindicator but no effects of the gases on the trees have thus far been detected by foliar analysis.

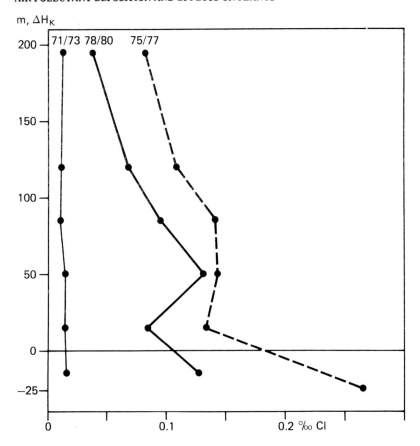

Figure 3: The effect of the height above (below) the chimney
mouth of a garbage incinerator on the chloride content of beech
foliage (each value is the average of 3-6 replicates).

Accumulation of S in young needles of spruce fumigated continuously over 105 days with low SO_2 concentrations is shown in
figure 4. The addition of a fertilizer (P) or a mycorrhizal in-
oculum (M) favored the growth of the seedlings in spite of increasing SO_2 uptake, manifested as higher S content. Apparently
the better fed spruces were able to make more efficient use of
the SO_2 taken up. It is not known, however, how long this plant
reaction would continue. Again no bioindication was used.

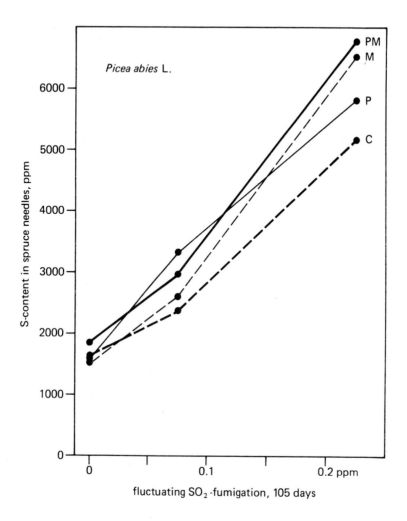

Figure 4: The effect of a prolonged SO₂ fumigation on the S content of spruce needles (each value is the average of 9 replicates).

In contrast to figures 3 and 4, figure 5 shows data on the
use of two bioindications in pine. The youngest needles of a pine
clone exposed to the gases of an aluminum smelter (F) and controls
were subjected periodically to measurements of peroxidase activity (4) and of ascorbic acid contents (5). The data show the
marked increases of peroxidase activity in F-exposed needles.
Peroxidases are a group of enzymes needed to oxidize toxic metabolic derivatives in the cell which occur due to ageing (controls)
or in consequence of stresses, such as air pollution. Ascorbic
acid contents on the other hand dropped, as is normally the case

when air pollution is involved (1,3). The difference was most
pronounced at the beginning when the sun was highest. Unfortunately solar radiation may severely interfere with this bioindication.

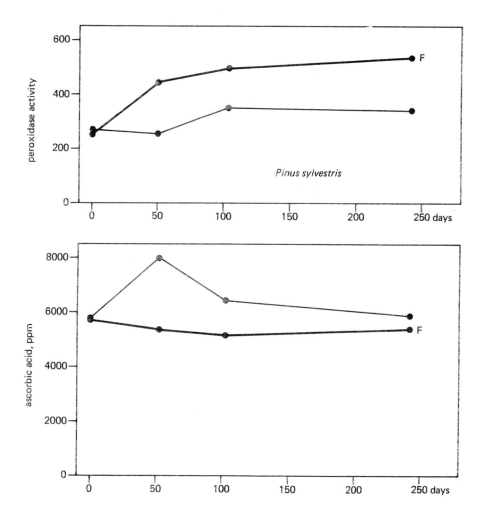

Figure 5: The effect of F-containing air on two bioindications
(peroxidase activity and ascorbic acid contents) in pine needles
over several months (each value is the average of 5 replicates).

Finally figure 6 gives the results of a 3-month winter fumigation of two spruce clones with SO_2. The S content of the
youngest needles by foliar analysis confirms the validity of
spruce as a bioindicator. It also demonstrates SO_2 uptake by

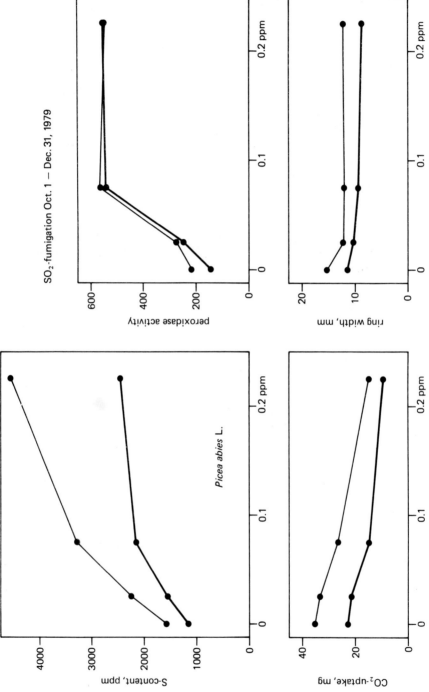

Figure 6: The effect of an SO2 fumigation on a bioindicator and on bioindications

plant organs outside the growing season. At any time of the year, therefore, needle analyses yield valuable information for the distribution of an SO_2 pollution. Figure 6 also includes data on the bioindications peroxidase activity and CO_2 uptake. Peroxidase activity shows a drastic increase at lower concentrations only because the highest SO_2 concentration killed many needles without initiating the formation of visual symptoms. These needles dropped, still green, after the experiment. The CO_2 uptake of the shoot shows a marked decline with increasing concentration in spite of the generally low activity outside the growing season. Since wood formation (width of annual ring) depends on CO_2 uptake this factor was also investigated. Ring width was measured for the growing season after fumigation. It reflects an after-effect which is not only statistically significant but also surprising because trees had been fumigated outside the growing season!

Summary

Plants may be efficient accumulators of air pollutants. The mere presence of an accumulated substance, however, usually does not indicate a toxic effect on biological matter. In addition, plants are often under stress without showing it through visible symptoms. Particularly when dealing with forests with important protective functions it is therefore necessary to use physiological, biochemical or ecological reactions to investigate whether a plant is affected in the invisible range by such accumulations ("latent injury").

Man often uses bioindicator plants as early warning monitors. The chosen strains of particular sensitivity may indicate a toxic action conveniently by visible symptoms. Often, however, they act as mere accumulators and make substances available for analysis which occur only temporarily or in such low concentrations that their detection in the air is very difficult. Not all substances, however, are suitable for such a foliar analysis.

Especially in the invisible range of injury, those bioindications should be used which show whether the plant is affected. Examples are given to demonstrate how such an effect may be detected by using enzymatic activities, CO_2 uptake or other plant reactions to air pollutants.

References

(1) Keller, Th., 1977: Begriff und Bedeutung der "latenten Immis-
    sionsschädigung". Allg. Forst- u. Jagdztg. 148 (6), 115-120
(2) Keller, Th., 1980: Der Nachweis einer Immissionsbelastung
    durch eine Müllverbrennungsanlage mit Hilfe der Blattanalyse
    auf Chlorid. Staub-Reinhalt. Luft 40 (3), 113-115
(3) Keller, Th., 1981: Folgen einer winterlichen SO_2-Belastung
    für die Fichte. Gartenbauwissenschaft 46 (4), 170-178
(4) Keller, Th., and Schwager, H., 1971: Der Nachweis unsichtba-
    rer ("physiologischer") Fluor-Immissionsschädigungen an Wald-
    bäumen durch eine einfache, kolorimetrische Bestimmung der
    Peroxidase-Aktivität. Eur. J. Forest Pathol. 1 (1), 6-18
(5) Keller, Th., and Schwager, H., 1977: Air pollution and ascor-
    bic acid. Eur. J. Forest Pathol. 7 (6), 338-350
(6) Schroeder, J., 1873: Die Einwirkung der schwefligen Säure auf
    die Pflanzen. Tharander forstl. Jahrbuch 23, 217-267

IUFRO-STUDIES ON MAXIMAL SO_2 EMISSIONS STANDARDS TO PROTECT FORESTS

Karl Friedrich Wentzel

Hessische Landesanstalt für Umwelt
Wiesbaden, Germany

This workshop deals with the 'effects of the accumulation of air pollutants in forest ecosystems'. This is a secondary part of the immission issue. In this framework I will try to delineate the present knowledge of the poisonous effects of air pollutants on the process of photosynthesis happening b e f o r e or during their accumulation in forest ecosystems. According to the long German research tradition in this field (beginning 1853 (5)) these effects are not any less important or convincing in explaining the reasons for the steadily increasing injuries caused by air pollution in forests and woodlands all across central Europe.

Only a small part of gaseous dry deposition penetrates into the foliage (9). The much greater part ist scavenged out of the air by the tree canopies, absorbed in water-films on the huge quantity of surfaces (2). This is carried as throughfall with precipitation to the ground (4). The poisonous effects however on the physiological processes in leaves depend on the concentrations of gaseous pollutants entering the stomata and the duration of their influence. Before being neutralised and deposited in the foliage the pollutants, having passed the stomata, affect the photosynthetic processes.

The accumulation of pollutants in the foliage cannot be considered a criterium of injury but only an index of injurious effects having already occured some time before. In 1898 already SO_2 and HF were defined as 'Poisons of Assimilation' (12). They decrease all metabolic

295

B. Ulrich and J. Pankrath (eds.), Effects of Accumulation of Air Pollutants in Forest Ecosystems, 295–302.
Copyright © 1983 by D. Reidel Publishing Company.

processes and enzymatic activity and increase transpiration by paralysing stomata-mechanism. Intensifying
eachother these effects reduce the vitality of trees
and the resistance to other dangerous factors while
causing premature senility and death. These results
are demonstrated by numerous investigations in polluted
areas and research experiments in fumigation chambers.

The International Union of Forestry Research Organisa-
tions (IUFRO) has created a subject group 'Air Pollution'
which has been involved in this research since 1957.
A principal field of study has been the dose-response
relationship between the immissions relative to concentration and duration of influence on one hand and the
response of the forests on the other. In 1980 I have
attempted to recapitulate the results of the last 15
years of research. This article has been published under
the title 'Maximale Immissionswerte zum Schutze der Wälder' (10). The attached diagram illustrates this. As
fumigation can only be carried out on young plants I
have included in this diagram only the results from
field measurements (solid lines).

The curves 1 to 8 show the SO_2 immission concentrations
of various central European forest-damage areas as
connecting curves between actual measured annual average
values (on the 50% line) and peak values (on the lines
of 95 or 97,5 percentile of 30 minute integrals).To the
numbersof the curves are noted the dose-response-rela-
tionships which are cited from the respective literature.
The measured data result in - as is evident from the
chart - a wide range of different concentrations between 12 µg and 150 ug for the average as well as 75
to 1000 µg for the peak value. This delineates the
range of concentrations which causes the so called
'chronic immission diseases' of the forests. Even higher
immission concentrations which cause 'acute immission
diseases' with speedy formation of necrotic tissue are
subsumed here. This was done as these concentrations
practically do not exist any more in the natural environment. In the last 30 years now gaseous emissions have
been vented through high stacks.

The range of 10 to 25 µg of the annual average can be
called the range of invisible, latent, or hidden injury.
Our subject group of the IUFRO issued the resolution on
air quality standards for the protection of forests in
1978 (3). Here, for practical reasons, the sensitivity
of Norway spruce as a guiding species was used. Abies
Alba in central Europe shows a higher sensitivity than

SO$_2$ Field Measurements in European Immission Areas

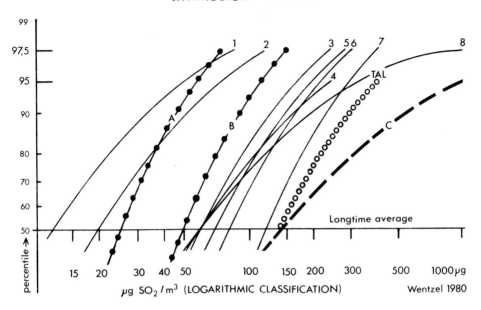

1. MATERNA 1973 (13): High ranges of 'Erzgebirge' = resistance diminished to frost and other secondary injuries.
2. MATERNA 1972 (14): Wood increase - loss 20% in 'Erzgebirge'.
3. MATERNA a.o. 1969 (15): Dieback of whole Norway spruce stands in the Bohemian part of the 'Erzgebirge'.
4. WENTZEL 1979 (16): Good sites in Rhein-Main area sufficiently protected.
5. STEIN a. DÄSSLER 1968 (17): Moderate injury in the Saxonian part of the 'Erzgebirge'.
   LUX 1976 (18): Situation comparable in 'Niederlausitz'.
6. KNABE 1970 (19); WENTZEL 1971 (20): Strong decrease of growth in the Ruhr district.
7. KNABE 1972 (21): Economical forestry with Norway spruce and Scotch pine made impossible in the Ruhr district.
8. GUDERIAN a. STRATMANN 1968 (22): Immission type 'Bierstorf' = single immission source in mountain valley. Heavy growth damage to nearly all species (young plants).

Norway spruce; it dies already at annual average concentrations of 20 μg SO_2. It therefore was not considered in this tabulation because of doubts on climatic sensitivity cooperating.

Also indicated in the diagram are the conclusions as
deriving from this knowledge. We have recommended to
the governments of our countries to strive for SO_2 concentrations which are indicated by the dotted lines A
and B. This would fairly assure the protection of the
forests. The line A is to be used for extreme (poor)
sites, line B for average (good) sites.

Also included in the diagram is the requirement for
protection of the human health (line C) as determined
by German medical research. Additionally the line 'TAL'
marks the requirement for siting of new industrial installations in the Federal Republic of Germany. It is
clearly evident from the diagram that the line 'TAL'
protects human health and avoids 'acute' injury to the
forests but it does not protect them from the widespread chronic injury. Practically all elevated sites
of German mid-range mountains are located within the
range of annual average concentrations from 10 to 25 μg
SO_2. This is a result of the residence time for SO_2
in the atmosphere on an average of 1 to 3 days (1). It
has been proven that the sensitive Abies alba cannot
tolerate this concentration (11). Additionally as a result of recent research it can be assumed that photo-
oxydants are affecting these regions just as SO_2 does.

I am convinced that the poisonous effect of these low
level immissions which affect the foliage directly are
the most important primary cause for the phenomenon of
long range injury through air-pollution in the central
European forests. Their poisonous effect which remains
hidden over long periods of time weakens the vitality
of the trees. It also gradually lessens their resistance
to other damaging agents. This is the triggering factor
in a progressing disease which in time will take on a
more and more complicated character. Only over a long
incubation period will this become externally evident.
The heavier the dose and the longer the duration the
more serious the effects of the accumulation of air-
pollutants in the ecosystem will be. This is especially
valid for the gradual acidification of the soil and the
accumulation of heavy metals (6, 7, 8).

Resolution on Air Quality Standards for the Protection
of Forests

Subject Group S2.09-00, Air Pollution, considers as
its most important assignment to set up effective limiting values of pollutant concentrations producing
no injury to forest stands on the basis of the latest
scientific knowledge and silvicultural experiences.
These air quality standards should serve to the responsible authorities as a basis for measures to obtain clean air which is necessary for the protection
of the forests.

At the X. meeting in Ljubljana, Yugoslavia, the air
quality standards for sulphur dioxide and hydrogen
fluoride - necessary for the protection of the forests -
were thoroughly discussed. For practical reasons it is
useful to fix only one or two standards. It is of advantage to use Norway spruce (Picea abies) as a guiding species since it is most common in Central and
Northern Europe and its requirements against air pollution include the protection of most of the other tree
species.

According to the current state of research in pollution ecology and forest science the protection of Norway spruce forests is guaranteed with high probability
if the following air quality standards are being observed. If they are exceeded a reduction of vitality,
growth performance, and resistance against biotic and
abiotic influences must be expected. These limits are
valid for separate occurrence of the respective pollutants. In combined occurrence of different air pollutants synergetic effects have to be taken into consideration.

The restriction of pollutant concentrations to these
air quality standards cited next page is only sufficient to prevent the forests from direct effects, but
not from all other gradual modifications of the soils
by acid precipitation with combined effects.

Members are invited to inform their governments and
responsible authorities about the results, and to ask
for promotion of research activities on various air
pollutants in sensitive forests, including acid precipitation and combined effects.

IUFRO Air quality standards 1978 (3)
- confirmed 1980 (10) -

Sulphur dioxide in micrograms per cubicmeter of air

annual average	average of 24 hrs	97,5 percentile of 30 minutes values in period of vegetation
50	100	150 /µg

(full production on most sites; the average of 24 hours may be exceeded 12 times during a period of six months)

| 25 | 50 | 75 /µg |

(necessary to maintain full production and environmental protection e.g. against erosion, avalanches, in higher regions of mountains, in boreal zones, extreme sites etc.).

Hydrogene fluoride in /µg per cubicmeter of air

| 0,3 | | 0,9 /µg |

(full production on most sites; further research necessary for additional figures as above)

References

1. Georgii, H.W. 1978: Die Verteilung der Schwefelver-
   bindungen in der nicht verunreinigten Atmosphäre.
   VDI-Berichte Nr.314, pp. 57-61.

2. Höfken, K.D., Georgii, H.W. a. Gravenhorst, G.1981:
   Untersuchungen über die Deposition atmosphärischer
   Spurenstoffe an Buchen- und Fichtenwald. Ber. d.
   Inst. f. Meteorologie u. Geophysik d. Universität
   Frankfurt, Nr. 46, 141 pages.

3. IUFRO-Fachgruppe Air Pollution 1978: Resolution über
   Maximale Immissionswerte zum Schutze der Wälder.
   IUFRO-News 25 (1979).

4. Künstle, E., Mitscherlich, G. a. Rönicke, R. 1981:
   Untersuchungen über Konzentration und Gehalt an
   Schwefel, Chlorid, Kalium und Calcium sowie den
   pH-Wert im Freilandniederschlag und Kronendurchlaß
   von Nadel- und Laubholzbeständen bei Freiburg/Br.,
   Allg. Forst- u. Jagdzeitung 152, pp. 147-165.

5. Stoeckhardt, A. 1853: Untersuchung junger Fichten
   und Kiefern, welche durch den Rauch der Antons-Hütte
   krank geworden. Thar. Forstl. Jhb.9, pp. 169-172.

6. Ulrich, B., Mayer, R. a. Khanna, P.K. 1979: Deposition von Luftverunreinigungen und ihre Auswirkungen
   in Waldökosystemen im Solling. Schrift.d.Forstl.Fak.
   d.Univ. Göttingen,Vol. 58.

7. Ulrich, B. 1980: Die Wälder in Mitteleuropa: Meßer-
   gebnisse ihrer Umweltbelastung, Theorie ihrer Gefähr-
   dung, Prognose ihrer Entwicklung. Allg.Forstzeit-
   schrift, pp. 1198-1202.

8. Ulrich, B. 1981: Gefahr für das Waldökosystem durch
   saure Niederschläge. Sonderheft d.Landesanstalt für
   Ökologie Nordrhein-Westfalen, pp. 9-25.

9. Wentzel, K.F. 1982: Foliar Analysis and Air Purification. Eur. J. Forest Path., in print.

10. Wentzel, K.F. 1981: Maximale Immissionswerte zum
    Schutze der Wälder. Mitt. d. Forstl. Bundesversuchs-
    anstalt Wien, Vol. 137/II, pp. 175-180 a. 327-328.

11. Wentzel, K.F. 1980: Weißtanne - immissionsempfindlich-
    ste einheimische Baumart. Allg.Forst-Z., pp. 373-374.

12. Wislicenus, H. 1898: Resistenz der Fichte gegen
    saure Abgase bei ruhender und tätiger Assimilation.
    Thar.Forstl.Jhb. 48, pp. 152-173.

Literature cited on SO$_2$ Field Measurements

13. Materna, J. 1973: Kriterien zur Kennzeichnung einer
    Immissionswirkung auf Waldbestände. Proceedings
    3. Intern. Clean Air Congr.Düsseldorf, pp. 121-123.

14. Materna, J. Jirgle, J. a. Kučera, J. 1969: Ergebnis-
    se von Messungen der SO$_2$-Konzentration im Erzgebirge.
    Ochrana ovzduši 6, pp.84-92.

15. Materna, J. 1972: Beziehungen zwischen der SO$_2$-Kon-
    zentration in der Luft und der Beschädigung von Fich-
    ten. Manuscript IUFRO-Fachtagung Sopron/Ungarn.

16. Wentzel, K.F. 1979: Die Schwefel-Immissionsbelastung
    der Koniferenwälder des Raumes Frankfurt/Main.
    Forstarchiv 50, pp. 112-121.

17. Stein, G. a. Dässler, H.G. 1968: Die forstliche
    Rauchschadengroßraumdiagnose im Erz- und Elbsand-
    steingebirge 1964/67. Wiss.Z.TH Dresden, pp.1397.

18. Lux, H. 1976: Ergebnisse der Rauchschadengroßraum-
    diagnose 1973 in der Niederlausitz. Wiss.Ztschr.
    d. TU Dresden 25, pp. 663-668.

19. Knabe, W. 1970: Kiefernwaldverbreitung und SO$_2$-Im-
    missionen im Ruhrgebiet. Staub 30, pp. 32-35.

20. Wentzel, K.F. 1971: Habitus-Änderung der Waldbäume
    durch Luftverunreinigung. Forstarchiv, pp. 165-172.

21. Knabe, W. 1972: Immissionsbelastung und Immissions-
    gefährdung der Wälder im Ruhrgebiet. Mitt. d. Forstl.
    Bundesversuchsanstalt Wien, Vol. 97/I, pp. 53-87.

22. Guderian, R. a. Stratmann, H. 1968: Freilandver-
    suche zur Ermittlung von Schwefeldioxidwirkungen
    auf die Vegetation. Forschungsberichte Nordrhein-
    Westfalen Nr. 1920, Westdeutscher Verlag, Köln u.
    Opladen.

LONGTERMED FLUORIDE POLLUTION OF A FOREST ECOSYSTEM:
TIME, THE DIMENSION OF PITFALLS AND LIMITATIONS

Hannes Flühler

Swiss Federal Institute of Forestry Research,
CH-8903 Birmensdorf

Abstract

This is a synthesis of various forest damage investigations
carried out in the Swiss Rhone Valley from 1977 - 81. Meteorologically the valley is an enclosed system. The fluoride pollution
covers a time span of 74 years which makes it possible to focus
on slowly progressing phenomena such as fluoride accumulation in
soils and forest damage development.
The role of field research as a tool for problem recognition
and analysis is discussed.

INTRODUCTIVE SUMMARY

The pine forests in the Swiss Rhone Valley are severely
damaged. Air pollutants, fluorides among others, are considered
to be one of the prime causes. The air pollution history of this
valley is old. It started in 1908 when the first two aluminum
smelters were built and possibly ended in 1981 when the F-emissions were drastically reduced.
In this contribution we focus on some draw backs and limitations of investigating this particular case which is fairly
well defined not only in the sense of knowing its early beginning
and development but also well defined in a geographical, meteorological, social and political context. The objective of this
paper is to illustrate the following statements:
. Our awareness of an environmental problem and the attitude
of different segments of society toward any measures for its
control are changing with time.
. Our perceptive capabilities are often restricted to a spectrum

303

B. Ulrich and J. Pankrath (eds.), Effects of Accumulation of Air Pollutants in Forest Ecosystems, 303–317.
Copyright © 1983 by D. Reidel Publishing Company.

within the limits of the available analytical and experimental methods.

. Even consistent and convincing field experiment data are seldom (if ever) a proof of cause and effect.

. Dealing with slowly progressing phenomena our experimental evidence is unavoidably a momentary picture or at best a short sequence of pictures in a long movie.

These statements are neither original nor absolute. Keeping them in mind provides a sounder base for argumentation.

AIR POLLUTION IN THE SWISS RHONE VALLEY - TODAY

In March 1978 samples of pine needles were collected in the Rhone Valley from Lake of Geneva 120 km upstream. The samples were analyzed for sulfur, chloride and fluoride (1). The S-, Cl-, and F-contents were averaged per site. The site averages shown in Figure 1 were smoothed with a cubic spline regression (2) in order to illustrate their regional distribution along the valley axis. Despite the fact that the soils vary remarkably from site to site the S-contents decrease systematically upstream. However, they are not alarmingly high. Their obvious trend very likely reflects the atmospheric SO_2-load. The shaded bands cover the range of those values observed in an unpolluted control area. The Cl-contents exhibit local peaks in the vicinity of the relatively large chemical plants in Visp and Monthey but most measurements are within the range of the natural background values of the control area. The spatial pattern of F-contents differs from those of the two other components. The aluminum smelter in Chippis is apparently the most significant F-emission source, that of Steg ranks second and the smelter in Martigny raises the level of F-contents hardly above the natural background. The most seriously F-polluted area around Chippis and Visp was sampled more densly with approximately four sampling plots per km of the valley axis (cf. the map on page 408 of (1)). The F-pollution in the central valley in Chippis and to a lesser degree in Steg was apparently intense. However, in the lower valley in Martigny it was almost not detectable at the time of that survey.

The air volume within the narrow and meteorologically enclosed valley is small and prone to air pollution. Frequent and relatively stable temperature inversions aggravate the problem. Preliminary results from other studies hint at the possibility that other air pollutants, especially oxidants, may also play a certain role.

After evaluating the actual situation we focus on the historical dimension of the problem which makes fluoride become the main topic.

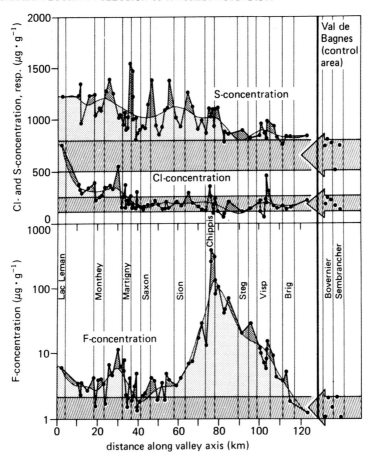

Figure 1. S-, Cl-, and F-contents of one year old 1977 pine needles in the Rhone Valley sampled in March 1978. The samples from Val de Bagnes can be considered as natural background.

THE HISTORICAL VIEW

At the onset on the twentieth century the incentive to industrialize this valley was great. Labor was cheap and hydroelectric energy available. In 1907/8 the two aluminum smelters in Martigny and Chippis went into operation producing approximately 2000 and 11000 t Al annually (1916), respectively (9800 and 28000 t Al/y in 1981). For years and decades the F-emissions were of minor concern and environmental control an unthinkable criterion. The change of attitude toward industrial air pollution presents itself in many superficial or even anecdotic details: Years ago the smoking stack on the letter head of the aluminum company in Chippis was a symbol of progress and prosperity. Our generation

inherited the socio-economic benefits of the technological euphoria
along with the environmental side effects of a continuous production. Today, we are inclined to relate the smoking stack much
more to the latter than to the former consequences of industrial
activity.

The analytical abilities and the awareness of environmental
problems were linked together - as they still are today -. Faes
(1921) and Wille (1922), the first environmental consultants in
that area published air analyses of samples taken at the nearby
nunnery on the Geronde cliff. This site is located in Chippis,
100 m from the smelter and some 30 m above its roof tops. The
F-concentrations in the air collected on the roof tops were then
at the lower limit of resolution. The "0.1 mg F per 1000 litres of
air" were then found to be negligibly small. Nowadays such concentrations of 100 $\mu g \cdot m^{-3}$ (roof top sample) and 20 - 30 $\mu g \cdot m^{-3}$
(Geronde cliff samples) are considered to be excessively high and
hazardeous for plant and animal life. Misjudgments like the one
just quoted are only worthwile to be mentioned if we also run
the risk of drawing erroneous conclusions from imperfect experimental evidence, a risk which is immanent in the theme of this
workshop. Looking backwards over a time span of 74 years we have
a chance to learn not only from scientific facts but also from
the way they were interpreted in the past.

As mentioned above two of the smelters were founded in 1907
and 1908, respectively. In the thirties a third one was built
close to the one in Martigny which was remodeled in 1965. The
new location was meteorologically unfavorable, amidst the main
stream of the prevailing winds. The old smelter closed much later
in 1956. Both together produced only 3000 to 5000 t Al per year.
However, the F-emissions must have been substantial since no
filters had been installed then. In 1962 a 48000 t Al/y smelter
was constructed in Steg equipped with wet scrubbers which at that
time represented the latest and most efficient filter technology.
The smelter in Martigny installed the same equipement in 1965,
when its capacity was doubled. In the seventies, after decades
of complaints by farmers and after numerous studies on the impact
of atmospheric fluoride on agricultural crops have been completed
the fluoride problem got political momentum. The heated political
discussions and violent actions calmed down when the F-emission
standards (1.5 kg total F per t Al produced) were established and
enforced (Oct. 1978). As a consequence the dry absorption filter
technique was introduced in all three smelters. Today, the F-emissions are supposedly within safe limits.

The take home lessons of this historical review are trivial
but unfortunately quite symptomatic for such problems:
. More than only scientific evidence is needed to resolve such
  relatively clear cut problems. In addition to the scientific
  diagnosis it takes time, much time and it requires the in-
  sisting activity of pressure groups in order to converge toward
  an environmental compromise.

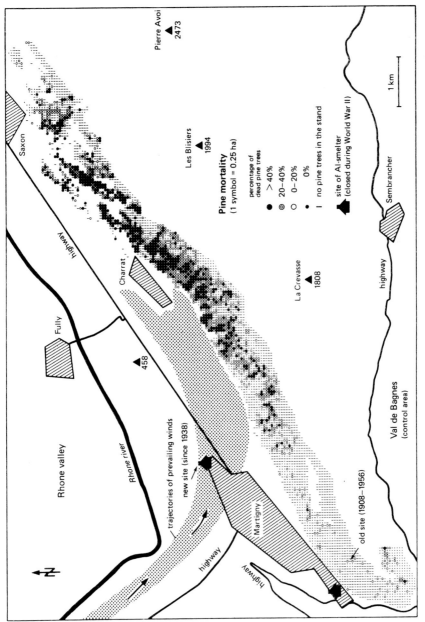

Figure 2. Pine mortality rated on 1:13000 colored infrared aerial photographs (data from Graf et al. 1982). Wind pattern and trajectories (Environmental Protection Agency, State of Wallis).

- The handling of environmental risks is a mirror for the ethical and socio-economic base of society. In this case - and very likely in many other cases as well - the awareness of the public for such phenomena grows more slowly than the problem itself.
- Even when we assume that this problem has been resolved it leaves us with a new question, that of the resilience of a disturbed system. With other words, we should know the time span of recovery for the damaged habitats. This leads us back on scientific grounds.

SIGNATURES OF AIR POLLUTION IN FOREST STANDS

The plant tissue analyses of the survey 1978 (Fig. 1) suggest that the level of atmospheric SO_2 and Cl in the lower Rhone Valley were slightly elevated but that of F insignificant. Earlier records indicate that the F-contents were much higher in the past. Trees turning brown and F-contents being low is a perfect motive for disregarding F as a possibly damaging agent and for postulating other hypotheses. However, using other survey techniques such as mapping pine mortality and using tree ring analysis we obtained conflicting evidence.

The spatial distribution of pine mortality is shown in Figure 2. Colored infrared aerial photographs were sampled systematically. Each sample plot (0.25 ha) was rated according to various site, stand and damage characteristics (5, 6, 7). The most severely damaged stands are located on the rugged slopes above Saxon and Charrat Northeast of Martigny. These sites are exposed toward the prevailing strong winds upstream (Westwind conditions). Virtually no damages and intact pine stands were found out of reach of the main wind trajectories. The spatial pattern of damage and wind distribution might be coincidental but leaves ample space for speculation. The data are consistent and convincing but definitely no proof of cause and effect.

The history of forest damage development was reconstructed by means of tree ring analysis (8). The cores were taken from both damaged and healthy looking pine trees in the stands of Saxon. More than 80 % of the cores exhibit a sharp discontinuity in ring widths as shown at the top of Figure 3. The year of this growth shock as well as those of the first and last ring of the individual cores were crossdated using the narrow rings formed during extremely dry years. Half of the trees with that ring anomalie were damaged between 1941 and 1947 irrespective of age, site and stand type. These were the years of the most pronounced periods of drought after the second smelter in Martigny was built at the new location. Again, this might be mere coincidence.

A tree ring model suggests that climate is not the prime cause of the growth retardation. This model makes use of the "response functions", which are used in connection with a multiple

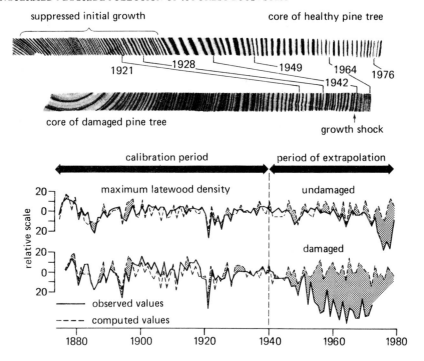

Figure 3. X-ray pictures and densitograms from pine cores of a damaged and a healthy tree (data and model computations from Kienast 1982).

regression *(9)*. The model was adapted by Bräker, Swiss Federal Institute of Forestry Research, and applied by Kienast *(10)*. It relates maximum latewood density and ring width to monthly rainfall and temperature data. The time series of tree rings and climatic variables from 1874-1940 were used for calibration. A shorter calibration period (1874 until 1920) yields similar model parameters. Using these parameters and the climatic records after 1940 the model repredicts the tree ring sequence of the last 40 years. The simulated tree ring series agree fairly well with those observed on healthy trees but deviate systematically from those of damaged trees indicating that the growth conditions changed after 1940. Apparently, the damages we observe today were induced some 20 to 40 years ago. Tacit and explicit conclusion drawn from such data can easily be killed by raising additional questions. Such evidence yields rather a base for new and more precise hypotheses than a proof of cause and effect.

FLUORIDE DEPOSITION AND ACCUMULATION

   For topographical reasons most of the fluoride emitted since
1908 must have been deposited within the valley. This thought
evokes a whole complex of questions about the materials balance
of this pollutant in the environment. The data presented below
relate only to the questions of how much F enters a forest ecosystem and where it is stored. The experiments were conducted
in the vicinity of the smelter of Chippis (11). The F-contents
(F_i) of the six months old pine needles (1977) were used to define
the F-exposure (p_i) of the various sites (i). The F_i are expressed
relative to the average F-content of the most intensely contaminated sites within an area of less than 800 m from the smelter
($F_{max} \approx 255$ µg F·g^{-1})

$$p_i = (F_i - F_o) / (F_{max} - F_o) \text{ with } o \leq p_i \leq 1.0 \tag{1}$$

where F_o is the F-content of pine needles collected at uncontami-
nated remote sites outside of the main valley ($F_o \approx 5$ µg F·g^{-1}).
   Various plant and soil materials were sampled in two pine
stands (experimental site B) located 1.8 km from the smelter where
the F-exposure was p = 0.24. The concentrations and the quantities
of fluoride stored in these compartments per hectare are given
in Table 1. Due to the uncertainties of estimating the mass of
these compartments and due to the natural variability of the F-concentrations the F-storage data indicate the order of magnitude
only. The most significant portion of the fluoride pool at this
site is the water-insoluble fraction of soil F (88 %) which can
be considered to a large extent as being the soilborne F of the
mineral constituents (12). The least significant portion of F is
stored in the wooden tissue of the tree trunks and roots. Even
the amount of F dissolved in the soil solution of the root zone
(30 cm of depth) is apparently greater than that stored in the
biomass above ground.
   The concentrations and areal quantities in Table 1 are a
merely static description and hence not conclusive for assessing
the current state of contamination. It is crucial to know whether
the system already reached a quasi steady state after the 74 years
of continuous contamination or whether the observed level of
accumulation is a momentary picture of a transient change. If
the rate of deposition and the fluxes of fluoride from one compartment into others were known the system's state could be
evaluated. The available information is still far from being
sufficient for such conclusions.
   Four independent schemes were used to estimate the F-deposition rate:
. gross mass balance of the total F-emissions in the valley
. F-accumulation in a contaminated soil profile
. F-accumulation in standard soil columns
. direct measurements with Bergerhoff gauges.

Compartment		mass of compartment** t·ha⁻¹	F-concentration* [μg F·g⁻¹]		storage [kg F·ha⁻¹]	storage [% of total F]
main root zone	F in soil solution (A)	700*	7	(A)	4.9	.151
	waterextractable F (B) minus (A x water content)	3500*	105	(B)	367.	11.3
	insoluble F (= total F minus (A+B)) (C)		820	(C)	2870.	88.4
litter		8	163		1.3	0.040
roots	cortex	3	25		0.075	0.002
	wooden tissue	8	6		0.042	0.001
soil vegetation (sparse)		0.5	100		0.500	0.015
pine trees	trunk bark	3.9	35		0.136	0.004
	wood	22.1*	3		0.066	0.002
	branches bark	8.7	19		0.165	0.005
	wood	26.2	10		0.262	0.008
	needles ½ 1975	0.8	156		0.720	0.022
	1976	3.4	123			
	1977	3.4*	52			

Table 1 Approximate estimates of fluoride storage within various compartments of a pine stand (site B, cf. FERLIN et al. 1982, POLOMSKI et al. 1982)

* measured data   ** data from literature

The results are summarized in Table 2

Gross Mass Balance of the Total F-Emissions in the Valley

Conservative estimates of the fluoride emitted in total (E_{tot}) since 1908 are in the order of 11000 t. The official estimate of the annual F-emission in 1978 was 579 t $F \cdot y^{-1}$. These quantities were deposited within an area of roughly 430 km^2. As a first approximation we assume that the rate of deposition (D_i) is proportional to the F-exposure (p_i) as defined in Eq. 1. The area with clearly elevated F-contents in pine needles (F_i > 15 µg $F \cdot g^{-1}$) is only 33 km^2. It is obvious that some fluoride must have been deposited outside of this zone. The residual F-content F_r observed within 92 % of the 430 km^2 area in the valley varies mostly between F_o (natural background) and this arbitrary limit of 15 µg $F \cdot g^{-1}$. The average F_r is assumed to be 10 µg $F \cdot g^{-1}$. The hypothetical residual deposition rate D_r is obtained from a weighted average mass balance:

$$E_{tot} = D_r \sum_i \frac{p_i}{p_r} \cdot A_i \quad \text{which yields} \quad D_r = \frac{E_{tot} \cdot p_r}{\sum_i p_i \cdot A_i} \tag{2}$$

where A_i is the area of the zone with a F-exposure p_i and p_r the residual F-exposure defined by F_r and Eq. 1 ($p_r = 0.02$). The deposition rate expected at site i is

$$D_i = D_r \cdot (\frac{p_i}{p_r}) \equiv E_{tot} \frac{p_i}{\sum_i p_i A_i} \quad \text{or equally} \quad D_i = D_r \cdot \frac{F_i - F_o}{F_r - F_o} \tag{3}$$

D_i and E_{tot} refer either to annual rates or to the total emission since 1908. Accepting the inaccuracies and limitations of this scheme we may speculate how much F was deposited at the two experimental sites (cf. Table 2c).

F-Accumulation in a Contaminated Soil Profile

Figure 4 shows the total and water extractable F-contents of the soil profile located at a distance of 1.0 km from the smelter. ($p_i = 0.80$). The corresponding values of comparable soils in the zone of the residual deposition ($p_r = 0.02$) were similar to the F-contents in the subsoil of this profile (z > 150 cm). We assume without having further evidence that the F-content in the subsoil approximately equals the natural background value. Integrating the total F-contents over depth and subtracting the assumed soil-borne quantity yields an estimate of the total input since 1908 (Table 2b). For reasons given below it is unlikely that much fluoride was leached out of the profile: The concentration of soil solution samples varied with time but did not exceed 8 µg $F \cdot ml^{-1}$ which corresponds with the CaF_2 solubility product. The total mass of F accumulated in the profile and this maximum con-

		site A	site B
a) site characteristics	distance from smelter [km]	1.0	1.8
	F-content of pine needles 1977 (Nov.) (F_i) [μg F$\cdot$g^{-1}]	205	65
	F-exposure (p_i) (Eq. 1)	0.80	0.24
b) experimental estimates	accumulation within a soil profile 0 - 150 cm since 1908	970 g F$\cdot$m^{-2} (D_i = 13.1)	–
	accumulation within soil columns 1978/79 (D_i)	–	7.7 $\pm$ 3.5
c) computed from gross mass balance Eq. 1-3	E_{tot} (1908-82)= 11000 t F, D_r = 19.4 g F$\cdot$m^{-2}	776 g F$\cdot$m^{-2} (D_i = 10.4)	–
	E_{tot} (1978) = 580 t F$\cdot$y^{-1}, D_r = 1.02 g F$\cdot$m$^{-2}\cdot$y^{-1}	–	12.3
d) Bergerhoff gauges	average D_i (1974-78)	0.7	0.5

Table 2 F-accumulation and F-deposition D_i [g F$\cdot$m$^{-2}\cdot$y^{-1}] at two grass covered sites

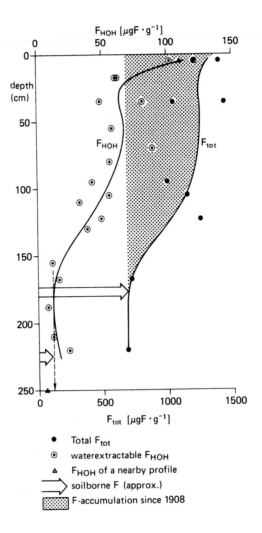

Figure 4. Concentrations of total and water-extractable F in
a soil profile located 1 km from the Al-smelter.

centration in the solution are inconsistent results. Assuming
that the entire rainfall since 1908 (∿ 4300 cm of water) contained
the 8 µg F·ml⁻¹ and infiltrated without any evaporation loss only
one third the actually observed accumulation could be explained.
Laboratory experiments indicate that even in such calcareous soils
the F-concentration in a rapidly infiltrating solution may considerably exceed the CaF_2 equilibrium concentration because the
kinetics of the CaF_2 precipitation is slow as compared with the
rapid pore velocity (13).
   The fact that F accumulates in soils in measureable quantities
was overlooked until quite recently (14, 15, 16). The pool of soil-

borne F is large compared with the shorttermed input from the atmosphere. The sometimes highly variable mineral composition of the soil might have masked the accumulation of the airborne contaminant in many cases. Experiments of excessive duration would be required to demonstrate the accumulation process under such conditions.

F-Accumulation in Standard Soil Columns

Columns filled with a standard soil material of a low F-content and known adsorption characteristics were exposed for one year (1978-1979) at site B (1.8 km from the smelter, Polomski et al. 1982). The net increase of F in the columns represents the actual F-deposition. Amazingly, this value is in close agreement with the previous estimates despite the small cross section of the colums (100 cm^2) and the considerable variability between the four replicates (cf. Table 2b). However, none of the three schemes presented so far will withstand any criticism. The underlying assumptions, the spatial variability of the measurements and the analytical inaccuracies make these estimates prone to error. In addition, the time consuming experiments provide only isolated point informations which can hardly be extrapolated in space.

Direct Measurements with Bergerhoff Gauges

The 1000 ml polyethylene beakers placed 2.0 m above ground were sampled in two week intervals from 1974-78. These data were obtained and made available by the Alusuisse Company in Chippis. This technique is a widely accepted routine procedure. These measurements are closely related with the F-contents of plant samples (Ferlin et al. 1982). Hence the relative differences from site to site are consistent with the survey data of 1978 and hence with the F-exposure p_i (Eq. 1). However the maximum values of the F-depositions determined in this manner are even less than the hypothetical mean deposition rate calculated for the entire valley (($\overline{D} = E_{tot}$ / 430 km^2 ≈ 1.3 g F·m^{-2}·y^{-1}, cf. Table 2d). Apparently, such measurements greatly underestimate the actual deposition. Once again, the experimental technique per se limited or mislead the investigator when he concluded that the F-deposition was negligibly small.

CONCLUDING REMARKS

From our point of view the ultimate goal of this workshop is to become better prepared for future tasks in this field which we will undoubtedly be confronted with. One way to learn is to avoid those mistakes we have a chance to know about. Studying longtermed slowly progressing phenomena is a confusing and some-

times frustrating experience. There are some major difficulties
specific for such cases. Some of them are illustrated in this
text and summarized in the introduction as axiomatic statements.
They could be rephrased more imperatively provided the reader
tolerates such stylistic means in a scientific contribution:
- The public needs more and more frequent information even when
  it is still incomplete.
- We have to admit that certain phenomena are experimentally
  not yet or no longer accessible.
- Repredictions of past events or projections into the future
  should not rely on a single approach.
- The limits of interpretation of field data should be clearly
  stated. The widespread believe or expectancy that scientists
  will or should come up with clear cut answer should be
  cautioned. More than only hard facts can be useful. Conclusions
  by analogy, plausibility explanations, a sound base for hypotheses or just the scientist's opinion might be valuable enough
  to be put forth.

References

(1) Flühler, H., Keller, Th., and Schwager, H., 1981: Die Imis-
    sionsbelastung der Föhrenwälder im Walliser Rhonetal. Mitt.
    eidg. Anst. forstl. Vers'wes. 57/4, pp. 399-414.

(2) Kimball, B.A., 1976: Smoothing data with cubic splines.
    Agron. J. 68, pp. 126-129.

(3) Faes, H., 1921: Les dommages causés aux cultures par les
    usines d'électro-chimie. 107 pp., Lausanne/Paris, Payot.

(4) Wille, F., 1922: Die Rauchschadenfrage der Aluminiumfabriken
    mit besonderer Berücksichtigung der Aluminiumfabrik Chippis.
    Parey, Berlin, 66 pp.

(5) Scherrer, H.U., Flühler, H., Mahrer, F., and Bräker, O.U.,
    1980: A sampling technique to assess site, stand, and damage
    characteristics of pine forests on CIR aerial photographs.
    14. Congr. Int. Soc. Photogrammetry, Hamburg, Proc. XXIII B8,
    Comm. VII, pp. 804-811.

(6) Scherrer, H.U., Flühler, H., and Mahrer, F., 1981: Alterna-
    tive Verfahren für die Interpretation von Föhrenschäden
    (Pinus silvestris L.) auf mittelmasstäblichen Infrarot-
    Farbaufnahmen. Mitt. eidg. Anst. forstl. Vers'wes. 57/4,
    pp. 433-452.

(7) Graf, H., Flühler, H., and Scherrer, H.U., 1982: Waldschaden-
    kartierung Martigny/Saxon. (Interner Bericht, Eidg. Anst.
    forstl. Versuchswesen, Birmensdorf.)

(8) Kienast, F., Flühler, H., and Schweingruber, F.H., 1981:
Jahrringanalysen an Föhren (Pinus silvestris L.) an immissionsgefährdeten Waldbeständen des Mittelwallis (Saxon,
Schweiz). Mitt. eidg. Anst. forstl. Vers'wes. 57/4,
pp. 415-432.

(9) Fritts, H.C., 1971: Multivariate techniques for specifying
tree-growth and climate relationships and for reconstructing
anomalies in paleoclimate. J. appl. Meteorology 10, pp. 845-864.

(10) Kienast, F., 1982: Jahrringanlysen in immissionsgefährdeten
Waldschadengebieten des Mittelwallis (Saxon). Diplomarbeit,
Geograph. Inst. Univ. Zürich, 112 pp.

(11) Ferlin, P., Flühler, H., and Polomski Janina, 1982: Immis-
sionsbedingte Fluorbelastung eines Föhrenstandortes im Unteren
Pfynwald. Schweiz. Z. f. Forstw. 133, pp. 139-157.

(12) Polomski, Janina, Flühler, H., and Blaser, P., 1981: Konta-
mination des Bodens durch Fluorimmissionen. Mitt. eidg. Anst.
forstl. Vers'wes. 57/4, pp. 479-499.

(13) Flühler, H., Polomski Janina, and Blaser, P., 1982: Movement
and retention of fluoride in soils. J. environ. Qual. (in
press).

(14) Sidhu, S.S., 1979: Fluoride levels in air, vegetation and
soil in the vicinity of a phosphorous plant. J. of Air
Pollution Control Assoc. 29, pp. 1069-1072.

(15) Groth, E., 1975: An evaluation of the potential for ecological
damage by chronic low level environmental pollution by fluoride
Fluoride 4, pp. 224-240.

(16) Polomski, Janina, Flühler, H., and Blaser, P., 1982: Accumulation of airborne fluoride in soils. J. environ. Qual. (in
press).

THE PROBLEM OF DETERMINING GROWTH LOSSES IN NORWAY SPRUCE
STANDS CAUSED BY ENVIRONMENTAL FACTORS

S. ATHARI and H. KRAMER

Institut für Forsteinrichtung und Ertragskunde der
Universität Göttingen, Büsgenweg 5, D 3400 Göttingen

Abstract

Several reasons make it difficult to carry out accurate measurements to define the amount of reduction of increment in
Norway spruce stands which is originated by air pollution.
The main problems are described
- as difficulties to find undamaged stands for comparison
  purposes
- as possible mistakes resulting from high variability of
  results from yearring-analyses of single trees;
- climatic and soil parameter beside air pollution may influence yearrings
- as necessity to synchronize each yearring to be measured
  in order to avoid mistakes caused by missing yearrings; up
  to 24 missing yearrings were found in disks at breast hight.

Part of the difficulties could be overcome by analyzing disks
both in breast hight and in higher parts of the trunk. Comparison of yearring diagrams resulting from analyses of limed
and unlimed tree groups in an area which is characterized by
acidified soils, justified losses of increment in the unlimed
stand.

Earlier research has shown that air pollution has a negative
influence on the growth of a forest. The direct pollution uptake damages the stomata and results in defoliation. This
process reduces the vitality of the forest trees and causes
a decrease in growth increment and, in late phases, mortality.
Air pollution, however, can also lower the growth increment

319

B. Ulrich and J. Pankrath (eds.), Effects of Accumulation of Air Pollutants in Forest Ecosystems, 319–325.
Copyright © 1983 by D. Reidel Publishing Company.

of trees indirectly. Soil acidification upsets the nutrient
supply and a precipitation deficiency at the same time results
in a defoliation of the trees.

The determination of reduced growth can be obtained by comparing damaged and undamaged trees. The growth development of
the undamaged trees is considered to be the normal growth.
The assumption is that the damaged trees would have the same
growth increment as the undamaged trees if they had not been
damaged. This assumption is only correct if the compared trees
have the same growth conditions with the exception of the
damage-causing factor. This can hardly be realized in the field.

Another problem is that not all of the trees in a stand are
damaged to an equal extent. The growth of the trees of the
damaged stand can be increased due to lower stand density which
results from the higher mortality. The growth information can,
therefore, only be evaluated if the competition factors for
damaged and undamaged trees are known.

A one time stand inventory is normally not sufficient. Conti-
nual observation of stands provides better information about
the growth development of compared stands. With the establish-
ment of permanent sample plots in areas with air pollution, we
would be able to obtain the necessary information for the
growth study and use this information in evaluating the effect
of the damaging factor.

The growth evaluation of single sample trees can give mislea-
ding information since the growth of each tree depends on specific conditions and the tree's genetic composition. The analysis
of a great number of sample trees, on the other hand, requires
much time and effort.

The growth pattern of single trees and stands can be obtained
through yield research. Multiple influences and interactions
of factors can increase or decrease growth. The growth losses
related to air pollution are the result of a disease complex.
The determination of certain growth patterns, therefore, require close cooperation with researchs in other fields, such
as Soil Science, Botany, and Meteorology.

Air pollution impairs the vitality of the trees and results
in a disorder of the year-ring formation. Under extremely bad
conditions, it is possible that no year-rings are formed. By
synchronisation of year-ring series of a severely damaged
dominant tree, ATHARI (1981) registered up to 19 missing year-
rings on a single radius of a disk at breast height (1,3 m).
The year-rings which were missing at 1,3 m existed in disks
from the crown area. The problem of determining the growth of

Stem - analysis of an damaged Norway spruce with missing annual rings

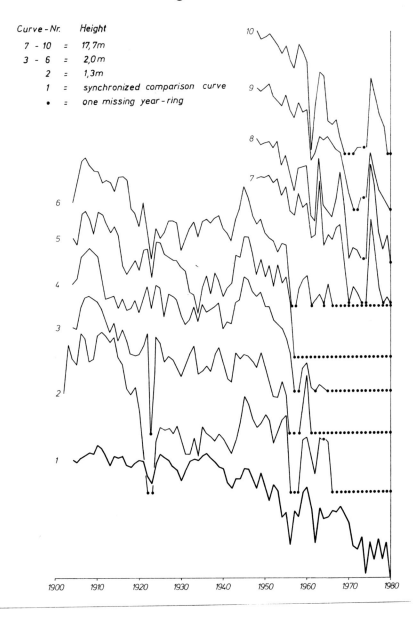

Curve - Nr.	Height
7 - 10 =	17,7m
3 - 6 =	2,0m
2 =	1,3m
1 =	synchronized comparison curve
• =	one missing year - ring

Figure 1: Damage related losses of year-rings

trees damaged by air pollution, insect attacks, etc. is, that some living trees have not developed visible year-rings in certain years. In order to determine the yearly growth, the measured year-rings must be dated. The number of missing year-rings can supply information about the degree of damage of the analysed trees. The missing of year-rings are rare in the crown area. By comparison of annual growth curves from the upper and the lower part of the trunk, and with the help of a previously synchronized comparison curve, it is possible to date the year-rings.

Figure 1 shows the growth curves at different heights of an 80 year old Norway Spruce from Forstamt Neuhaus (Solling). In order to avoid interference between curves, the origins of the logarithmic ordinates are shifted upwards for the different disks. This vital looking tree apparently had gone through periods without growth in the last 24 years. This tree has the highest number of missing year-rings known to us in central Europe. It is the first indication that severely damaged trees might not form year-rings in the crown area.

In order to obtain growth curves at different tree heights, a stem analysis is essential. The limitation of research to the lower trunk sections would lead to an overestimation of growth losses because it would leave the real growth in the higher trunk sections unconsidered.

In the following, preliminary results of stem analysis of a Spruce-Liming-Experiment in the Staatlichen Forstamt Neuenheerse (Nordrhein-Westfalen) are presented. The trees of the limed stand are compared with those of an unlimed stand. This research supplies meaningful information about the effect of liming to acid soil on growth. The liming was done in April of 1964 with 5000 kg/ha "Hüttenkalk".

The growth determination was initially limited to four stem analysis per plot. Dominant and predominant trees were taken as sample trees. To compare the soil conditions of both plots, soil samples were taken near each sample tree and the pH of the soil was measured. The unlimed plot had a mean pH (CACL2) of 3.5, the limed plot a pH of 3.9.

After measuring the radial increment of the disks, each increment was dated and measurement errors were corrected. Only one tree in the unlimed plot had 2 missing year-rings. The development of radial growth of the limed and unlimed trees at a height of 1,3m is shown in figure 2.

It is clear that both groups of trees have very similar growth variation. In the 22 year period before the liming (1942 -

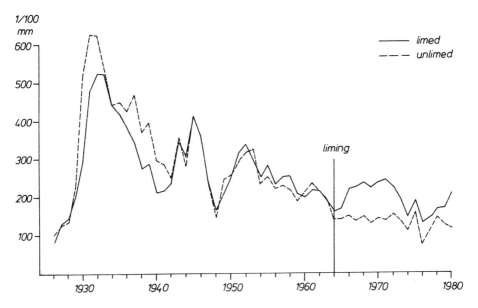

Figure 2: Yearring diagrams of the limed and unlimed tree
         groups at a height of 1,3 m

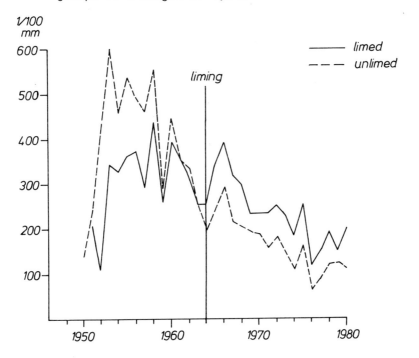

Figure 3: Yearring diagrams of the limed and unlimed tree
         groups at a height of 14,3 m

1963), both groups had equivalent growth. After the liming in 1964, the limed trees showed increased growth up to the end of the observed period in 1980.

In figure 3 the growth development of both groups of trees is shown at a trunk section of 14,3 m height. The positive effect of the liming on the growth at 14,3 m is clear and does not show a tendency to fall off even 17 years after the liming. This effect of liming could also be observed at other heights. This concurs with the results from JOHANN (1980). The better growth can be observed at every height of the trunk.

The growth development of the limed and unlimed trees have the same age trend and the growth (fig. 2) is very similar before liming in 1964. The assumption that the radial growth of both tree groups should remain the same if no new influencing factor appears, seems justified. The obvious difference in the growth development before and after liming can, therefore, be attributed to the liming. A relative comparison of both tree groups can illustrate this.

The relative increment of limed and unlimed trees was calculated as percent of the mean annual increment before liming (1959 - 1963). Figure 4 shows the percentage of these relative

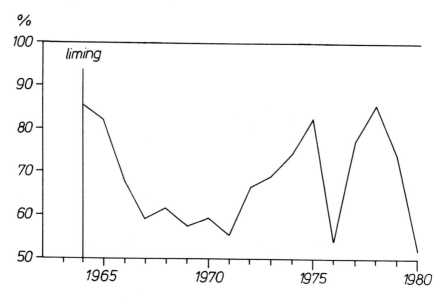

Figure 4: Relative radial increment of the unlimed trees at 1,3 m

increments unlimed to limed trees. The unlimed trees show great losses of increment in comparison to the limed trees. The effect of the liming is already visible in the first year after liming.

The mean annual increment of the unlimed trees since 1964 is 69% of the mean annual increment of the limed trees in the lower trunk section.

When comparing the mean basal area increment of the same trees the effect of the liming is also clear. In the period 1964 - 1980 the unlimed trees had 65% of the basal area of the limed trees at a height of 1,3 m. The increase in the volume of the limed and unlimed trees should be similar to that of the basal area growth.

From this preliminary research, we can suppose that the growth losses are often higher than can be expected from the outward appearance of a forest. An exact assertion demands precise stand - and stem analysis. For the future it is advisable to establish permanent sample plots.

REFERENCES

ATHARI, S. 1981: Jahrringausfall,ein meist unbeachtetes Problem bei Zuwachsuntersuchungen in rauchgeschädigten und gesunden Fichtenbeständen. Mitt.d.Forstl.Bundes-Versuchsanstalt Wien, 139, 7-27

JOHANN, K. 1980: Zuwachs in verschiedenen Schafthöhen ausgewählter Analysebäume nach Düngung. Mitt.d.Forstl.Bundes-Versuchsanstalt Wien, 130, 109-127

FIRST INFORMATION ABOUT INVENTORY OF EMISSION-DEPENDING DAMAGES
ON NORWAY SPRUCE IN LOWER SAXONY/FED. REP. GERMANY

Dr. J. B. Reemtsma

Forest Experiment Station of Lower Saxony
3400 Göttingen/Federal Republik of Germany

ABSTRACT
Paper shortly deals with projects of the experiment station
on immission study as inventory as investigation. Regional
distribution is explained and a first rough estimation of
the extend of probably immission dependend damages on spruce
forests is given.

Due to a more or less dramatic increase of forest damages
during the last years, our experiment station has the order
to survey as the type of damage attributed to immissions or
of unknown cause as the area of its occurrence in Lower
Saxony. Because we are quite short of personel and there is
around one million hectars of forests to cover, the survey
runs at various levels of intensity and accuracy. Projects
1-4 are mainly designed to provide quick information for the
forest administration, the government, the parliament and
the public. All work will be done mainly on spruce since we
don't know clear and visible symptoms on pine and broadleave
trees up to now.

1st project: Going by car into the forest in different
parts of the whole area, we try to get a
general idea of extend and distribution of
damages. First results will be given later.

2nd project: In addition to the normal inventory work an
area of around 40.000 ha of forests in the
southern and high Harz mountains will be
evaluated by the inventory staff. Each spruce

B. Ulrich and J. Pankrath (eds.), Effects of Accumulation of Air Pollutants in Forest Ecosystems, 327–330.

stand of more then 40 years of age will be
classified for the degree of damages and for
the portion of damaged trees in the classes of
predominants and dominants. To do it in a
simple way, we classified into 9 groups as
shown in table 1.

Table 1: Classification for degree of damage
and portion of damaged trees

degree of damage: loss of needles		portion of damaged trees in predominants and dominants	
41	<30%	1	<30%
42	30-60%	2	30-60%
43	>60%	3	>60%

First results will be available next winter.

3rd project: In the Hils-mountains - covered by around
10.000 ha of forests - infra-red-pictures are
taken by plane and compared to a ground-inven-
tory as described before. Pictures will be
evaluated by Prof. KENNEWEG of the Institute
of Inventory of the forest faculty, ground
classification will be carried out by the
experiment station.

Until winter we hope to know if we can work
better or faster by aerial photography than by
ground inventory. This work seems most important
for the private forests where no forest maps
are available but whose portion is around half
of all forests. In addition: it is the only way
of documentation of the actual situation for
comparison with later investigations and
developments.

4th project: Randomly selected areas distributed over the
whole state will be evaluated by ground classification as described before. The accuracy of
this level is between the 1st evaluation by
car and the complete inventory of project 2
and 3. From this project we hope to get a rough
estimate of degree and extend of damages on an
area basis. Because of our lack of experience
I can't estimate the time needed to get realistic
results.

5th project: At randomly selected places sample trees are
uprooted and investigated for damages on
needles, stems, roots, for pH in different soil
layers of the main rooting zone, for soil
description etc. in form of a check list.
Depending on the funds, samples for analyses
of needles, ring-width and soil will be taken.
The type of analyses is in discussion with
Dr. ULRICH. This project should provide parameters for differentiation of unspecific
symptoms or by other biotic and abiotic influences.

6th project: Simple experimental plots will be installed to
compare limed and unlimed plots on a basis of
5 t/ha Calzium carbonate or silicate, to
various tree species, soil conditions and
immission. The distribution of these plots
depends in the hilly and mountanous area of
the south of Lower Saxony on the probably
given different immission intensity. In the
northern lowlands it depends on a systematic
distribution. We are planning ca. 20 experimental areas x 4 trees species x two treatments, all together around 160 plots.

Up to now there is no precise experiment design.
These plots are open also for all other people
interested. Possible investigations are for
example input, acidification, leaching, development of damages, growth, silvicultural
treatments. Oberservation time f.e. three
decades.

That is, what we started so far or we will start soon.

The symptoms are explained by some other authors in this
meeting. They are not quite specific. They are starting on
the best, the predominant trees. Nowhere has been written
up to now that best trees are dying first in a stand, except
by air pollution. On the other hand: stands with worst
damages we find on poor and exposed sites, esp. on western
borderings of stands. This is not very specific but logical.

The country wide distribution seems to be as follows:
The southern portion of Lower Saxony south of line Osnabrück-
Hannover-Berlin is a hilly and mountainous area. There are
2/3 of all spruce stands. This area is strongly influenced
by southwestern and western wind-directions from the Ruhr-
gebiet, the highest industrialised area of Germany,

additionally by a lot of industries in Ost-Westfalen and
local. Tops and ridges of nearly all mountains and higher
hills are medium or highly damaged. The middle and lower
elevations mostly have lower damages, and only protectet
sites show no visible symptoms. There is a strikt relation
to geomorphology rather then to site quality.

In the middle of the country between the line Hannover-
Berlin and line Bremen-Hamburg there seem to be mostly low
or zero grades of damage, except in local exposed positions
and rests of storm-thinned stands which have middle grades.
There seems to be a trend of decreasing grades from west to
east.

In the north of the line Bremen-Hamburg damages are increasing probably because of industries on the lower Weser-
river. In this area middle grades of damage are found much
more often then in the middle area.

In the lowlands of the northwest the situation is still
unclear because there are few forests and a small portion
of spruce in it. Additionally most of the old stands we
lost by a storm 1972 and in the new plantations symptoms
are quite uncertain. But we know it is not a healthy world.

For spruce stands of more than 40 years of age my personal
first rough estimate at the present, very incomplete level
of visitation, gives the following picture:

zero grade: probable damage on soil   10-20%
                     without actual symptoms
                     on stands,

lower grades: up to 60% in grade 1 or   55-70%
                     up to 30% in grade 2 of
                     Tab. 1,

middle grades: more than 60% in grade 1,   10-20%
                     30 to 60% in grade 2 or
                     below 30% in grade 3,

worst grade: it must be exspected that   5-10%
                     stands will dy latest in
                     the next decade, if things
                     continue to develop like in
                     the last years.

DIE-BACK OF RED SPRUCE, ACID DEPOSITION,

AND CHANGES IN SOIL NUTRIENT STATUS - A REVIEW

G.H. Tomlinson II

Domtar Inc., Montreal

Die-back and death of red spruce is now occurring on non-calcareous soils in the Adirondack Mountains, the Green Mountains of Vermont, and the Tremblant area of the Laurentian Mountains of Quebec. The die-back syndrome involves a slowing of growth over a period of a number of years with a loss of foliage, initially at the crown, and subsequently moving down the tree. Examination of the roots shows death of the fine feeder roots, and particularly the absence of the short branch roots. Trees of all ages are affected.

A literature review indicates strong support for a hypothesis, largely based on Ulrich's studies in Germany [1], which, through a series of intermediate stages, links SO_2 emissions with this spruce die-back. Ulrich has called attention to the significance of the fact that only a portion of the SO_2 entering the atmosphere returns to the earth as sulfuric acid in rain. The balance, remaining in the gaseous state, is "dry deposited" on various surfaces. The tree canopy provides an important sink for gaseous SO_2. As it is carried in the wind from its many sources, it is adsorbed and absorbed on the bark, leaves and needles, oxidized to sulfuric acid, and, with or without partial neutralization from cations present in the canopy, is washed to the forest floor with rain. Air-borne dust and acid aerosols are also deposited and washed from the trees in a similar manner. The combined effect of acid already contained in the rain and that formed on the trees, and then washed from them, results in a series of changes after entering the soil. As the soil acidifies, its buffering capacity is reduced, nutrient cations are leached from the rooting zone, and aluminum ions, which are toxic to tree roots, are formed from the minerals present, and enter the soil solution.

B. Ulrich and J. Pankrath (eds.), Effects of Accumulation of Air Pollutants in Forest Ecosystems, 331–342.

These changes are superimposed on the normal acidification - deacidif- ication phases which occur in the soil, i.e. those resulting from nitrogen cycling. When protein in the humus breaks down, nitric acid is formed. However, since nitric acid is assimilated by plant and tree roots to form the protein required in their growth, no net change in acidity takes place when these are in balance. However, humus break-down which is a biotic reaction is very dependent on temperature and cyclic weather patterns. During hot dry years, the excessive quantity of nitric acid formed from the humus results in a major "acidification push" which is followed by a de-acidification phase in subsequent cool moist years when humus accumulates. Seasonal acidity changes during the course of a year resulting from different periods of maximum decomposition and growth, and even major weather-induced uncouplings, which occur at several- year intervals, can apparently be accommodated in a well buffered soil. However, the continuing superimposed one-way acidification resulting from acid deposition, with resultant reduced buffering of the soil, makes natural recovery more difficult.

Ulrich (2) has demonstrated that with Norway spruce, a soil solution containing a molar ratio of Ca^{2+}/Al^{3+} of less than 1 at a pH of less than 4.0 can result in damage to the fine feeder roots, as indicated in Fig. 1. Unless new feeder roots are established during succeeding years, the stressed tree will show reduced growth, lose foliage and eventually die. During the period of reduced growth, the rate of assimilation of nitric acid is also reduced, this resulting in an additional acidification factor resulting in further loss of nutrients from the rooting zone. There is considerable evidence that this sequence is now occurring in North America, and that this syndrome is showing with red spruce in acid impacted non-calcareous areas.

On Camel's Hump Mountain in Vermont, the basal area of red spruce was measured in 1965, and re-measured in 1979, showing a reduction from 6.7 m^2/ha to only 3.7 m^2/ha. In 1979, the basal area of dead trees was more than twice that of those still alive (3).

Figure 2 illustrates the changes in rates of growth measured by Johnson (4) on 100 to 220 year old red spruce on Camel's Hump. Stress, as indicated by the rapid reduction in growth rate which started in the mid sixties, followed a period characterized by unusually hot dry weather. The pattern of reduced growth occurred during the same years for all trees, regardless of their age, indicating an environmental effect, rather than one related to age alone.

A further important series of experiments has been reported by Raynal et al (5) and these will be reviewed in relation to Ulrich's studies. They also observed a decrease in growth rate, starting in the mid sixties, of red spruce in the Huntington Forest of the Adirondack Mountains of New York State where spruce die-back is now occurring. Table 1 shows data on the soil in this site. The pH of the humus layer and the A Horizon of the mineral soil are low, being 3.4 and 3.6 respectively. In the A

Horizon, the exchangeable calcium and total exchangeable base values are only 0.08 and 0.17 meq/100 g respectively. Soils having a pH of about 3.5 are particularly vulnerable to continuing SO_4^{2-} and H^+ inputs since these inputs result in a decreased inventory of nutrient cations such as Ca^{2+} and Mg^{2+} and an increased inventory of toxic Al^{3+}.

Raynal et al (5) measured inputs to the soil from bulk precipitation and throughfall, and also measured seepage losses using lysimeters set at various depths in the soil. The total sulfate deposited on the forest floor amounted to 61.8 kg/ha/year, the quantity in throughfall being 1.6 times that in bulk precipitation. These investigators found that SO_4^{2-} is not retained in the mineral soil, and that this ion carries with it substantial amounts of Ca^{2+} and Mg^{2+} from the rooting zone.

It is of considerable interest to estimate the loss of exchangeable calcium as a percentage of its total inventory above various depths in the soil. Table 2 illustrates the successive stages in the estimation. It will be noted from Column B that the quantity of calcium deposited on the forest floor was considerably greater in the throughfall measured under the conifers than in the bulk precipitation measured in the open. Previous studies of Abrahamsen et al (6), indicate that throughfall washes "dry deposits" of SO_2, aerosol, and mineral dust from the canopy, and that metabolites can be absorbed or leached from it. However, to avoid overstating the net loss of calcium from the soil horizons of the Huntington Forest (Column C), it was assumed that all of the calcium in throughfall was derived from the atmosphere. Also, because the 105 day test period was relatively short, the loss, calculated to 365 days, was reduced to 75% of its value, again to avoid overstating the loss (Column E). The annual loss of exchangeable calcium from each of the horizons is over 10% of the total inventory in that horizon plus the quantity in the horizon or horizons above it. This is an exceedingly high loss rate, particularly in view of the low calcium inventory of the soil.

Cowell et al (7), referring to the role of the upper horizons in relation to soil sensitivity stated, "A soil depth of 25 cm is considered important because this is the zone of maximum nutrient cycling and uptake by root systems. ---- Exchangeable base loss is considered a better measure of sensitivity than cation exchange capacity." These investigators further state, referring to soils having a pH of less than 4.5 that, "In this situation, any further loss of cation is considered significant, however small the loss may be." These authors in their Table 13, indicated that a loss of 25% of exchangeable base in 25 years, or an annual loss of 1% of the inventory, would be a sign of a sensitive soil. On the basis of their criteria, the Huntington Forest soil with a loss of over 10% of exchangeable calcium per year, is indeed endangered.

The reasons why these low pH soils have such special sensitivity as contrasted with those of higher pH can be best understood from a comparison of the specific separate effects of acid inputs on soils of various pH levels, as described by Ulrich (8).

Calcareous soils having a pH above about 6.5 are in the "Carbonate Buffering Range", and are largely buffered by $CaCO_3$. The effect of acid, either H_2CO_3, resulting from reaction of CO_2 with water, or H_2SO_4, results in the release of Ca^{2+} as follows:

(a) $2 CaCO_3 + 2H^+ \longrightarrow 2 Ca^{2+} + 2 HCO_3^-$

Soils having a pH in the range of about 5.0 to 6.5 are in the "Silicate Buffering Range" and are buffered by feldspars such as orthoclase ($K_2O.Al_2O_3.6SiO_2$), anorthite ($CaO.Al_2O_3.2SiO_2$), etc. The action of acid, either H_2CO_3 or H_2SO_4, results in the release of nutrient cations, as indicated by the following typical reaction:

(b) $CaO.Al_2O_3.2SiO_2 + 2H^+ + H_2O$

$\longrightarrow Ca^{2+} + Al_2O_3.2SiO_2.2H_2O$
    (kaolinite)

Soils having a pH in the range of about 4.2 to 5.0 are in the "Cation Exchange Buffering Range". The negatively charged kaolinite (clay) attracts, and holds in storage for subsequent release, the positively charged cations, namely Ca^{2+}, Mg^{2+}, K^+, Na^+, NH_4^+, Al^{3+} and H^+. The clay acts in a manner similar to ion exchange resins such as those used in water softeners. As H_2SO_4 is added to the soil, the ratio of H^+ to other cations increases, and the H^+ ion is exchanged for Ca^{2+} or other cations held by the clay, and the released Ca^{2+}, together with the SO_4^{2-}, is leached from the rooting zone of the soil. Fig. 3 illustrates the effect of increased H^+ ion input on the base saturation of soil. At pH 5.0, 60% of exchange sites would be occupied by the base cations such as Ca^{2+}, Mg^{2+}, etc. As increasing inputs of H^+ enter the soil, the Ca^{2+} is replaced by H^+ and at pH 4.2 only about 8% of these sites are occupied by the base cations. If the pH of the soil is subsequently raised, base cations, resulting from breakdown of litter and humus, percolating through the soil will again be retained on sites previously occupied by H^+, only to be lost from the ecosystem as a result of subsequent acid inputs.

When the clay is held over long periods in the Cation Exchange Buffering Range, hydroxy aluminum compounds are formed from the kaolinite:

(c) $Al_2O_3.2SiO_2.2H_2O \longrightarrow 2 AlOOH + 2SiO_2 + H_2O$

Soils having a pH value of less than 4.2 are in the "Aluminum Buffering Range", in which Al^{3+} is released by reaction of H^+ with aluminum hydroxy compounds as indicated:

(d) $AlOOH + 3H^+ \longrightarrow Al^{3+} + 2H_2O$

In this range, a decrease in acidity would result in conversion of Al^{3+} into one of the many aluminum hydroxy compounds, while subsequent increased acid inputs would again result in formation of Al^{3+} which is toxic to feeder rootlets as indicated in Fig. 2.

As is apparent from the above, calcium and other nutrient cations are contained in soil minerals such as limestone and feldspars, and can be released in the so-called "weathering reaction" by the action of carbonic acid (H_2CO_3) formed from CO_2 and H_2O. However, at pH values below about 5.0, carbonic acid breaks down to H_2O and CO_2, with CO_2 leaving as a gas as indicated:

$$(e) \quad H_2CO_3 \rightleftharpoons H^+ + HCO_3^- \longrightarrow H_2O + CO_2\uparrow$$

For this reason, loss of nutrients in the Cation Exchange Buffering Range, at pH values below 5.0, cannot result from the action of carbonic acid. In contrast, input of sulfuric acid not only accelerates the release of nutrients in soils at pH above 5.0, but unlike carbonic acid, will reduce the pH into the Cation Exchange Buffering Range, resulting in a major reduction of the inventory of the nutrient cations.

At the still lower pH range of 4.2 or less in the Aluminum Buffering Range, feldspars and other nutrient cation-containing minerals become essentially exhausted from the fine soil structures, as a result of continuing SO_4^{2-} inputs and nutrients become increasingly less available from this soil. Moreover in granitic rock, the concentration of calcium is substantially less than that of aluminum, and thus the release of cations resulting from the action of continuing sulfuric acid inputs cannot restore a favorable cation balance. Nutrient cations released by mineralization of the litter and humus, unless re-absorbed in active feeder roots, will be leached by the SO_4^{2-} through the rooting zone from the nutrient pool of the ecosystem. Thus, as sulfuric acid inputs continue with the passage of time, the molar ratio of Ca^{2+}/Al^{3+} will continue to decrease in soil solutions in the Aluminum Buffering Range. As can be seen from Fig. 1., under these conditions, the fine roots in the mineral soil become increasingly stunted and less able to take up the Ca^{2+} and other nutrients that must be extracted from the mineral soil to make up for the leaching losses from the litter and humus layers.

In the Huntington Forest, the molar ratio of Ca^{2+}/Al^{3+} in the soil solution of the A Horizon was found to be 0.84, with a pH of 3.6, while the ratio in the combined B Horizons is 0.70 with a pH of 4.2 in the upper B 21h Horizon and 4.7 in the lower B 23 Horizon (9). Raynal et al (5) in their Table 3-2 noted that the fine roots were missing from the mineral soil, and subsequent examination of the roots in the humus layer showed obvious damage.

In an actively growing forest, nitric acid atmospheric inputs to the soil are normally largely assimilated by the roots for conversion to protein needed for growth of the trees. Such inputs, resulting from oxidation of

NO_x, act as a fertilizer. However, as noted previously, during hot dry weather spells, nitric acid formation from the humus can exceed the trees' ability to reassimilate this and an "acidification push" takes place. If a non-calcareous soil is in the Silicate Buffering Range, which would be the normal situation in pre-industrial times with low atmospheric acid input, where release of nutrient cations from the soil results from only carbonic acid weathering, little damage would be expected. The soil acidity would be later reduced in cool moist years and any loss of Ca^{2+} from the rooting zone would be made up from weathering of minerals and by the calcium in rain and dust carried by the wind from calcareous areas. Unfortunately, in impacted zones, such as the Huntington Forest, calcium inputs from rain and dust are now pre-acidified as indicated by the acidity of rain and the dust-containing throughfall under the trees, and thus are not retained in the soil.

When an "acidification push" occurs with soil in the Aluminum Buffering Range, the non-assimilated NO_3^- will carry with it the nutrients released from litter and humus from the upper soil horizons. At the same time, the H^+ releases Al^{3+} from the hydroxy aluminum compounds present in the soil. Under these conditions, the roots in the mineral soil may become permanently damaged. The absence of active fine roots in the mineral soil eliminates the upward movement of calcium and other nutrients, released by weathering, which would otherwise cycle through the tree, foliage and litter to the humus. Future growth of the stressed tree will thus be controlled by the rate of assimilation, through roots remaining above the mineral soil, of those nutrients not leached by rain, from the decomposing previously formed humus. This appears to be the situation with the moribund spruce in the Huntington Forest, where the rate of tree growth has decreased, and the trees are dying.

Weetman et al (10) calculated that in a red spruce and balsam fir forest in Quebec, the calcium content of the trees, including foliage and roots, was 530 kg/ha. Approximately this amount should obviously be present in the soil to allow regeneration of a new viable forest. Since the potential rooting zone of the A Horizon in the Huntington Forest has a pH of 3.6 and is in the Aluminum Buffering Range, the original stores of calcium in the fine soil structure have been largely exhausted from the minerals present and from ion exchange sites. The calcium formed from breakdown of litter and humus is being rapidly leached from the already low inventory in the organic horizons as can be seen from Table 2. Only 61 kg/ha of calcium remain in the humus layer and, quite clearly, insufficient calcium is present to allow regeneration of a spruce forest to replace the moribund and dying trees at this location.

In addition to the inorganic acids, organic acids, formed during the decomposition of litter and humus, are also present in the soil. Humic acids, which have a relatively high molecular weight, are less damaging in the soil than the inorganic acids, since they convert aluminum to a chelated form which is less toxic to roots. The lower molecular weight organic acids are largely broken down to CO_2 and H_2O within the normal rooting zone and therefore do not have the same leaching action as the "mobile" sulfate ion.

Unfortunately, atmospheric inputs of sulfuric acid have a year by year accumulative effect on the chemical composition of the soil. For instance, at the present rate of input in the Huntington Forest, over 6 tons of sulfuric acid per hectare will enter the soil during a 100 year span in the lifetime of a tree.

It is important that the significance of these changes, and the damage that can result for present and future forestry, become more generally recognized.

BIBLIOGRAPHY

1.   Ulrich, B., "Die Walder in Mitteleuropa: Messergebnisse Ihrer Umweltbelastung, Theorie Ihrer Gefahrdung, Prognose Ihrer Entwicklung".   Algemeine Forstzeitschrift, Vol. 44, 1980, (Translation available, G.H. Tomlinson).

2.   Ulrich, B., "Destabilisierung von Waldökosystemen durch Akkumulation von Luftverunreinigungen", Der Forst und Holzwirt, 36, 525-532, 1981.

3.   Siccama, T.G., Bliss, M., and Vogelmann, H.W., "Decline of Red Spruce in the Green Mountains of Vermont".   Manuscript for Publication (Available from H.W. Vogelmann, Dept. of Botany, Un. Vermont, Burlington, VT).

4.   Johnson, A.H., "Assessing the Importance of Aluminum as a Link Between Acid Precipitation and Decreased Forest Growth", Presented at USEPA Acidic Deposition Effects Program, Raleigh, N.C., Feb. 1982.

5.   Raynal, D.J., Leaf, A.L., Manion, P.D. and Wang, G.J.K., "Actual and Potential Effects of Acid Precipitation in the Adirondack Mountains".   Dec. 1980, 56/ES-/HS/79.   N.Y. State Energy Research and Development Authority.

6.   Abrahamsen, G., Horntvedt, R., and Tveite, B., "Impacts of Acid Precipitation on Coniferous Ecosystems".   Proc. of 1st Int. Symposium on Acid Precip. and the Forest Ecosystem, Dochinger and Seliga, U.S.D.A. Forest Service, Report NE 23, Upper Darby, Penn.

7.   Cowell, D.W., Lucas, A.E., and Rubec, C.D.A., "Development of Ecological Sensitivity Rating for Acid Precipitation Impact Assessment".   Working Paper No. 10, Land's Directorate Environment Canada, Catalogue No. EN 13-4/10E Ottawa, March, 1981.

8.   Ulrich, B., "Gefahren für das Waldökosystem durch Saure Nieder-
     schläge" in Immissionsbelastungen von Waldökosystem, published by
     Landesanstalt für Okologie Landschaftsentwicklung und Forst-
     planung Nordrhein-Westfalen.   4350 Recklingshausen, Germany,
     1982.

9.   Molitor, A.V., and Raynal, D.J., "Acid Precipitation and Ionic
     Movements in Adirondack Forest Soils". Soil Sci. Soc. of Amer. J.,
     46, 137-141, 1982.

10. Weetman, G.F., and Webber, B., "The Influence of Wood Harvesting
     on the Nutrient Status of Two Spruce Stands". Can. Journ. For.
     Res. 2, 351-360, 1972.

11. Coote, D.R., and Wang, C., "The Significance of Acid Rain to
     Agriculture in Eastern Canada". Land Resource Research Institute
     Contribution No. 19, Agriculture Canada, Ottawa.

TABLE 1. SOIL CHARACTERISTICS AND EXCHANGEABLE CALCIUM CONTENT OF SOIL IN HUNTINGTON FOREST OF THE ADIRONDACKS DATA OF RAYNAL ET AL (5)

Horizon	Depth cm	pH	Organic Matter %	Cation Exchange Capacity meq/100 g	Exchangeable Ca^{2+}		
					meq/100 g	Saturation % of CEC	kg/ha
O1 (litter)	20-16	4.3	92.2	61.4	2.32	3.8	9.3
O2 (humus)	16-0	3.4	83.8	108.9	1.07	1.0	51.4
A2	0-15	3.6	1.2	3.8	0.08	2.1	26.4
B21h	15-20	4.2	14.8	29.0	0.14	0.4	15.4
B21hir	20-34	4.9	15.9	38.2	0.02	0.05	6.2
B23	34-40	4.7	6.2	15.6	0.03	0.2	4.0
C IX	40-90	4.7	5.1	5.1	0.05	1.0	55.0

TABLE 2. FLUX AND CALCULATED NET LOSS OF EXCHANGEABLE CALCIUM FROM EACH HORIZON AT THE HUNTINGTON FOREST CONIFER SITE, BASED ON AUGUST 8TH TO NOVEMBER 21, 1979 (105 days). DATA FROM RAYNAL ET AL (5) TABLE 13-3

	Total depth including over-lying horizons	105 day transfer from eco-system strata	105 day net loss based on 2.6 kg/ha through-fall input	Calculated net loss for 365 days	Estimated loss, 75% of Column D	Inventory including overlying horizons	Annual loss as % of inventory
	cm	kg/ha	kg/ha	kg/ha/365 days	kg/ha/365 days	kg/ha	
	A	B	C	D	E	F	G
Deposited on Soil							
Bulk Precipitation		0.5					
Throughfall		2.6					
Seeping from Soil							
O Horizon	20	5.2	2.6	9.0	6.75	60.7	11.1
A Horizon	35	8.5	5.9	20.5	15.38	87.1	17.7
B Horizon	60	6.8	4.2	14.6	10.95	102.5	10.7

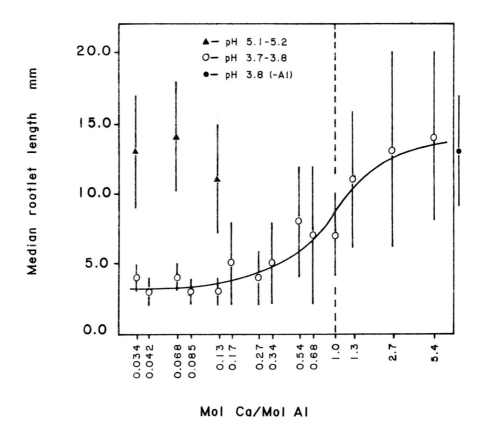

FIG. 1. Effect of pH and molar ratio of Ca/Al in soil solution
on length of branch rootlets of Norway spruce (From
Ulrich (2))

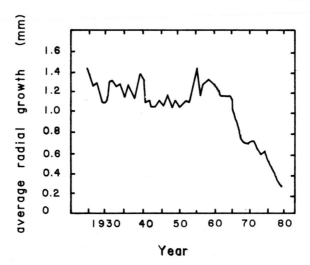

FIG. 2.   Growth rate of Picea Rubens in the Boreal forest zone on
          camels hump. Line represents the mean of the 10 oldest
          trees cored in June 1980. Tree ages ranged from 100 to
          220 years. (From Johnson (4))

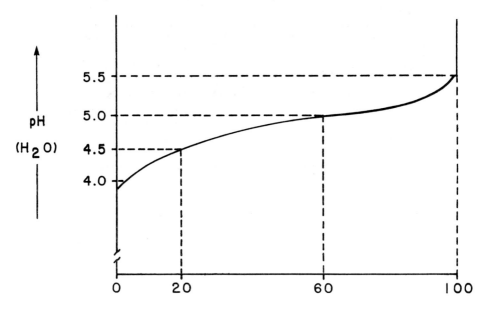

FIG. 3.   Relationship between soil pH (H_2O) and percent base
          saturation (From Coote et al. (11))

SCOTS PINE-DYING WITHIN THE NEIGHBOURHOOD OF AN INDUSTRIAL AREA

Karl Kreutzer, Andreas Knorr, Franz Brosinger,
Peter Kretzschmar
Lehrstuhl für Bodenkunde, University of Munich

ABSTRACT

A Scots pine dying close to an industrial area with brass processing plants was investigated. Comparative studies concerning soil
properties and nutrition status of selected pine stands inside and
outside of the damage area indicated that Cu and/or Zn immissions
are the most probable cause. Within the damage area a strong enrichment of HNO_3 - extractable Cu and Zn could be found in the surface humus and in the mineral top soil layer. The contents amounted in the F + H layer up to 3000 µg/g, in the uppermost mineral
soil up to around 90 µg/g. Likewise the Cu and Zn contents in the
needle dry matter were unusually high (Cu 22-80 µg/g, Zn 130-247
µg/g). Threshold ranges for toxicity could not be worked out because a part of the Cu and Zn in the needles seems to be physiologically not active.

The symptoms of the damage were discribed as well as the course
of the diameter growth. Obviously before external symptoms occur
the diameter growth is decreasing. Sound appearing trees within
the damage area have a reduced diameter growth as well as raised
Cu and Zn contents in the needles.

1. INTRODUCTION

In the following a report is given on the results of a preliminary
investigation of Scots pine-dying which occurred in the vicinity
of an industrial area near Nuremberg. The damages were observed
for the first time in the early seventies.

B. Ulrich and J. Pankrath (eds.), Effects of Accumulation of Air Pollutants in Forest Ecosystems, 343–357.
Copyright © 1983 by D. Reidel Publishing Company.

The investigations, performed from 1974 to 1976, should bring out
a first approach to the causes of the damages by comparing soil,
growth and needle data from inside and from outside of the damage
area. The aim was to give some shortterm recommendations to the
forest practice and to the plant managers in order to avoid further damages on the pines at this place, which was possible and
somewhat satisfying, though the mechanisms of the destructive factors are not understood completely in detail. Therefore some new
investigations are intended, also in other places, where similar
symptoms occur.

2. THE DAMAGE AREA

Fig. 1 shows the result of a mapping of the area.[*] The damaged
forest stands are close to the industrial area, with several brass
processing plants. The main damaging substances emitted from the
plants were supposed to be metallic dusts, containing copper and
zinc, besides the emissions of SO_2, HF, NO_x, Calciumcarbide and
organic compounds as acetylene. An exact list of the emissions was
not available.

Along the highways (Autobahn) road-side effects (caused by deicing
salts, tire abrasion substances or exhaust gases) might have inter-
fered, but could not be made responsible for the whole damage in
the area as the map is showing a good orientation of the damages
to the industrial area.

3. SYMPTOMS

The trees are slowly dying, shedding the needles and little twigs.
For mapping so-called "heavily damaged trees" were considered to
have lost the needles from more than one third of the crown, whether in one piece, mostly from the top, or scattered all over the
crown, retaining until next season only a part of the needles from
the last year. But it is somewhat surprising that heavily damaged
trees are mixed with those looking quite healthy, in the following
called as "sound appearing trees in the damage area".

The buds of the obviously damaged trees are breaking regularly.
Within the first weeks they do not differ in colour from the needles of healthy trees, but they are smaller. During the summer chlorotic stains and spots occur on them: The needles become yellow-
green mottled or yellow-tipped with yellow edges without sharp demarcations between green and chlorotic. These needles have also a
water content than healthy ones.

[*] The mapping was performed by Forstrat Kienzler, Oberforstdirektion Ansbach

In late summer necrotic spots occur in the centres of the chlorotic
stains and tips, often a yellowing is observed at the base of the
needle.

During the winter the necrosis is spreading. A part of the needles
is dying. The surviving ones often develop sharp demarcations between green and necrotic spots, cross stripes or bases. But towards
the end of their second season nearly all of them are dying and shed.

Unlike of the needles of these damaged trees those of the "sound
appearing ones" remain green or become only to a much less extent
chloroticly mottled in the first season, or weakly necrotic spot-
ted in the second (and third) one.

These symptoms correspond to damages in the root system: In the
early season about 40 to 60 % of the fine roots of heavily damaged
trees are dead, whereas sound appearing trees only have about 20 %
dead fine roots.

4. THE COMPARING STUDIES

4.1 Sites and Stands

Within the damage area mainly three site units occur ("Sand",
"Sandstone", "Loam"). On each site unit three older stands, rather
similar in their phenotype, were selected for investigation. The
same selection was performed outside of the damage areas (five to
six kilometers away) on the same site units using stands of similar age and similar management as within. Thus a paired comparison
could be achieved for each site unit between the presumably charged
and uncharged area.

> Site unit "Sand": Rather flat podsols derived from di-
> luvial middle and coarse sands, very poor in silicates.
> Raw humus layer rather thin, because of former litter
> utilisation. pH see Fig. 2.

> Site unit "Sandstone": Likewise flat podsols, derived
> from coarse grained sandstone, with similar humus form
> and acidity.

> Site unit "Loam": Mainly podsol-pseudogley soils derived
> from a layered substratum: Loamy sand above clayey layer,
> intermittend with stagnant water table. The deeper sub-
> soil in the clayey layer is marly. Raw humus, similar to
> that before, as well as acidity in the topsoil.

The mean annual precipitations amount to 750 mm, the mean annual
temperature to 8,4 °C.

4.2 Methods

a) Soil investigations. At several depths of the soil profile the
   HNO_3-extractable fraction of Cu, Zn, Ca, K, Na and Mg was determined according to Westerhoff 1955 (2); supposing that this
   fraction is also containing the main part of secondary enriched
   amounts especially of heavy metalls, as it is proved for Cu (4).
   The pH-values were measured in a 1 m KCl solution.

b) Growth investigations. The course of the diameter growth was
   investigated from 1950 to 1974. The different grwoth levels of
   the stands caused by differences in site, age or stand density
   were eliminated by relating the mean growth of the stands to an
   average growth during a period before an impact was supposed.
   As within the damaged area the trees differed very significantly
   in their health appearance, for growth investigation two groups
   of trees from each damaged stand were selected: a) heavily damaged and b) sound appearing ones. Each group per stand consisted of 10 trees, thus 270 trees were   analyzed by measuring the
   yearrings in two directions (1).

c) Needle analysis. In the needle dry matter the contents of N, P,
   K, Na, Ca, Mg, S, Si, Al, F, Mn, Fe, Cu and Zn were determined
   according to proved methods of needle analysis. As well the needle dry weight was measured. Before drying, the needles were
   washed three times with destilled water. Within the damaged
   stands the same two groups of trees as for the growth investigation (heavily damaged and sound appearing) were separately
   investigated (1).

4.3 Results of soil investigations

Within the same site unit no essential differences could be found
between inside and outside of the damage area for the pH and for
the Na and K contents in all investigated soil layers. The Ca contents, however, showed within the damage area a tendency to higher
values   in the F+H layer (surface humus) and in the upper mineral
soil down to 10 cm depth, partly with statistically significant
differences, presumably due to Ca input. But the absolute differences are rather small. In the deeper layers no essential differences occurred. The Mg-contents tended to smaller values within
the damage area,possibly due to enhanced leaching.

Unlike these findings very large and statistically high significant differences were found for the contents of Cu and Zn in the
F+H layer and in the mineral top soil (see Fig. 3 and 4). Within
the damage area the soils of the sample plots are obviously enriched with Cu down to about 20 cm depth, with Zn down to about
50-60 cm. The deeper distribution of Zn may be due to a relatively
better mobility in the soil.

But even in the undamaged area the surface humus shows some raised
Cu and Zn contents in comparison with values from literature (3),
indicating that there a small secondary enrichment by immissions
has probably taken place, even when the biogenetic enrichment in
the organic substance is drawn into consideration (5, 6).

4.4 Results of the investigation of the growth

The relative diameter growth of the stands from the site unit
"Sand" is delineated in Fig. 5 as an example, being very similar
to the growth on the other site units. The 100 % line is representing the mean of the comparison trees in the undamaged stands. One
can see that in 1974 the heavily damaged trees were only 20 to 40%
as efficient in diameter growth as the comparison trees from outside of the damage area. But the sound appearing trees seem to be
suffering too, as their productivity is only 60 to 80 % of those
outside. The decrease of diameter increment is rather continuous,
starting in S II and S III around 1960, indicating a long suffering time. Both stands are located very near to the industrial
area, whereas S I with a later beginning is located farther away
in the outer zone of the damage area.

4.5 Results of the needle analysis
Tab. 1 shows those elements, whose contents are very   s i m i l a r
in the needles of the stands inside and outside of the damage area:

    N   very low supply, probably due to former litter utilization.
    P   low supply, but with respect to N rather sufficient.
        The N/P-ratio lies near 8.
    K   rather sufficient.
    Al values as usual.
    S   no indice of critical charging, but boundary range.

Tab. 2.1 is representing those elements which are in the most cases
significantly   h i g h e r   concentrated in needles of the stands
within the damage area:

    Ca possibly due to Ca-containing immissions.
    F   possibly due to HF-immissions; the values lie, however,
        beyond the usual level of toxicity, when not combined with
        other stress factors.
    Cu, Zn very high values, especially for Cu on the site unit
        "Sand"; the sound appearing trees have also rather high con-
        tents; presumably a strong enrichment is given in the cell
        walls without being physiologically active. The raised con-
        tents may be due not only by root uptake but also by de-
        positions on the needle surface.
    Mn no significant enrichment in the needles from the site unit
        "Sand", but very significant on the site unit "Loam"; "Sand-
        stone" lies between.

Table 2.2 shows those elements, which are (in the most cases) significantly l o w e r concentrated in the needles of the stands inside the damage area.

Mg   possibly due to enhanced Mg losses, but rather speculative.
Si   unclear.

Table 3 shows some difficulty to explain relations, namely, a tendency that the heavily damaged trees have higher contents of sodium and iron than the sound looking ones. Possibly there exists a concentration effect due to less needle mass. Concerning sodium one might suppose that in the heavily damaged trees the selective uptake of ions is disturbed.

Well understandable is the fact, that the sound appearing trees have three times as heavy needles as the heavily damaged trees.

5. DISCUSSION AND CONCLUSIONS

It seems to be rather probable that in the vicinity of the industrial area the investigated forest stands were charged with Cu and Zn immissions. Specific symptoms for Cu and Zn toxicity could not be worked out as well as toxicity thresholds of Cu and Zn in the Scots pine needles, because Cu and Zn may be partly stored up in the needle cell walls without being physiologically active. The findings of soil and needle analysis, however, are pointing to a Cu and Zn poisoning of the trees, possibly combined with other stress factors. One of these stress factors might be the low N or Mg supply, but there must be drawn other ones into consideration as immission of Cd, Pb or root distructions by Al poisoning a.s.o., which were not investigated.

The clear orientation of the damage zones to the industrial area and the relative small extent of the damage area suggested that the Cu and Zn enrichment was caused by dust immissions. This suggestion was proved, since some years after the installation of a high efficient dust filter in one of the plants, the damages in some parts of the area were significantly reduced and the Cu and Zn contents of the needles decreased (Braun, unpublished data). This again is indicating that the high Cu and Zn contents from the 1974 needles are mainly caused by direct deposition, as the amounts in the soil are still the same.

Supposing that a Cu and Zn intoxication had taken place as well as a nitrogen and/or magnesium deficiency, liming with Mg-chalk plus phosphate and nitrogen fertilization was recommended. The liming combined with phosphate brought an additional effect in contrast to N-fertilization. The doses were 20 dt Magnesiachalk (Ca, Mg-carbonat) and 5 dt Hyperphos (raw phosphate) per ha.

LITERATURE

1   Knorr, A. and F. Brosinger: 1975, Diplomarbeit, Forstwiss. Fakultät München

2   Kretzschmar, P.:   1977, Diplomarbeit, Forstwiss. Fakultät München

3   Mayer, R.:   1981, Göttinger Bodenkundliche Berichte No. 70

4   Rieder, W., and U. Schwertmann: 1972, Landwirtschaftliche Forschung No. 25

5   Schlichting, E.:   1955, Acta Agric. Scandinavia 5 (4)

6   Schlichting, E.:   1965, Zeitschr. Pflanzenern., Düng., Bodenk. No. 110

7   Westerhoff, H.:   1955, Landw. Forschung No. 7, pp. 190-193

350

K. KREUTZER ET AL.

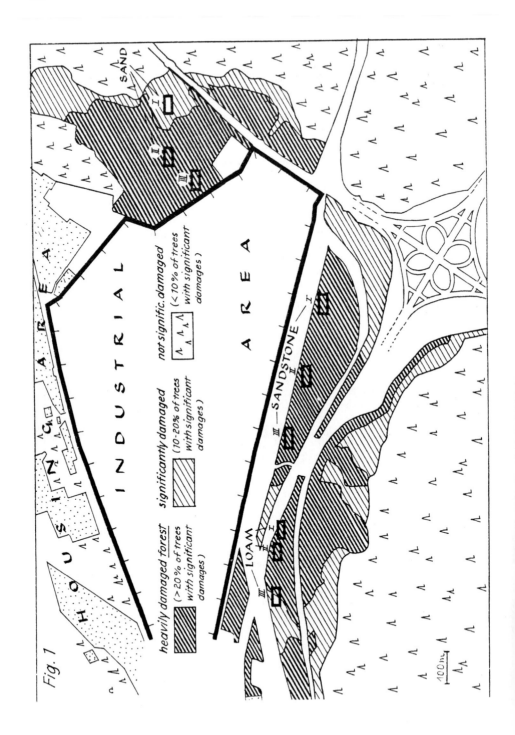

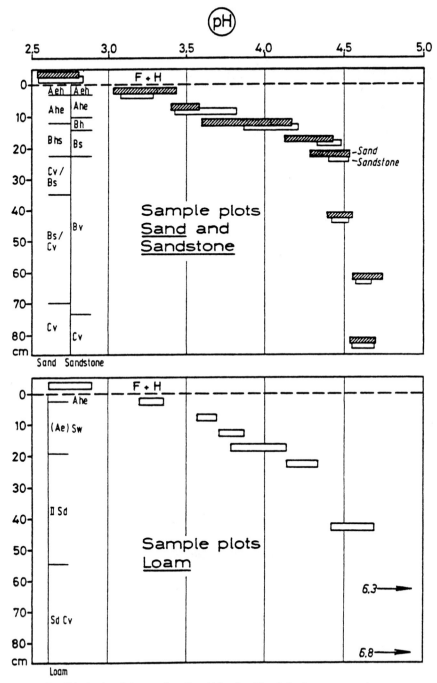

Fig. 2 Distribution of pH with depth (whole ranges)

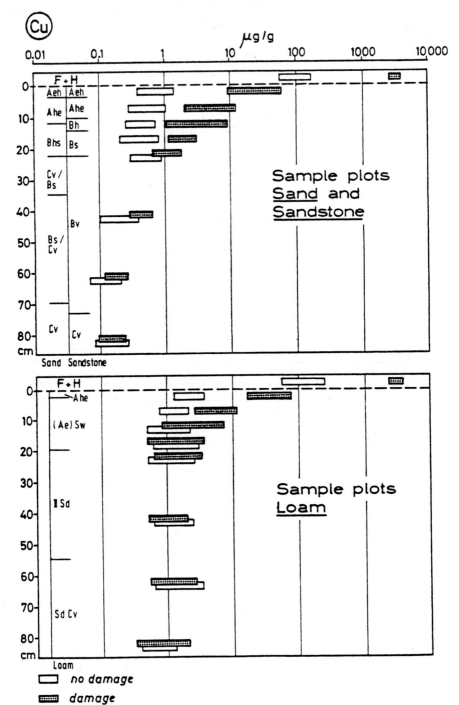

Fig. 3 Distribution of HNO$_3$-extractable Copper with depth

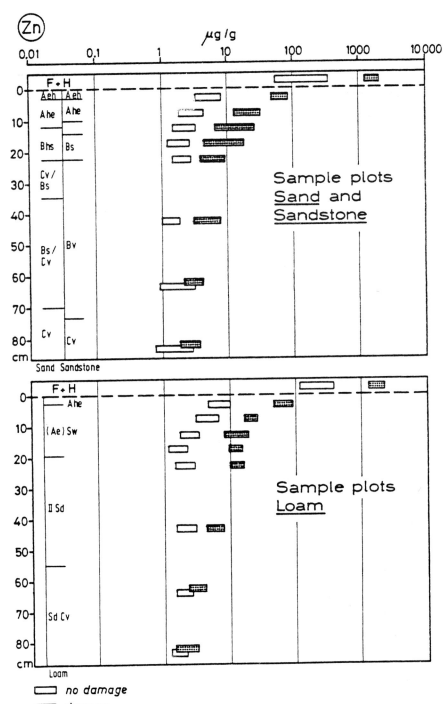

Fig. 4 Distribution of HNO$_3$-extractable Zinc with depth

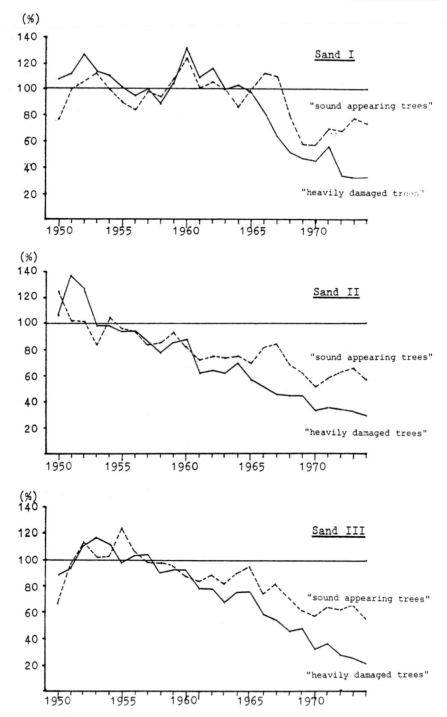

Fig. 5 Relative diameter growth

Comparison of the needle analyses

d = heavily damaged trees from within the damage area
s = sound appearing trees   "   "   "   "   "
C = Comparison trees from outside the damage area
Figures in brackets mean deviation

Table 1 Statistically not significant differences between the
        "damage area" and the "comparison area"

		Sand		Sandstone		Loam	
N	d	10,3	(1,0)	10,3	(0,4)	10,2	(0,2)
mg/g	s	11,6	(0,3)	11,4	(0,7)	10,6	(0,2)
	C	10,9	(0,5)	10,9)	(0,9)	12,0	(1,1)
P	d	1,18	(0,15)	1,26	(0,05)	1,23	(0,04)
mg/g	s	1,23	(0,09)	1,24	(0,06)	1,19	(0,08)
	C	1,25	(0,02)	1,17	(0,17)	1,25	(0,11)
K	d	4,82	(0,45)	4,80	(0,11)	4,50	(0,78)
mg/g	s	5,25	(0,19)	4,65	(0,10)	4,34	(0,85)
	C	4,35	(0,22)	4,32	(0,26)	4,51	(0,02)
Al	d	0,20	(0,03)	0,23	(0,01)	0,24	(0,02)
mg/g	s	0,23	(0,03)	0,24	(0,01)	0,27	(0,01)
	C	0,29	(0,01)	0,28	(0,04)	0,27	(0,08)
S	d	1,02	(0,09)	0,84	(0,08)	1,03	(0,10)
mg/g	s	0,91	(0,30)	0,74	(0,09)	0,85	(0,11)
	C	0,69	(0,08)	0,82	(0,07)	1,08	(0,09)

Table 2   In the most cases statistically significant differences between the "damage area" and the "comparison area"

2.1 "Damage area" has higher values than "comparison area"

		Sand		Sandstone		Loam	
Ca	d	3,42	(0,32)	4,10	(0,11)	3,60	(0,49)
mg/g	s	4,00	(0,05)	4,28	(0,75)	3,83	(0,36)
	c	2,56	(0,24)	2,60	(0,39)	2,74	(0,22)
F	d	5,0	(1,4)	4,1	(0,3)	4,0	(1,6)
µg/g	s	3,7	(1,1)	4,2	(0,6)	3,4	(0,2)
	c	2,7	(0,2)	3,1	(0,2)	1,7	(0,8)
Zn	d	257	(35)	196	(26)	146	(45)
µg/g	s	247	(54)	192	(39)	130	(12)
	c	100	(8)	93	(2)	63	(5)
Cu	d	90	(7)	24	(3)	22	(7)
µg/g	s	54	(5)	17	(2)	22	(4)
	c	10	(3)	7	(1)	8	(1)
Mn	d	550	(130)	580	(72)	1550	(475)
µg/g	s	580	(110)	450	(112)	2010	(123)
	c	610	(135)	310	(175)	910	(262)

2.2 "Damage area" has deeper values than "comparison area"

		Sand		Sandstone		Loam	
Mg	d	0,52	(0,28)	0,61	(0,12)	0,70	(0,17)
mg/g	s	0,60	(0,15)	0,67	(0,06)	0,71	(0,09)
	c	0,90	(0,08)	0,96	(0,11)	1,10	(0,07)
Si	d	0,95	(0,06)	1,10	(0,10)	0,75	(0,11)
mg/g	s	1,22	(0,11)	1,25	(0,19)	0,91	(0,30)
	c	1,95	(0,16)	1,85	(0,29)	1,40	-

Table 3   In the most cases statistically significant differences
          between "heavily damaged trees" and "sound appearing
          trees" within the damage area

        3.1 "Damaged" ones have higher values than "sound appearing"
            ones

		Sand		Sandstone		Loam	
Na	d	55	(6)	53	(12)	85	(6)
µg/g	s	25	(15)	31	(1)	33	(9)
	c	20	(3)	23	(6)	29	(5)
Fe	d	153	(18)	103	(9)	137	(15)
µg/g	s	105	(27)	82	(8)	99	(9)
	c	81	(2)	76	(3)	88	(7)

        3.2 "Damaged" ones have deeper values than "sound appearing"
            ones

		Sand	Sandstone	Loam
needle	d	2,74 (0,37)	3,58 (0,17)	3,23 (0,44)
dry	s	7,66 (0,61)	6,54 (0,34)	6,25 (1,18)
weight				
g/500	c	7,42 (0,10)	7,28 (0,36)	7,69 (0,17)

GROWTH PATTERNS, PHLOEM NUTRIENT CONTENTS AND ROOT CHARACTERISTICS OF BEECH (FAGUS SYLV.L.) ON SOILS OF DIFFERENT REACTION

K.E. Rehfuess, H. Flurl, F. Franz and E. Raunecker

Chair of Soil Science and Chair of Forest Growth Science
University of Munich

ABSTRACT

A pilot study was run in 6 older beech stands in Lower Franconia growing either on Terrae fuscae and Parabraunerde soils of high pH and rich in bases (A) or on acid to podzolic brown forest soils (B). Intensive growth analyses of 2-4 representative trees per stand revealed different levels, but similar time patterns of diameter and volume increment. The extremely dry early summer season of 1976 depressed growth on both acid and highly buffered soils. Beech phloem sampled at breastheight contained more K, Mn, Zn and Al, but less Ca on (B) than on (A), whereas contents of N, P, Mg and Cu were similar. Density of living roots of selected trees in the topsoil (0-50 cm) attained the same order of magnitude on both groups of substrata; but less roots were detected in the subsoil of acid sites. Living fine roots of trees on (A) showed evidence of containing more Ca and less Mn than trees on (B); N, P, K, Mg, Fe, Cu, Zn and Al however varied on similar levels.

2 selected trees growing on a podzolic soil and suffering from bark necrosis exhibited, as compared with unaffected neighbours, similar development of diameters from 1940-75 and the same abrupt decrease of radial increment in 1976, but no subsequent regeneration. They had exceptionally high N, P and Zn, but very low K contents in the phloem. The root system of one of those trees was strikingly shallow and covered by Armillaria mellea rhizoids. The fine roots didn't differ in element contents from those of adjoining healthy trees.

B. Ulrich and J. Pankrath (eds.), Effects of Accumulation of Air Pollutants in Forest Ecosystems, 359–375.
Copyright © 1983 by D. Reidel Publishing Company.

1. INTRODUCTION

Beech (Fagus sylv.L.) occurs in Germany both on acid and alcaline
substrata. Under comparable moisture and temperature conditions
it grows better on soils well provided with bases and ranging in
pH from 5 to 8 than on soils with an acid reaction. On acid sites,
liming seems to improve the growth of beech more efficiently than
that of Norway spruce (Picea abies L.; 3). This superiority of
growth on soils with a high pH level is generally attributed to
an optimal supply of N, P, Ca and Mg. The absence of toxic levels
of Al and Mn may also be a main factor contributing to this phenomenon.

According to a theory forwarded by B. Ulrich and coworkers (4),
vast areas of forest land in Central Europe suffer nowadays from
a rapid decrease of pH due to interactions between acid precipitation and the production of mineral acids within the soils themselves (e.g. HNO_3 via mineralization and nitrification). This acid
formation, they suggest, is promoted by warm and dry weather conditions. Ulrich and coworkers (4) expect this rapid acidification
to reduce the vitality and growth of forests mainly by releasing
toxic amounts of Al and Mn into the soil solution, by enhancing
the leaching of bases and by inducing the accumulation of nutrients in an inactive organic floor. They fear, that this complex of damaging events is already causing premature shedding of
leaves and dieback in stands of different species and should
finally lead to a degradation of forests into acid heathlands
(4, 5).

If this fast reduction of soil pH together with its suggested negative effects is already taking place, one would expect, that
beech stands on acid sandy soils exhibit a decline in growth and
Al- or Mn- induced damages within the root system, whereas comparable trees on highly buffered loamy or clayey sites with high
pH are not yet affected.

These considerations and similar research approaches in Scandinavia initiated the following pilot study which compares growth,
nutritional status and root characteristics of selected beech
trees on soils of different textural and chemical properties.
The study was undertaken in order to detect, if possible, site-
specific deviations in behaviour of those trees indicating a rapid deterioration of soil fertility as suggested by B. Ulrich.

2. MATERIAL AND METHODS

Six older beech stands were chosen in Lower Franconia at 5-60 km
distance from Würzburg in a NNE direction. Their soils were derived either from triassic limestones (Muschelkalk in one case

mixed with basalt) or from triassic old red sandstone (Buntsandstein). Both types of sedimentary rock are partly covered by loess
deposits. Table 1 describes in detail the sampling sites and soils.
Climatic conditions within the study area are rather similar with
the only exception being site "Kalkofen" (2).

The area under investigation is situated rather far away from big
urban or industrial agglomerations, but nevertheless seems to have
been affected by soil acidification during the past two decades
(Wittmann 1981, personal comm.).

On each sampling site a homogenous plot of 900 or 1600 m² was
established. The diameters at breastheight were recorded for all
living trees. After computation of the total basal area per hectare, 2-4 beech trees per plot, showing the mean diameter and the
average height of the 100 largest stems per hectare (top height
trees), were cut. For these trees, characterized in Table 2, we
measured total height, crown width and length and determined age.
Afterwards the stems were divided into ten sections. For each section one representative stem slice was sampled. An additional cut
was made at breastheight. With the help of a special program
(Kawabata a. Schiibajashi 1977, Kennel 1977, Flurl 1981) we were
able to compute and to plot complete stem analyses and the altera-
tion with time of diameter, height, basal area and volume increments for the whole life span of each individual tree.

Due to technical circumstances, the study had to be started in
March 1980, when no foliage was present. Therefore we used the
phloem analysis (1) for an evaluation of the nutritional status
of sample trees, analysing the fresh inner bark sampled at breastheight.

At 1 m distance from the stumps of 4 additional beech trees on
different soils, pits were dug out 3 m wide and 1 m deep. After
careful preparation of the front wall, all living and dead roots
were identified and classified with regard to diameter and then
counted. Afterwards representative samples of living roots of the
diameter classes "finest roots" (< 1 mm) and "fine roots" (1-2 mm)
were cut, carefully cleaned with distilled water, dried and ground
(woody and bark material together) for chemical analysis. From
the rooting space, soil was sampled at 10 cm intervals.

Plant tissue analyses for N, P, K, Ca, Mg, Fe, Mn, Cu, Zn and Al
and soil pH determinations were performed using standard laboratory procedures (2).

Table 1: Description of sampling sites
Br = Braunerde, Tf = Terra fusca, Pb = Parabraunerde, Ps = Pseudogley, Pd = Podsol

Characteristic	Sampling area					
	Kalkofen	Dachsbau	Dianenlust	Reuterlein	Brückengraben	Kleinhahn
Forest District	Steinach	Bad Neustadt	Münnerstadt	Würzburg	Steinach	Steinach
Compartment	V 3b	III 3b	XIX 2a	II 1b	XII 3f	I 2b
Growth district	Hohe Rhön	Nördl.Fränk. Platte	Nördl.Fränk. Platte	Südl.Fränk. Platte	Vorrhön	Vorrhön
Mean elevation (m)	800	370	380	320	350	350
Aspect	NW	SW	-	-	-	SE
Inclination	steep	gentle	-	-	-	steep
Mean annual temp. (°C)	5,7	7,5	8,0	8,0	8,0	8,0
Mean annual precipitation (mm)	1050	620	600	600	650	650
Parent rock	mixed slope deposits of Basalt and Muschelkalk	Muschelkalk	Loess covering Muschelkalk	Loess covering Muschelkalk	Loess covering old Red Sandstone	old Red Sandstone
Soil type	Br.-Tf	Tf	Pb.-Tf	Pb.-Tf	Ps.-Pb	Pd.-Br.
Texture	clayey loam	loamy clay	silty loam/ loamy clay	silty loam/ loamy clay	silty loam/ clayey sand	loamy sand
Range of pH (H$_2$O)	6,1-7,8	5,5-7,6	5,1-7,1	4,8-7,1	4,0-6,5	3,3-4,7
Plant association	Asperulo-Fagetum	Galio-Carpinetum	Galio-Carpinetum	Melico-Fagetum	Luzulo-Fagetum	Luzulo-Fagetum

Table 2: Characteristics of healthy sample trees

Characteristic	Sampling area					
	Kalkofen	Dachsbau	Dianenlust	Reuterlein	Brückengaben	Kleinhahn
n	3	4	3	1	3	2
age	87-102	77-82	71-76	120	104-111	109-111
DBH (cm)	28,8-33,2	33,0-35,3	38,0-42,2	50,0	32,9-35,4	32,4-33,5
height (m)	25,5-28,8	24,6-26,5	26,3-27,6	28,7	26,3-28,0	24,2-27,2
Top height site class (according to yield table SCHOBER 1967; moderate thinning)	I,3-II,7	I,3-II,1	0,8-I,3	II,4	II,5-II,9	II,7-III,5

3. RESULTS AND DISCUSSION

3.1 Growth characteristics

Figure 1 demonstrates the time patterns of annual radial increment

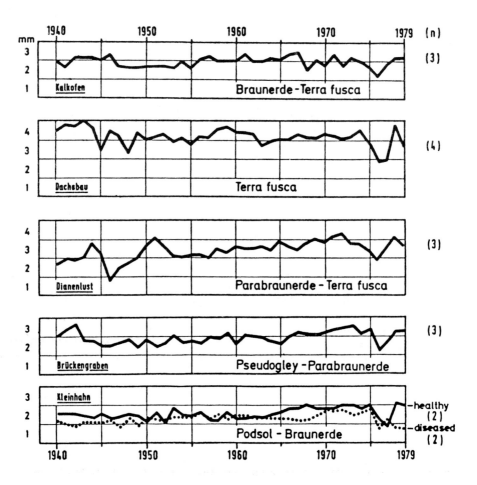

Figure 1. Mean year ring width at breastheight of 2-4
sample trees at five locations differing in
soil reaction

at breastheight during the past 40 years. In correspondance with
the differences in site classes (Table 2) its level is generally
lower on acid substratum than on soils well provided with bases.
After a pronounced reduction of year ring width in the late for-
ties, the majority of sample trees exhibited a slight, but steady
increase of radial growth until 1973/74. This tendency was even

more obvious for stands on acid soils (Brückengraben, Kleinhahn)
as compared with beech trees growing on base-rich substratum
(Kalkofen, Dachsbau), although the former stands were older. Another sharp decline of radial growth occured in 1975/76, followed
in most cases by rapid recovery. This depression, coinciding with
extremely low precipitation mainly in 1976 (Figure 2) was observed

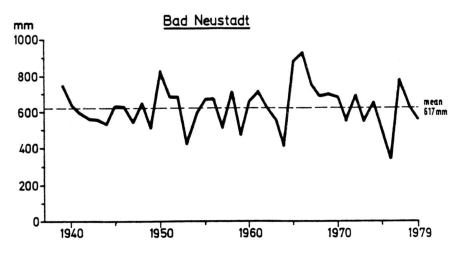

Figure 2. Annual precipitation during the period 1939-
1979. Forest District Bad Neustadt/Saale

both on acid sandy sites and on well buffered Braunerde-Terra fusca
at higher elevation. These observations indicate that Al or Mn toxicity was not a major factor in causing this chronosequence. Two
sample trees at the site "Kleinhahn" suffering from bark necrosis
revealed, as compared with unaffected neighbours, after similar
development of diameter increment from 1940-75, the same significant decline in 1976, but no subsequent rise.

Table 3, which summarizes the mean current annual diameter increment data of healthy sample trees during three subsequent periods,
prevailingly exhibits a growth increase on sandy acid soils as
well as on substrata high in pH in 1970-74. This phase was followed by a remarkable decline of diameter increments from 1975
onwards. The mean annual current volume increment (Table 4) rised
even more pronouncedly during 1970-74 on all sites under study and
was generally depressed in the following period including unusually
dry years.

3.2 Element contents of beech phloem

Phloem sampled at breastheight from healthy trees generally contained less Ca, but more K, Mn, Zn and Al on acid than on base-rich

Table 3: Comparison of current annual diameter increment of healthy sample trees during three
subsequent periods

Sampling site	tree number	mean current annual diameter increment (mm)			Difference (mm)	
		(a) 1965-69	(b) 1970-74	(c) 1975-79	(b)-(a)	(c)-(b)
Muschel-kalk						
Kalkofen	1	5,8	4,6	3,9	-1,2	-0,7
	2	4,3	4,0	3,1	-0,3	-0,9
	3	3,3	3,1	2,7	-0,2	-0,4
Dachsbau	12	6,3	6,5	5,8	+0,2	-0,7
	13	6,3	5,1	4,1	-1,2	-1,0
	14	4,5	5,4	5,0	+0,9	-0,4
	15	5,9	6,1	4,4	+0,2	-1,7
Dianenlust	9	7,0	6,5	5,2	-0,5	-1,3
	10	6,1	6,7	11,6	+0,6	+4,9
	11	5,2	5,8	6,8	+0,6	+1,0
Buntsand-stein						
Brücken-graben	6	2,5	2,9	3,0	+0,4	+0,1
	7	4,1	4,4	4,4	+0,3	-
	8	5,4	6,8	4,0	+1,4	-2,8
Kleinhahn	4	3,4	3,5	3,4	+0,1	-0,1
	5	4,0	4,1	3,4	+0,1	-0,7

Table 4: Comparison of current annual volume increment of healthy sample trees during three subsequent periods

Sampling site	tree number	mean current annual volume increment (dm^3)			Difference (dm^3)	
		(a) 1965-69	(b) 1970-74	(c) 1975-79	(b)-(a)	(c)-(b)
Muschel-kalk						
Kalkofen	1	27,4	30,2	30,0	+2,8	-0,2
	2	29,9	33,4	30,4	+3,5	-3,0
	3	21,1	26,9	26,6	+5,8	-0,3
Dachsbau	12	27,7	36,9	38,6	+9,2	+1,7
	13	31,6	31,9	31,0	+0,3	-0,9
	14	27,0	35,1	32,9	+8,1	-2,2
	15	33,2	44,4	41,0	+11,2	-3,4
Dianenlust	9	45,2	55,2	39,6	+10,0	-15,6
	10	46,0	53,2	41,9	+7,2	-11,3
	11	43,8	48,5	43,9	+4,7	-4,6
Buntsand-stein						
Brücken-graben	6	23,6	28,1	33,3	+4,5	+5,2
	7	27,4	35,1	34,0	+7,7	-1,1
	8	24,3	37,3	29,4	+13,0	-7,9
Kleinhahn	4	26,7	24,4	23,6	-2,3	-0,8
	5	26,5	31,8	27,8	+5,3	-4,0

soils, whereas contents of N, P, Mg and Cu varied on similar levels (Tables 5 and 6). Annual diameter and volume increments were
positively correlated with Fe contents in phloem, but inversely
related to K. The first relationship mainly originates from su-
perior Fe supply of the most vital stand "Dianenlust" on Para-
braunerde-Terra fusca. The negative correlation of both growth
parameters to K is due to the fact, that slowly growing stands
on acid sandy soils are well provided with this element, whereas
the most vigorous trees on Pb.-Terra fusca had only moderate K
contents in phloem. The phloem levels of Al and Zn reflected increased solubility of these elements in acid soils derived from
old red sandstone. The best growing trees on Pb.-Terra fusca, however, had only slightly less Al in phloem. The Ca/Al ratios (on
an equivalent basis) of all trees far exceeded the value of 1.
Trees affected by bark necrosis didn't differ significantly in
phloem contents of Al and Mn from neighbouring healthy trees, but
contained more N and Zn and much less K.

3.3 Root densities

For 4 selected beech trees the densities of living roots were determined as root numbers per area unit (Table 7). In the top soil
these densities attained the same order of magnitude for all trees
which were comparable in age and diameter irrespective of soil
chemical status and health conditions.

On the other hand there is evidence that less living roots were
present in the subsoil of acid substrata. The root system of one
diseased tree growing on the most acid site covered by this study
was remarkably shallow and contained many Armillaria mellea rhizoids and dead tissues.

3.4 Nutrient contents of roots

The same four trees used for studies on root distribution were
analyzed for the nutrient and aluminum contents of living finest
and fine roots (Tables 8 and 9). Healthy trees on base-rich soils
contained more Ca, but less P and Mn in both root fractions as
compared with beech trees on soils derived from old red sandstone.
N, K, Mg, Fe, Zn, Cu and Al contents, however, varied on similar
levels. The Ca/Al ratios of fine and of finest roots on acid soils
varied from 0,9 to 4,4 and from 0,9 - 2,2 respectively. Roots of
the diseased tree did not differ significantly in element contents from those of its healthy neighbour and generally exhibited higher Ca/Al ratios.

Table 5: Macronutrient contents of beech phloem at breastheight. Ranges and means.

Sampling area	number of trees	N	P	K	Ca	Mg
				mg/g dry matter		
Muschelkalk						
1.Kalkofen	3	7,5-7,7/7,6	0,33-0,34/0,33	2,1-2,5/2,3	64-100/78	0,41-0,46/0,42
2.Dachsbau	4	5,2-6,4/5,9	0,27-0,35/0,32	2,0-2,6/2,3	70-82 /77	0,34-0,42/0,39
3.Dianen-lust	3	6,5-6,7/6,6	0,29-0,35/0,32	1,8-2,0/1,9	33-50 /44	0,37-0,43/0,39
Buntsand-stein						
5.Brücken-graben	3	5,4-6,1/5,8	0,24-0,25/0,24	2,3-2,6/2,4	26-28 /28	0,42-0,42/0,42
6.Kleinhahn -healthy trees	2	4,1-6,9/5,5	0,34-0,34/0,34	2,3-3,0/2,7	29-35 /32	0,36-0,45/0,41
-diseased trees	2	6,1-10,6/8,4	0,29-0,68/0,49	0,3-0,8/0,5	31-39 /35	0,44-0,44/0,44

Table 6: Micronutrient and aluminum contents of beech phloem at breastheight. Ranges and means.

Sampling area	number of trees	Fe	Mn	Zn	Cu	Al
			µg/g dry matter			
Muschel-kalk						
1.Kalkofen	3	16-17 /17	640-1230 /907	3-4 /3	2-3 /3	Tr*)
2.Dachsbau	4	15-18 /17	350- 450 /418	3-3 /3	3-4 /4	Tr*)-15 /7
3.Dianen-lust	3	18-24 /22	640-1170 /887	3-4 /3	3-4 /3	25-29 /27
Buntsand-stein						
5.Brücken-graben	3	12-17 /15	1660-2620 /2063	5-7 /6	2-3. /3	29-37 /33
6.Kleinhahn						
-healthy trees	2	15-19 /18	1390-1650 /1520	4-4 /4	2-3 /3	27-35 /31
-diseased trees	2	19-20 /20	1450-2740 /2095	9-10 /10	3-5 /4	33-38 /36

*) Traces only

Table 7: Mean number of living roots per dm² rootable area (a) and percentage of dead roots (b). Rootable area = total area of the vertical cross-section of a particular soil layer minus area occupied by stones > 5 cm diameter

			Sampling site		
Feature	Kalkofen	Reuterlein		Kleinhahn	
				healthy tree	diseased tree
Soil type	Br.-Terra fusca	Parabraun- erde		Pods.-Braun- erde	Pods.-Braun- erde
Range of pH(H₂O)	6,1-7,8	5,2-7,1		3,9-4,7	3,3-3,9

$pH(H_2O)$ range: Kalkofen 6,1-7,8; Reuterlein 5,2-7,1; Kleinhahn healthy 3,9-4,7; Kleinhahn diseased 3,3-3,9

Sample tree:	Kalkofen	Reuterlein	Kleinhahn healthy	Kleinhahn diseased
age	87	120	100	80
DBH (cm)	29,5	50,0	25,0	29,0
height (m)	20,1	27,2	18,7	19,4

Roots per dm²

0-50 cm	a	b	a	b	a	b	a	b
< 2 mm	13,4	<5	8,6	<5	13,2	<5	13,8	11
2-5 mm	0,7	<5	1,3	6	1,4	<5	0,8	34
> 5 mm	0,5	<5	0,9	<5	0,7	<5	0,4	10

50-100 cm	a	b	a	b	a	b	a	b
< 2 mm	5,2	<5	3,6	<5	2,6	<5	1,4	59
2-5 mm	0,3	<5	1,0	<5	0,2	<5	-	86
> 5 mm	0,2	<5	0,5	<5	0,1	<5	-	0

0-100 cm	a	b	a	b	a	b	a	b
< 2 mm	10,2	<5	6,1	<5	7,9	<5	7,6	11
2-5 mm	0,5	<5	1,1	<5	0,8	<5	0,4	43
> 5 mm	0,4	<5	0,7	<5	0,4	<5	0,2	9

Table 8: Nutrient and aluminum contents of living fi-
         nest roots (diameter < 1 mm). Ranges of values
         in topsoil (ts) and subsoil (ss)

Element	Sampling site				
	Kalkofen	Reuterlein	Kleinhahn		
			healthy tree	diseased tree	
	ts ss			ts ss	
N mg/g	10 - 6	11 -5	7 -9	8 -5	
P mg/g	0,8- 1,0	0,9-0,7	1,5-1,3	0,7-1,0	
K mg/g	2 - 2	2 -2	3 -3	2 -1	
Ca mg/g	6 -16	2 -4	1 -1	2 -1	
Mg mg/g	0,6-0,7	0,8-1,2	0,7-0,6	0,8-0,7	
Fe mg/g	3- 2	1- 1	2- 2	1- 2	
Mn µg/g	100-64	93-84	374-227	455-309	
Cu µg/g	9- 8	8- 7	6- 5	6- 5	
Zn µg/g	41-28	90-20	59-50	66- 90	
Al µg/g	190-512	334-395	504-523	410-519	

———————— no clear depth function

Table 9: Nutrient and aluminum contents of living fine
roots (1-2 mm). Ranges of values in topsoil
(ts) and subsoil (ss)

Element	Kalkofen	Reuterlein	Kleinhahn healthy tree	Kleinhahn diseased tree
	ts ss			ts ss
N mg/g	7 - 5	8 -5	5 -5	8 -3
P mg/g	1,0- 0,7	0,9-0,8	1,5-1,2	1,7-1,0
K mg/g	2 - 2	2 -2	2 -3	2 -3
Ca mg/g	8 -16	5 -1	2 -1	3 -1
Mg mg/g	0,6- 0,5	0,8-1,5	0,8-0,7	1,2-0,8
Fe mg/g	2- 1	1- 2	1- 2	1- 1
Mn μg/g	76- 39	288-47	453-150	490-131
Cu μg/g	1- 6	7- 5	4- 4	6- 3
Zn μg/g	31- 17	100-11	64- 42	90- 43
Al μg/g	139-512 *)	399-527 *)	370-508 *)	489-297 *)

*) maximum always in A_1 or B_{v1}

——— no clear tendency within the profile

4. CONCLUSIONS

We were not able to detect any recent growth decline specific for beech trees on acid soils.

The obviously long-lasting differences in site class between beech stands on acid soils and those on base-rich substrata could not be explained satisfactorily by our analyses. Differences in stand age, in availability of water and in climatic site conditions to some extent may have obscured the relations between element supply and growth. The Ca nutrition should be taken into special consideration during further investigations.

Phloem as well as root analysis partly reflected the increased solubility and uptake of Al, Zn and Mn on acid soils. The differences in tissue levels between the most vital trees and slowly growing beeches however were generally small. Therefore these elements are assumed not to exert any toxic effects in the stands investigated. It is not probable, that Al or Mn toxicity were responsible for the decline in growth observed in dry periods.

All trees which were sampled and were comparable in age, exhibited similar root density in the topsoil (0-50 cm), but there was evidence of less roots being present in the subsoil (50-100 cm) on acid sites. This shallower root system, the reasons of which are not yet clear, is an important disadvantage during dry periods.

Bark necrosis of selected sample trees coincided with an unusually high percentage of dead roots, with shallow root system and with the occurence of Armillaria mellea. The extremely low contents of K and the very high levels of Zn in the phloem of those trees need further attention.

ACKNOWLEDGEMENTS

We gratefully acknowledge the technical support by the Bavarian State's Forest Administration (Oberforstdirektion Würzburg, FA. Bad Neustadt, FA. Münnerstadt, FA. Steinach and FA. Würzburg) during the field investigations and a financial grant of the Bayerische Düngekalk-Gesellschaft mbH, 8400 Regensburg.

5. REFERENCES

HOHENADL, R., M. ALCUBILLA and K.E. REHFUESS: 1978, Z. Pflanzenern., Bodenkde. 141, pp. 687-704.

RAUNECKER, E.: 1981, Dipl.Arbeit Univ. München

SEIBT, G. und J.B. REEMTSMA: 1977, Schriftenreihe Forstl. Fak. Göttingen 50, pp. 89-298

ULRICH, B., R. MAYER und P. KHANNA: 1979, Schriftenr. d. Forstw. Fak. d. Univ. Göttingen u.d. Niedersächs. Forstl. Vers.Anst. 58.

ULRICH, B.: 1981, Forstwiss. Centralbl. 100, pp. 228-236.

Address of the authors: Amalienstr. 52, D 8000 Munich, Germany

BIOLOGICAL ALTERATIONS IN THE STEM AND ROOT OF FIR AND SPRUCE DUE TO POLLUTION INFLUENCE

Josef Bauch

Institute of Wood Biology, University Hamburg

Wood biological investigations were carried out on healthy and diseased trees of 70 firs (Abies alba Mill.) and 33 spruce (Picea abies Karst.) from six sites in Germany. The tree-ring analyses reveal that trees even from the same site begin to respond to pollution in different calendar years. Stress due to drought, as in 1976, advances growth reduction, but is not its major cause. At some sites the decrease of tree vitality can be traced back for more than 30 years. Serious alterations could be observed on a cellular level with regard to the occurrence of elements in the fine roots. In acidic soils of about pH 4 or less the fine roots are deficient in the macro nutrients calcium and magnesium. Aluminium, normally present in healthy trees, had not been accumulated in the cortex of diseased trees.

1. INTRODUCTION

Numerous investigations on trees influenced by pollution reveal that the plant can be damaged directly in the crown (e.g. 9, 11, 12, 19) and indirectly through the acidified soil (e.g. 17, 18). Many alterations in the needles and leaves were described whereas analyses of the wood (2, 13) and in particular of the fine roots (10, 16) are few. In the following investigation the first question is directed to the analysis of the growth increment of fir and spruce from several sites in order to determine where in a tree the severest damage occur and how far this damage can be traced back to the year of beginning. The second question refers to alterations in the fine roots. The occurrence and distribution of elements in the tissue of healthy and diseased trees were investigated on a cellular and partly sub-cellular level. Special

B. Ulrich and J. Pankrath (eds.), Effects of Accumulation of Air Pollutants in Forest Ecosystems, 377–386.
Copyright © 1983 by D. Reidel Publishing Company.

emphasis was given to the uptake of macro and micro nutrients
through the cortex, and the presence of aluminium.

2. MATERIAL AND METHODS

2.1 Trees investigated

On account to the past experience with diseased firs (2, 6) discs
of 70 firs and 33 spruce trees from six sites were selected for
wood biological studies (table 1). Additionally, seven year old
fir and spruce plants were available from the institute's nursery.
Morphological studies and microanalyses of the fine roots were
carried out on 10 trees listed in table 2. For these trees also
a description of the habit, the sulphur content of the needles
and pH-values of the surrounding soil are presented (table 2).

2.2 Growth-ring analyses

Growth-ring measurements were carried out on all trees listed in
table 1. For this purpose an EKLUND device connected to an
electronic data processor was used. Due to the various missing
rings in diseased trees four radii per disc were measured. In
general 3-5 discs from different tree height were available.

Table 1. Selection of discs of fir and spruce trees.

SITE	FIR				SPRUCE			
	OLD GROWTH TREES	AGE	YOUNG GROWTH TREES	AGE	OLD GROWTH TREES	AGE	YOUNG GROWTH TREES	AGE
BAVARIA:								
FA BODENMAIS	5	80-115	6	20-50	10	80-140	-	--
FA KELHEIM	11	55-115	20	20-45	5	50-100	-	--
FA WÖRTH/D.	13	90-120	-	-	5	95-105	-	--
BADEN-WÜRTTEMBERG:								
FA BAD HERRENALB	12	85-175	-	-	5	110-120	5	35-50
HAMBURG:								
BFH - HAMBURG	-	-	FIELD EXPERIMENTS	7	-	-	FIELD EXPERIMENTS	7
HAUSBRUCH-HAMBURG	-	-	-	-	1	130	-	-
SCHLESWIG-HOLSTEIN:								
FA EUTIN	3	50-60	-	-	2	50-60	-	-

2.3 Micro-analyses

Fine roots from altogether ten trees (table 2) were analysed for the localization of elements in the individual cells. Tips of fine roots from different depths were freeze-dried and embedded in methacrylate without any fixation. The localization of elements within individual cells was carried out with a Laser-Microprobe-Mass-Analyzer (LAMMA 500) and a X-ray dispersive analysis (EDXA) in combination with a transmission electron microscope (TEM). Fine root cross sections of 1 μm in thickness were placed in a laser microscope connected to a mass spectrometer. With this instrument areas of about 1 μm^2 can be analysed (3). X-ray dispersive analysis was applied to 0.5 μm cross sections, and results obtained individually for cell walls and lumina.

Table 2. Selection of fine roots of fir and spruce trees.

TREES	SITE	AGE (YEARS)	DESCRIPTION OF THE TREES	pH-VALUE OF THE SOIL (10 AND 30 CM DEPTH)
FIR 1 AND 2	BFH-HAMBURG	7	HEALTHY YOUNG FIRS FINE ROOTS WELL DEVELOPED	6,8 / 6,8
FIR 3 AND 4	FA BAD HERRENALB	100	HEAVILY DISEASED FIRS SMALL CROWN, GROWTH INCREMENT REDUCED, PARTIALLY MISSING RINGS S-CONTENT OF THE NEEDLES 10% HIGHER THAN THE CONTROLS FINE ROOTS GREATLY REDUCED	3,4 / 3,6
FIR 5, 6, 7	FA EUTIN	50	SLIGHTLY DISEASED FIRS SMALL CROWN, GROWTH INCREMENTS SLIGHTLY REDUCED FINE ROOTS MODERATELY DEVELOPED	4,4 / 5,2
SPRUCE 8	FA HAMBURG	130	DISEASED SPRUCE SMALL CROWN, GROWTH INCPEMENTS REDUCED, S-CONTENT OF THE NEEDLES 10% HIGHER THAN THE CONTROLS FINE ROOTS GREATLY REDUCED	3,2 / 3,3
SPRUCE 9 AND 10	FA EUTIN	50	SLIGHTLY DISEASED SPRUCE SMALL CROWN, GROWTH INCREMENTS ONLY SLIGHTLY REDUCED MODERATELY DEVELOPED FINE ROOTS	4,1 / 5,2

3. RESULTS

3.1 Decrease of growth in fir and spruce

The growth-ring analysis reveals that, as a rule, the severest reduction in growth occurs near the butt of a diseased tree. The growth ring pattern of a diseased fir (FA Kelheim) may illustrate this gradient from bottom to top (figure 1): At 0.5 m tree height no xylem was developed at all in the years 1976-1978, at 10 m height tree growth was still severely retarded while near the crown at 19 m no negative influence could be observed.

In fig. 2 the sensitivity of spruce and fir is presented jointly growing at the FA Wörth site (comp. table 1). Most of the fir trees are diseased or even dried out while only few are still healthy. A part of the former have apparently suffered already for about thirty years. Some trees exhibit a complete growth stop at the butt for several years. This observation could be confirmed at the site FA Herrenalb. It is remarkable that at FA Wörth spruce does not suffer so far.
From the comparison of growth ring patterns of all trees investigated, it can be concluded that trees from the same site begin to respond to environmental stress in different calendar years. Physiological stress due to drought, as in 1976, advances this reaction but it is not its major cause.

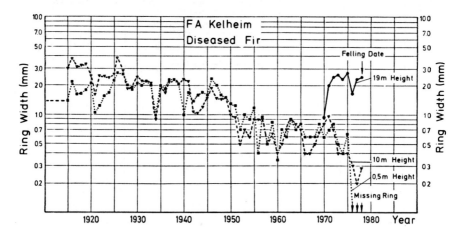

Fig. 1. Growth-ring patterns of three different heights of a
        diseased fir tree (FA Kelheim).

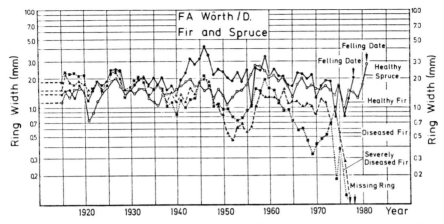

Fig. 2. Growth-ring pattern of a healthy spruce; for comparison
         those of a healthy, a diseased and a severely diseased
         fir (FA Wörth/D.).

3.2 Localization of elements in the fine roots

For the search of a primary damage in the fine roots of diseased
fir and spruce the occurrence and distribution of elements in
individual cells was analysed (comp. table 2). Mass-spectrometry
was carried out with a series of cross sections from the cortex
to the primary xylem. Figure 3, for example, demonstrates the
individual measurement covering the wall of an inner cortex cell
of a healthy fir. Sodium, magnesium, aluminium, potassium and
calcium were identified. Figure 4 illustrates the situation for
a corresponding measurement of a diseased fir. Sodium, aluminium
and potassium are present as in a healthy tree. However, the cell
wall of the diseased tree does not contain calcium and magnesium.
        In order to obtain more meaningful results the average values
encompassing several trees were computed and listed in table 3.
They underline the severe shortage of the macro nutrients calcium
and magnesium in the cortex of diseased fir and spruce. It is
remarkable that the aluminium content in the cortex of the diseased
firs does not differ much from that of the healthy plants.
        Comparative measurements of the cell walls of the primary
xylem within the vascular cylinder reveals the presence of the
same elements. In healthy trees the aluminium content in the wall
of primary xylem cells is low with 1.0 - 1.7 rel. units, in those
of diseased firs the amount is about double. In the diseased
spruce no aluminium was found.
        These results were confirmed by X-ray microanalysis (3).

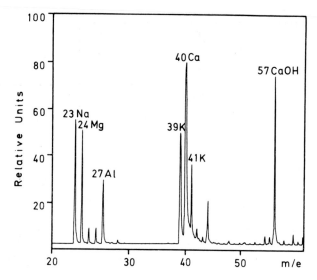

Fig. 3. Positive mass spectrum of a cell wall in the root cortex
of a healthy fir.

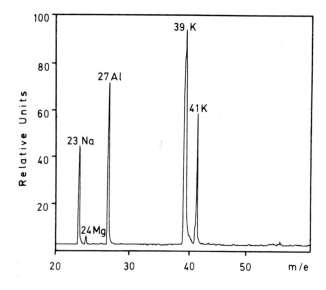

Fig. 4. Postitive mass spectrum of a cell wall in the root cortex
of a diseased fir.

Table 3. Determination of elements in cells of the fine roots (data in rel. units).

TREE	CELL TYPE, POINT OF MEASUREMENT	NA	MG	AL	K	CA	CAOH
HEALTHY FIRS 1 AND 2	CELL WALL OF THE CORTEX:						
	OUTER ZONE (CELL SERIES 1-3)	7,2	1,6	7,0	9,3	7,2	6,6
	INNER ZONE (CELL SERIES 4-6)	6,1	1,5	3,9	8,9	6,7	4,6
	CELL WALL OF PRIMARY XYLEM:						
	S 2 -WALL	4,7	1,0	1,0	9,6	6,6	4,6
	ML/P-WALL	5,3	0	1,7	10,6	6,8	6,0
DISEASED FIRS 3 AND 4	CELL WALL OF THE CORTEX:						
	OUTER ZONE (CELL SERIES 1-3)	4,9	<0,1	7,1	12,9	0	0,1
	INNER ZONE (CELL SERIES 4-6)	3,9	0,1	7,3	12,0	0	0
	CELL WALL OF PRIMARY XYLEM:						
	ML/P AND S 2-WALL	5,9	3,5	3,5	13,8	1,0	1,5
DISEASED SPRUCE 8	CELL WALL OF THE CORTEX:						
	OUTER ZONE (CELL SERIES 1-3)	3,5	1,4	4,8	12,8	<0,1	<0,1
	CELL WALL OF PRIMARY XYLEM:						
	ML/P AND S 2-WALL	2,8	<0,1	0	14,3	0	0

4. DISCUSSION

The biological analysis of the wood in diseased conifers should
contribute to localize the alterations in terms of initiation
and extension, in these trees. Previous investigations of the
pathological wetwood formation (2, 14) revealed, as a rule, that
the severest damage occurred at the butt and in the roots of the
trees. Yet there is evidence that the bacterial disease already
represents a secondary effect following a reduction of the tree's
vitality by pollution of the environment (6). The growth-ring
analyses corroborate these findings and demonstrate that the
severest loss in cambial activity may also be initiated in the
butt part of the tree (comp. 1). The relation between growth
depression and dry years was not very pronounced at the six sites.
It is striking that the growth ring depression of individual
trees begins in different years even within the same site.
Slight damages can be traced back in some firs for about thirty
years, preferably in those from sites with higher elevation. Dry
years as e.g. in 1976 enhance the decline but are not the main
cause for it. Similar dendroclimatological investigations as
they were previously carried out for trees in urban areas (8)
agree with these results (7).

From the biological alterations in the wood of diseased fir
and spruce it may be assumed that the damages are more likely
to have their origin in the roots. A subsequent study on fine
roots geared to the identification of elements in the root cells
supports this possibility.

An acidified soil apparently leads to an insufficient supply
of calcium and magnesium. The cortex of the fine roots is in
direct contact with the soil substrate. In an acidified system
the increased number of hydrogen ions is in competition with these
nutrients for the uptake by the root cells. As the cortex transports
the ions to a large degree within the cell walls, this system is
liable to be altered with the change of the pH-value of the soil
in a fashion similar to an ion exchanger.

The deficiency in calcium may have multiple significance.
The efficiency of the cortex cell walls with regard to water
transport could become insufficient. Another affect may lead to
a destabilization of enzyme systems. Finally calcium is important
for the structural development of cell wall; even the lignification
depends on this element. The consequence would be a severe reduction
in growth.

The deficiency in magnesium which belongs to the most
enzyme-active elements will also affect the tree's vitality (4).

The distribution of aluminium in the fine roots of healthy
and diseased trees indicates that this element might not play
the key role as the toxic element as stated previously indicated
(18), because its concentration in diseased roots does not differ
very much from that in healthy roots. Nevertheless, the analyses
by Ulrich(16) on spruce roots could be confirmed, where the ratio

of Ca : Al is >1 in healthy trees and ≪1 in diseased ones,
however with the Al component remaining practically unchanged.
    An immediate toxic effect of aluminium to the roots has yet
to be verified experimentally. An important question is, wether
the mycorrhyza is damaged by these alterations in the soil. There
is some indication that this is indeed a viable possibility (5).
    As Schütt (15) and Ulrich (17) state the principal damage
observed in conifers may have its origin in the roots rather than
in the crown. The cellular analysis of elements in fine roots
support this statement.

5. SUMMARY

Growth ring measurements were carried out on discs of 70 firs
and 33 spruce trees selected from six different locations in
Germany known to have been subjects severe pollution. The growth
ring patterns reveal a latent loss in vitality on some sites
for about 30 years. As compared with fir spruce seems somewhat
less susceptible to pollution effects. The severest reduction
of growth occurs generally at the butt zone of the tree. In
trees from the same site a growth reduction is liable to start
in different years. The physiological stress in dry periods
enhances the growth reduction. However, drought is not the main
cause for the disease.
    Mass spectrometrical and X-ray dispersive analyses on a
cellular scale were carried out on cross-sections of fine roots
of 10 healthy and diseased fir and spruce trees. In acidified
soils the cortex of the fine roots does not contain calcium
and magnesium. Neither was aluminium accumulated in the cortex
as in common with the cells of healthy trees. There is no
indication that in the primary xylem of the vascular cylinder
in diseased trees the transport of elements is out of control.

6. REFERENCES

1.   Athari, S.: 1981, Mitteilungen der Forstl. Bundes-Versuchs-
     anstalt, Wien 139, pp. 7-27.
2.   Bauch, J., Klein, P., Frühwald, A., Brill, H.: 1979, Eur.
     J. For. Path. 9, pp. 321-331.
3.   Bauch, J., Schröder, W.: 1982, Forst. Cbl. 101, c. 5, in press
4.   Baumeister, W., Ernst, W.: 1978, G. Fischer Verlag Stuttgart-
     New York, 416 p.
5.   Blaschke, H.: 1981, Forst. Cbl. 100, pp. 190-195.
6.   Brill, H., Bock, E., Bauch, J.: 1981, Forst. Cbl. 100,
     pp. 195-206.
7.   Eckstein, D., Aniol, R.W., Bauch, J.: 1982, Eur. J. For.
     Path. 12, in press.
8.   Eckstein, D., Breyne, A., Aniol, R.W., Liese, W.: 1981,
     Forst. Cbl. 100, pp. 381-396.

9. Guderian, R., van Haut, H., Stratmann, H.: 1969,
    Forschungsbericht NW Nr. 1017, Köln und Opladen.
10. Hüttermann, A.: 1982, LÖLF-Mitteilungen, Landesanstalt für
    Ökologie, Landschaftsentwicklung und Forstplanung, Nordrhein-
    Westfalen, pp. 26-31.
11. Keller, T.: 1964, Schweiz. Z. Forstwes. 115, pp. 228-255.
12. Knabe, W.: 1982, LÖLF-Mitteilungen, Landesanstalt für
    Ökologie, Landschaftsentwicklung und Forstplanung, Nordrhein-
    Westfalen, pp. 43-57.
13. Schütt, P.: 1977, Forst. Cbl. 96, pp. 177-186.
14. Schütt, P.: 1981, Forst. Cbl. 100, pp. 174-179.
15. Schütt, P.: 1981, Forstl. Cbl. 100, pp. 286-287.
16. Ulrich, B.: 1981, Der Forst- und Holzwirt 36, pp. 525-532.
17. Ulrich, B.: 1982, LÖLF-Mitteilungen, Landesanstalt für
    Ökologie, Landschaftsentwicklung und Forstplanung, Nordrhein-
    Westfalen, pp. 9-25.
18. Ulrich, B., Mayer, R., Khanna, P.K.: 1979, Schriften Forstl.
    Fak. Univ. Göttingen 58, Sauerländer Verlag.
19. Wentzel, K.F.: 1966, Angew. Botanik 40, pp. 1-11.

INDEX

Acid Deposition   1, 106, 127,
    147, 195, 207, 219, 331
Activity products   161
Adirondack Mountains 331
Adsorption   235
Adsorption of SO$_4$ 181, 195
Aerosols   57, 65
Aggradation phase   1
Alkali cations   33
Aluminum   37, 47, 113, 127,
    147, 158, 171, 207, 222,
    347, 359, 277
Aluminum complexes   108
Aluminum buffer range   127,
    147, 334
Aluminumhydroxosulfate   128,
    152, 159
Aluminum smelters   289, 303
Al-toxicity   207, 219, 360, 384
Aminopeptidase   257
Ammonia   171
Ammonium   38, 57, 183
Ammonium sulfate   171
Ammonification   98
Animal manure   175
Armillaria mellea   359
Ascorbic acid contents   290

Bacterial disease   384
Basaluminate   109
Base neutralization capacity
    (BNC)   128
Basic aluminum sulfate   109
Batch experiments   164
Beech (Fagus silv.)   33, 48, 57,
    93, 147, 164, 359
β-glucosidase   257

Bioindicator   285
Biological activity   247
Biomass utilization   14, 106
Biotite   119
Birch (Betula pendula)   172, 18⟨
    207, 219
Brown earth   266
Buffer ranges   127, 147
Bulk precipitation   49, 65, 222

Cadmium   47, 57, 147, 233
Calcareous soils   171
Calcium   37, 122, 127, 171, 195,
    207, 347, 359, 377
Ca/Al ratio   207, 219
Calcium carbonate buffer range
    127, 180, 334
Canopy drip   148, 171, 220
Cation acids   158
Cation exchange   37, 164
Cation exchange buffer rabge
    127, 147, 334
Cation exchange capacity   129
Chemical equilibrium   159
Chemical soil state   1, 147
Chloride   33, 37, 57, 222, 288,
    304
Chlorinated hydrocarbons   65
Chromogenic substrates   268
Cloud droplets   33
CO$_2$ uptake   288
Cobalt   51
Collembols   249
Column lysimeters   195
Cortex   377
Copper   47, 233, 343, 359
Chromium   47

Debye-Hueckel law   161
Decomposition   94, 147, 177, 247
Dendroclimatological
    investigations   384
Deposition   285
Deposition of aerosols   33, 59
Deposition velocity   57, 85, 228
Desorption   235
Destabilization   1
Diameter growth   343
Drainage   183
Drought   380
Dry deposition   35, 47, 57, 83,
    171, 209, 219
Dryfall   65

Earth alkali cations   33
Earth worm activity   254
Enchytraeids   249
Epiphytic moss   65
Equilibrium soil   solution
    165, 222, 233
Exchangeable protons   159

Felspars   119, 179
Fir (Abies alba)   377
Fluoride   295, 303, 347
Flux balance   33, 48, 93, 147
Fog   33, 149
Foliar analysis   286
Forest canopy   33, 58, 183

Game   271
Garbage incinerator   289
Gas adsorption   33
Green Mountains   236
Growth retardation   308, 319
Growth development   319, 359,
    377

Heath (Calluna vulg.)   147, 184
Heavy metals   47, 51, 57, 147,
    158, 233, 271, 343, 359
Humic substances   233
Humus accumulation   108, 147
Humus disintegration   1
Humus forms   247
Humus layer   52, 94, 149

Impaction of aerosols   34

Interception deposition   33
Inventory of damages   327, 329
Ion cycle   3, 93, 154, 184, 227
Ion uptake   3, 93, 105, 149, 183,
    227
Iron   38, 48, 57, 122, 359
Iron buffer range   127
IUFRO   295

Jurbanite   109, 161

Lake acidification   108
Larix kaempheri   184
Laser-Microprobe-Mass-Analyser
    379
Latent injury   287
Laurentian Mountains   331
Leaching   50, 171, 197
Lead   47, 57, 233
Lignification   384
Litterfall   49, 148
Lower Saxony   327

Magnesium   37, 122, 147, 195,
    207, 359, 377
Manganese   37, 48, 57, 199, 347,
    359
Mathematical model   159
Mercury   271
Mineralization   3, 93, 105
Mycorrhyza   385

Nickel   47
Nitrate   38, 57, 171, 184, 222
Nitric acid   171
Nitrification   98, 171, 257
Norway Spruce   33, 47, 57, 93,
    147, 164, 207, 319, 327, 377

Oak (Quercus robur)   147, 172, 221
Organic acids   93, 227
Organic matter   233
Output   49, 148

Parabraunerde   359
Peroxidase activity   289
pH   106, 195, 220, 233, 379
Phosphatase   257
Phosphodiesterase   257
Physiology of the soil   257

Picea sitchensis 184
Pine needles 304
Pinus banksiana 195
Podzolization 1, 127
Polycyclic aromatic hydrocarbons
    65
Potassium 37, 122, 195, 347,
    359, 381
Precipitation deposition 33,
    147
Primary xylem 381
Proton consumption 93, 113, 147
Proton flux 105, 147, 183
Proton production 1, 93, 147,
    227

Rainwater 172
Resilience 1, 127
Rhone valley 303
Ring width 293
Roots 359, 368, 377

Scots pine 83, 147, 184, 207,
    221, 303, 343
Sedimentation 34
Silicate buffer range 127, 334
Silicate weathering 1, 113, 127
Simulated acid rain 197, 207
SO_2 33, 83, 108, 295
SO_2-adsorption 33, 42, 291
Sodium 37, 122, 195, 381
Soil acidification 1, 106, 127,
    171, 219
Soil alkalinization 1
Soil enzymes 257
Soil solution 147, 172, 207
Solling project 1, 33, 47, 57,
    93, 127, 157
Solubility product 165
Sorption 164
Stability 1, 93
Steady state 3, 93, 169
Stem analysis 322
Stemflow 148, 171, 220
Strain 1
Stress 1, 287
Strong acids 157
Sulfate 37, 57, 152, 171,
    195, 222
Surface resistance 83

Terra fusca 265, 359
Tillingbourne catchment study
    221
Titration 163
Toxicity 207, 219, 287, 360, 384
Transmission electron microscope
    379
Tree ring analysis 308
Tree vitality 33

Vascular cylinder 381

Weak acids 157
Weatherable minerals 179
Weathering balance 113
Wet deposition 34, 47, 58, 209
Wetfall 65
Wetwood formation 384

X-ray dispersive analysis 379

Zinc 47, 343, 359